Graduate Texts in Mathematics **118**

Springer
New York
Berlin
Heidelberg
Barcelona
Budapest
Hong Kong
London
Milan
Paris
Santa Clara
Singapore
Tokyo

Graduate Texts in Mathematics

continued after index

Gert K. Pedersen

Analysis Now

Springer

Gert K. Pedersen
Mathematics Institute
University of Copenhagen
Universitetsparken 5
DK-2100 Copenhagen Ø
Denmark

Mathematics Subject Classification (1991): 46-01, 46-C99

Library of Congress Cataloging-in-Publication Data
Pedersen, Gert Kjærgård.
    Analysis now / Gert K. Pedersen.
        p.   cm.—(Graduate texts in mathematics; 118)
    Bibliography: p.
    Includes index.
    1. Functional analysis. I. Title. II. Series.
QA320.P39   1988
515.7—dc19   88-22437

Printed on acid-free paper.

Typeset by Asco Trade Typesetting Ltd., Hong Kong.
Printed and bound by R.R. Donnelley & Sons, Harrisonburg, Virginia.
Printed in the United States of America.

9 8 7 6 5 4 3 2 (Corrected second printing, 1995)

ISBN 0-387-96788-5 Springer-Verlag New York Berlin Heidelberg
ISBN 3-540-96788-5 Springer-Verlag Berlin Heidelberg New York

For
Oluf and *Cecilie*,
innocents at home

Preface

Mathematical method, as it applies in the natural sciences in particular, consists of solving a given problem (represented by a number of observed or observable data) by neglecting so many of the details (these are afterward termed "irrelevant") that the remaining part fits into an axiomatically established model. Each model carries a theory, describing the implicit features of the model and its relations to other models. The role of the mathematician (in this oversimplified description of our culture) is to maintain and extend the knowledge about the models and to create new models on demand.

Mathematical analysis, developed in the 18th and 19th centuries to solve dynamical problems in physics, consists of a series of models centered around the real numbers and their functions. As examples, we mention continuous functions, differentiable functions (of various orders), analytic functions, and integrable functions; all classes of functions defined on various subsets of euclidean space $\mathbb{R}^n$, and several classes also defined with vector values. Functional analysis was developed in the first third of the 20th century by the pioneering work of Banach, Hilbert, von Neumann, and Riesz, among others, to establish a model for the models of analysis. Concentrating on "external" properties of the classes of functions, these fit into a model that draws its axioms from (linear) algebra and topology. The creation of such "supermodels" is not a new phenomenon in mathematics, and, under the name of "generalization," it appears in every mathematical theory. But the users of the original models (astronomers, physicists, engineers, et cetera) naturally enough take a somewhat sceptical view of this development and complain that the mathematicians now are doing mathematics for its own sake. As a mathematician my reply must be that the abstraction process that goes into functional analysis is necessary to survey and to master the enormous material we have to handle. It is not obvious, for example, that a differential equation, a system of linear equations, and a problem in the calculus of variations have anything in common. A knowledge of operators on topological vector spaces gives, however, a basis of reference, within which the concepts of kernels, eigenvalues, and inverse transformations can be used on all three problems. Our critics, especially those well-meaning pedagogues, should come to realize that mathematics becomes simpler only through abstraction. The mathematics that represented the conceptual limit for the minds of Newton and Leibniz is taught regularly in our high schools, because we now have a clear (i.e. abstract) notion of a function and of the real numbers.

When this defense has been put forward for official use, we may admit in private that the wind is cold on the peaks of abstraction. The fact that the objects and examples in functional analysis are themselves mathematical theories makes communication with nonmathematicians almost hopeless and deprives us of the feedback that makes mathematics more than an aesthetical play with axioms. (Not that this aspect should be completely neglected.) The dichotomy between the many small and directly applicable models and the large, abstract supermodel cannot be explained away. Each must find his own way between Scylla and Charybdis.

The material contained in this book falls under Kelley's label: What Every Young Analyst Should Know. That the young person should know more (e.g. more about topological vector spaces, distributions, and differential equations) does not invalidate the first commandment. The book is suitable for a two-semester course at the first year graduate level. If time permits only a one-semester course, then Chapters 1, 2, and 3 is a possible choice for its content, although if the level of ambition is higher, 4.1–4.4 may be substituted for 3.3–3.4. Whatever choice is made, there should be time for the student to do some of the exercises attached to every section in the first four chapters. The exercises vary in the extreme from routine calculations to small guided research projects. The two last chapters may be regarded as huge appendices, but with entirely different purposes. Chapter 5 on (the spectral theory of) unbounded operators builds heavily upon the material contained in the previous chapters and is an end in itself. Chapter 6 on integration theory depends only on a few key results in the first three chapters (and may be studied simultaneously with Chapters 2 and 3), but many of its results are used implicitly (in Chapters 2–5) and explicitly (in Sections 4.5–4.7 and 5.3) throughout the text.

This book grew out of a course on the Fundamentals of Functional Analysis given at The University of Copenhagen in the fall of 1982 and again in 1983. The primary aim is to give a concentrated survey of the tools of modern analysis. Within each section there are only a few main results— labeled theorems—and the remaining part of the material consists of supporting lemmas, explanatory remarks, or propositions of secondary importance. The style of writing is of necessity compact, and the reader must be prepared to supply minor details in some arguments. In principle, though, the book is "self-contained." However, for convenience, a list of classic or established textbooks, covering (parts of) the same material, has been added. In the Bibliography the reader will also find a number of original papers, so that she can judge for herself "wie es eigentlich gewesen."

Several of my colleagues and students have read (parts of) the manuscript and offered valuable criticism. Special thanks are due to B. Fuglede, G. Grubb, E. Kehlet, K.B. Laursen, and F. Topsøe.

The title of the book may convey the feeling that the message is urgent and the medium indispensable. It may as well be construed as an abbreviation of the scholarly accurate heading: Analysis based on Norms, Operators, and Weak topologies.

Copenhagen   Gert Kjærgård Pedersen

Preface to the Second Printing

Harald Bohr is credited with saying that if mathematics does not teach us to think correctly, at least it teaches us how easy it is to think incorrectly. Certainly an embarrassing number of mistakes and misprints in this book have been brought to my attention during the past five years. Also, more or less desperate students have pointed out many phrases and formulations that made little sense without further explanation. I am deeply grateful to Springer-Verlag for allowing the numerous corrections in this revised second printing, and hope that it will be of improved service to the fastidious mathematicians it was aimed for.

GKP

Contents

CHAPTER 1

General Topology

General or set-theoretical topology is the theory of continuity and convergence in analysis. Although the theory draws its notions and fundamental examples from geometry (so that the reader is advised always to think of a topological space as something resembling the euclidean plane), it applies most often to infinite-dimensional spaces of functions, for which geometrical intuition is very hard to obtain. Topology allows us to reason in these situations as if the spaces were the familier two- and three-dimensional objects, but the process takes a little time to get used to.

The material presented in this chapter centers around a few fundamental topics. For example, we only introduce Hausdorff and normal spaces when separation is discussed, although the literature operates with a hierarchy of more than five distinct classes. A mildly unusual feature in the presentation is the central role played by universal nets. Admittedly they are not easy to get acquainted with, but they facilitate a number of arguments later on (giving, for example, a five-line proof of Tychonoff's theorem). Since universal nets entail the blatant use of the axiom of choice, we have included (in the regie of naive set theory) a short proof of the equivalence among the axiom of choice, Zorn's lemma, and Cantor's well-ordering principle. All other topics from set theory, like ordinal and cardinal numbers, have been banned to the exercise sections. A fate they share with a large number of interesting topological concepts.

1.1. Ordered Sets

Synopsis. The axiom of choice, Zorn's lemma, and Cantor's well-ordering principle, and their equivalence. Exercises.

1.1.1. A binary relation in a set X is just a subset R of $X \times X$. It is customary, though, to use a relation sign, such as $\leq$, to indicate the relation. Thus $(x, y) \in R$ is written $x \leq y$.

An *order* in X is a binary relation, written $\leq$, which is *transitive* ($x \leq y$ and $y \leq z$ implies that $x \leq z$), *reflexive* ($x \leq x$ for every x), and *antisymmetric* ($x \leq y$ and $y \leq x$ implies $x = y$). We say that $(X, \leq)$ is an *ordered set*. Without the antisymmetry condition we have a *preorder*, and much of what follows will make sense also for preordered sets.

An element x is called a *majorant* for a subset Y of X, if $y \leq x$ for every y in Y. *Minorants* are defined analogously. We say that an order is *filtering upward*, if every pair in X (and, hence, every finite subset of X) has a majorant. Orders that are *filtering downward* are defined analogously. If a pair x, y in X has a smallest majorant, relative to the order $\leq$, this element is denoted $x \vee y$. Analogously, $x \wedge y$ denotes the largest minorant of the pair x, y, if it exists. We say that $(X, \leq)$ is a *lattice*, if $x \vee y$ and $x \wedge y$ exist for every pair x, y in X. Furthermore, $(X, \leq)$ is said to be *totally ordered* if either $x \leq y$ or $y \leq x$ for every pair x, y in X. Finally, we say that $(X, \leq)$ is *well-ordered* if every nonempty subset Y of X has a smallest element (a minorant for Y belonging to Y). This element we call the first element in Y.

Note that a well-ordered set is totally ordered (put $Y = \{x, y\}$), that a totally ordered set is a (trivial) lattice, and that a lattice order is both upward and downward filtering. Note also that to each order $\leq$ corresponds a reverse order $\geq$, defined by $x \geq y$ iff $y \leq x$.

1.1.2. Examples of orderings are found in the number systems, with their usual orders. Thus, the set $\mathbb{N}$ of natural numbers is an example of a well-ordered set. (Apart from simple repetitions, $\mathbb{N} \cup \mathbb{N} \cup \cdots$, this is also the only concrete example we can write down, despite 1.1.6.) The sets $\mathbb{Z}$ and $\mathbb{R}$ are totally ordered, but not well-ordered. The sets $\mathbb{Z} \times \mathbb{Z}$ and $\mathbb{R} \times \mathbb{R}$ are lattices, but not totally ordered, when we use the *product order*, i.e. $(x_1, x_2) \leq (y_1, y_2)$ whenever $x_1 \leq y_1$ and $x_2 \leq y_2$. [If, instead, we use the *lexicographic order*, i.e. $(x_1, x_2) \leq (y_1, y_2)$ if either $x_1 < y_1$, or $x_1 = y_1$ and $x_2 \leq y_2$, then the sets become totally ordered.]

An important order on the system $\mathscr{S}(X)$ of subsets of a given set X is given by inclusion; thus $A \leq B$ if $A \subset B$. The inclusion order turns $\mathscr{S}(X)$ into a lattice with $\emptyset$ as first and X as last elements. In applications it is usually the reverse inclusion order that is used, i.e. $A \leq B$ if $A \supset B$. For example, taking X to be a sequence (x_n) of real numbers converging to some x, and putting $T_n = \{x_k | k \geq n\}$, then clearly it is the reverse inclusion order on the tails T_n that describe the convergence of (x_n) to x.

1.1.3. The *axiom of choice*, formulated by Zermelo in 1904, states that for each nonempty set X there is a (choice) function

$$c: \mathscr{S}(X) \backslash \{\emptyset\} \to X,$$

satisfying $c(Y) \in Y$ for every Y in $\mathscr{S}(X) \backslash \{\emptyset\}$.

Using this axiom Zermelo was able to give a satisfactory proof of Cantor's *well-ordering principle*, which says that every set X has an order $\leq$, such that $(X, \leq)$ is well-ordered.

The well-ordering principle is a necessary tool in proofs "by induction," when the set over which we induce is not a segment of $\mathbb{N}$ (so-called *transfinite induction*). More recently, these proofs have been replaced by variations that pass through the following axiom, known in the literature as Zorn's lemma (Zorn 1935, but used by Kuratowski in 1922). Let us say that $(X, \leq)$ is *inductively ordered* if each totally ordered subset of X (in the order induced from X), has a majorant in X. *Zorn's lemma* then states that every inductively ordered set has a maximal element (i.e. an element with no proper majorants).

1.1.4. Let $(X, \leq)$ be an ordered set and assume that c is a choice function for X. For any subset Y of X, let $\mathrm{maj}(Y)$ and $\mathrm{min}(Y)$, respectively, denote the sets of proper majorants and minorants for Y in X. Thus $x \in \mathrm{maj}(Y)$ if $y < x$ for every y in Y, where the symbol $y < x$ of course means $y \leq x$ and $y \neq x$.

A subset C of X is called a *chain* if it is well-ordered (relative to $\leq$) and if for each x in C we have

$$c(\mathrm{maj}(C \cap \mathrm{min}\{x\})) = x. \qquad (*)$$

Note that $c(X)$ is the first element in any chain and that $\{c(X)\}$ is a chain (though short).

1.1.5. Lemma. *If C_1 and C_2 are chains in X such that $C_1 \not\subset C_2$, there is an element x_1 in C_1 such that*

$$C_2 = C_1 \cap \mathrm{min}\{x_1\}.$$

PROOF. Since $C_1 \backslash C_2 \neq \emptyset$ and C_1 is well-ordered, there is a first element x_1 in $C_1 \backslash C_2$. By definition we therefore have

(i) $$C_1 \cap \mathrm{min}\{x_1\} \subset C_2.$$

If the inclusion in (i) is proper, the set $C_2 \backslash (C_1 \cap \mathrm{min}\{x_1\})$ has a first element x_2, since C_2 is well-ordered. By definition, therefore,

(ii) $$C_2 \cap \mathrm{min}\{x_2\} \subset C_1 \cap \mathrm{min}\{x_1\}.$$

If the inclusion in (ii) is proper, the set $(C_1 \cap \mathrm{min}\{x_1\}) \backslash \mathrm{min}\{x_2\}$ (contained in $C_1 \cap C_2$) has a first element y. By definition

(iii) $$C_1 \cap \mathrm{min}\{y\} \subset C_2 \cap \mathrm{min}\{x_2\}.$$

However, if $y \leq x$ for some x in $C_2 \cap \mathrm{min}\{x_2\}$, then $y \in C_2 \cap \mathrm{min}\{x_2\}$, contradicting the choice of y. Since both x and y belong to the well-ordered, hence totally ordered, set C_2, it follows that $x < y$ for every x in $C_2 \cap \mathrm{min}\{x_2\}$. Thus in (iii) we actually have equality. Since both C_1 and C_2 are chains (relative to the same ordering and the same choice function), it follows from the chain condition $(*)$ in 1.1.4 that $y = x_2$. But $y \in C_1 \cap \mathrm{min}\{x_1\}$ while

$x_2 \notin C_1 \cap \min\{x_1\}$. To avoid a contradiction we must have equality in (ii). Applying the chain condition to (ii) gives $x_1 = x_2$ in contradiction with $x_1 \notin C_2$ and $x_2 \in C_2$. Consequently, we have equality in (i), which is the desired result. $\square$

1.1.6. Theorem. *The following three propositions are equivalent:*

(i) *The axiom of choice.*

(ii) *Zorn's lemma.*

(iii) *The well-ordering principle.*

PROOF (i) $\Rightarrow$ (ii). Suppose that $(X, \leq)$ is inductively ordered, and by assumption let c be a choice function for X. Consider the set $\{C_j | j \in J\}$ of all chains in X and put $C = \bigcup C_j$. We claim that for any x in C_j we have

$$C \cap \min\{x\} = C_j \cap \min\{x\}. \tag{**}$$

For if y belongs to the first (obviously larger) set, then $y \in C_i$ for some i in J. Either $C_i \subset C_j$, in which case $y \in C_j$, or $C_i \not\subset C_j$. In that case there is by 1.1.5 an x_i in C_i such that $C_j = C_i \cap \min\{x_i\}$. As $y < x < x_i$, we again see that $y \in C_j$.

It now follows easily that C is well-ordered. For if $\emptyset \neq Y \subset C$, there is a j in J with $C_j \cap Y \neq \emptyset$. Taking y to be the first element in $C_j \cap Y$ it follows from (**) that y is the first element in all of Y. Condition (**) also immediately shows that C satisfies the chain condition (*) in 1.1.4. Thus C is a chain, and it is clearly the longest possible. Therefore, maj$(C) = \emptyset$. Otherwise we could take

$$x_0 = c(\text{maj}(C)) \in \text{maj}(C),$$

and then $C \cup \{x_0\}$ would be a chain [(*) in 1.1.4 has just been satisfied for x_0] effectively longer than C.

Since the order is inductive, the set C has a majorant x_ω in X. Since maj$(C) = \emptyset$, we must have $x_\omega \in C$, i.e. x_ω is the largest element in C. But then x_ω is a maximal element in X, because any proper majorant for x_ω would belong to maj(C).

(ii) $\Rightarrow$ (iii). Given a set X consider the system M of well-ordered, nonempty subsets $(C_j, \leq_j)$ of X. Note that $M \neq \emptyset$, the one-point sets are trivial members. We define an order $\leq$ on M by setting $(C_i, \leq_i) \leq (C_j, \leq_j)$ if either $C_i = C_j$ and $\leq_i = \leq_j$, or if there is an x_j in C_j such that

$$C_i = C_j \cap \min(x_j) = \{x \in C_j | x <_j x_j\} \text{ and } \leq_i = \leq_j | C_i. \tag{***}$$

The claim now is that $(M, \leq)$ is inductively ordered. To prove this, let N be a totally ordered subset of M and let C be the union of all C_j in N. Define $\leq$ on C by $x \leq y$ whenever $\{x, y\} \subset C_j \in N$ and $x \leq_j y$. Note that if $\{x, y\} \subset C_i \in N$, then $x \leq_i y$ iff $x \leq_j y$ because of the total ordering of N, so that $\leq$ is a well-defined order on C. Exactly as in the proof of (i) $\Rightarrow$ (ii) one shows that if $x \in C_j$, then

$$C \cap \min\{x\} = C_j \cap \min\{x\} \tag{**}$$

(the result of 1.1.5 has been built into the order on M). As before, this implies that $(C, \leq)$ is well-ordered. The conclusion that $(C, \leq)$ is a majorant for N is trivial if N has a largest element (which then must be C). Otherwise, each $(C_i, \leq_i)$ has a majorant $(C_j, \leq_j)$ in N and is thus of the form $(\ast\ast\ast)$ relative to C_j; and, as $(\ast\ast)$ shows, also of the form $(\ast\ast\ast)$ relative to C. We conclude that $(C, \leq)$ is a majorant for N, which proves that M is inductively ordered.

Condition (ii) now implies that M has a maximal element $(X_\omega, \leq_\omega)$. If $X_\omega \neq X$, we choose some x_ω in $X \backslash X_\omega$ and extend $\leq_\omega$ to $X_\omega \cup \{x_\omega\}$ by setting $x \leq_\omega x_\omega$ for every x in X_ω. This gives a well-ordered set $(X_\omega \cup \{x_\omega\}, \leq_\omega)$ that majorizes $(X_\omega, \leq_\omega)$ in the ordering in M, contradicting the maximality of $(X_\omega, \leq_\omega)$. Thus $X = X_\omega$ and is consequently well-ordered.

(iii) $\Rightarrow$ (i). Given a nonempty set X, choose a well-order $\leq$ on it. Now define $c(Y)$ to be the first element in Y for every nonempty subset Y of X.   □

1.1.7. Remark. The subsequent presentation in this book builds on the acceptance of the axiom of choice and its equivalent forms given in 1.1.6. In the intuitive treatment of set theory used here, according to which a set is a properly determined collection of elements, it is not possible precisely to explain the role of the axiom of choice. For this we would need an axiomatic description of set theory, first given by Zermelo and Fraenkel. In 1938 Gödel showed that if the Zermelo–Fraenkel system of axioms is consistent (that in itself an unsolved question), then the axiom of choice may be added without violating consistency. In 1963 Cohen showed further that the axiom of choice is independent of the Zermelo–Fraenkel axioms. This means that our acceptance of the axiom of choice determines what sort of mathematics we want to create, and it may in the end affect our mathematical description of physical realities. The same is true (albeit on a smaller scale) with the parallel axiom in euclidean geometry. But as the advocates of the axiom of choice, among them Hilbert and von Neumann, point out, several key results in modern mathematical analysis [e.g. the Tychonoff theorem (1.6.10), the Hahn–Banach theorem (2.3.3), the Krein–Milman theorem (2.5.4), and Gelfand theory (4.2.3)] depend crucially on the axiom of choice. Rejecting it, one therefore loses a substantial part of mathematics, and, more important, there seems to be no compensation for the abstinence.

EXERCISES

E 1.1.1. A subset $\Re$ of a real vector space $\mathfrak{X}$ is called a cone if $\Re + \Re \subset \Re$ and $\mathbb{R}_+ \Re = \Re$. If in addition $-\Re \cap \Re = \{0\}$ and $\Re - \Re = \mathfrak{X}$, we say that $\Re$ generates $\mathfrak{X}$. Show that the relation in $\mathfrak{X}$ defined by $x \leq y$ if $y - x \in \Re$ is an order on $\mathfrak{X}$ if $\Re$ is a generating cone. Find the set $\{x \in \mathfrak{X} | x \geq 0\}$, and discuss the relations between the order and the vector space structure. Find the condition on $\Re$ that makes the order total. Describe some cones in $\mathbb{R}^n$ for $n = 1, 2, 3$.

E 1.1.2. Let α be a positive, irrational number and show that the relation in $\mathbb{Z} \times \mathbb{Z}$ given by

$$(x_1, x_2) \leq (y_1, y_2) \quad \text{if} \quad \alpha(y_1 - x_1) \leq y_2 - x_2$$

is a total order. Sketch the set

$$\{(x_1, x_2) \in \mathbb{Z} \times \mathbb{Z} \mid (x_1, x_2) \geq (0, 0)\}.$$

E 1.1.3. An *order isomorphism* between two ordered sets $(X, \leq)$ and $(Y, \leq)$ is a bijective map $\varphi: X \to Y$ such that $x_1 \leq x_2$ iff $\varphi(x_1) \leq \varphi(x_2)$. A *segment* of a well-ordered set $(X, \leq)$ is a subset of X of the form $\min\{x\}$ for some x in X, or X itself (the improper segment). Show that if X and Y are well-ordered sets, then either X is order isomorphic to a segment of Y (with the relative order) or Y is order isomorphic to a segment of X.

Hint: The system of order isomorphisms $\varphi: X_\varphi \to Y_\varphi$, where X_φ and Y_φ are segments of X and Y, respectively, is inductively ordered if we define $\varphi \leq \psi$ to mean that $X_\varphi \subset X_\psi$ and $\psi | X_\varphi = \varphi$. Prove that for a maximal element $\varphi: X_\varphi \to Y_\varphi$ either X_φ or Y_φ must be an improper segment.

E 1.1.4. The equivalence classes of well-ordered sets modulo order isomorphism (E 1.1.3) are called *ordinal numbers*. Every well-ordered set has thus been assigned a "size" determined by its ordinal number. Show that the class of ordinal numbers is well-ordered.

Hint: Given a collection of ordinal numbers $\{\alpha_j | j \in J\}$ choose a corresponding family of well-ordered sets $(X_j | j \in J)$ such that α_j is the ordinal number for X_j for every j in J. Now fix one X_j. Either its equivalence class α_j is the smallest (and we are done) or each one of the smaller X_i's is order isomorphic to a proper segment $\min\{x_i\}$ in X_j by E 1.1.3. But these segments form a well-ordered set.

E 1.1.5. Let $f: X \to Y$ and $g: Y \to X$ be injective (but not necessarily surjective) maps between the two sets X and Y. Show that there is a bijective map $h: X \to Y$ (F. Bernstein, 1897).

Hint: Define

$$A = \bigcup_{n=0}^{\infty} (g \circ f)^n (X \backslash g(Y)),$$

and put $h = f$ on A and $h = g^{-1}$ on $X \backslash A$. Note that $X \backslash g(Y) \subset A$, whereas $Y \backslash f(A) \subset g^{-1}(X \backslash A)$.

E 1.1.6. We define an equivalence relation on the class of sets by setting $X \sim Y$ if there exists a bijective map $h: X \to Y$. Each equivalence class is called a *cardinal number*. Show that the natural numbers are the cardinal numbers for finite sets. Discuss the "cardinality" of some infinite sets, e.g. $\mathbb{N}$, $\mathbb{Z}$, $\mathbb{R}$, and $\mathbb{R}^2$.

E 1.1.7. For the cardinal numbers, defined in E 1.1.6, we define a relation by letting $\alpha \leq \beta$ if there are sets A and B with $\text{card}(A) = \alpha$ and $\text{card}(B) = \beta$ (a more correct, but less used terminology would be $A \in \alpha$ and $B \in \beta$, cf. E 1.1.6), and an injective map $f: A \rightarrow B$. Show with the use of E 1.1.5 that $\leq$ is an order on the cardinal numbers. Show finally that the class of cardinal numbers is well-ordered.

Hint: For each set X, let $\omega(X)$ be the smallest ordinal number corresponding to well-ordered subsets of X with the same cardinality as X. Show that ω defines an order isomorphism between the class of cardinal numbers and a subclass of the ordinal numbers, and then apply E 1.1.4.

E.1.1.8. A set is called *countable* (or countably infinite) if it has at most (or exactly) the cardinality (cf. E.1.1.6) as the set $\mathbb{N}$ of natural numbers. Show that there is a well-ordered set $(X, \leq)$, which is itself un-countable, but which has the property that each segment $\min\{x\}$ is countable if $x \in X$.

Hint: Choose a well-ordered set $(Y, \leq)$ that is uncountable. The subset Z of elements z in Y such that the segment $\min\{z\}$ is un-countable is either empty (and we are done) or else has a first element Ω. Set $X = \min\{\Omega\}$. The ordinal number (corresponding to) Ω is called the first uncountable ordinal.

E.1.1.9. Let $\mathfrak{X}$ be a vector space over a field $\mathbb{F}$. A *basis* for $\mathfrak{X}$ is a subset $\mathfrak{B} = \{e_j | j \in J\}$ of linearly independent vectors from $\mathfrak{X}$, such that every x in $\mathfrak{X}$ has a (necessarily unique) decomposition as a finite linear combination of vectors from $\mathfrak{B}$. Show that every vector space has a basis.

Hint: A basis is a maximal element in the system of linearly independent subsets of $\mathfrak{X}$.

E 1.1.10. Show that there exists a discontinuous function $f: \mathbb{R} \rightarrow \mathbb{R}$, such that $f(x + y) = f(x) + f(y)$ for all real numbers x and y. Show that $f(I)$ contains arbitrarily (numerically) large numbers for every (small) interval I in $\mathbb{R}$.

Hint: Let $\mathbb{Q}$ denote the field of rational numbers and apply E 1.1.9 with $X = \mathbb{R}$ and $\mathbb{F} = \mathbb{Q}$ to obtain what is called a *Hamel basis* for $\mathbb{R}$. Show that f can be assigned arbitrary values on the Hamel basis and still have an (unique) extension to an additive function on $\mathbb{R}$.

E.1.1.11. Let X be a set and $\mathscr{P}(X)$ the family of all subsets of X. Show that the cardinality of the set $\mathscr{P}(X)$ is strictly larger than that of X, cf. E 1.1.7.

Hint: If $f: X \rightarrow \mathscr{P}(X)$ is a bijective function, set

$$A = \{x \in X | x \notin f(x)\},$$

and take $y = f^{-1}(A)$. Either possibility $y \in A$ or $y \notin A$ will lead to a contradiction.

1.2. Topology

Synopsis. Open and closed sets. Interior points and boundary. Basis and subbasis for a topology. Countability axioms. Exercises.

1.2.1. A *topology* on a set X is a system τ of subsets of X with the properties that

(i) every union of sets in τ belongs to τ.
(ii) every finite intersection of sets in τ belongs to τ.
(iii) $\emptyset \in \tau$ and $X \in \tau$.

We say that (X, τ) is a *topological space*, and that τ consists of the *open* subsets in (X, τ).

1.2.2. A *metric* on a set X is a function $d\colon X \times X \to \mathbb{R}_+$ (the distance function) that is symmetric $[d(x, y) = d(y, x)]$ and faithful $[d(x, y) = 0$ iff $x = y]$, and satisfies the triangle inequality $[d(x, y) \leq d(x, z) + d(z, y)]$. We declare a subset A of X to be open if for each x in A, there is a sufficiently small $\varepsilon > 0$, such that the ε-ball $\{y \in X \mid d(x, y) < \varepsilon\}$ around x is contained in A. It is straightforward to check that the collection of such open sets satisfies the requirements (i)–(iii) in 1.2.1, and thus gives a topology on X, the *induced topology*. Conversely, we say that a topological space (X, τ) is *metrizable* if there is a metric on X that induces τ.

1.2.3. Remark. It is a fact that the overwhelming number of topological spaces used in the applications are metrizable. The question therefore arises: What is topology good for? The answer is (hopefully) contained in this chapter, but a few suggestions can be given already now: Using topological rather than metric terminology, the fundamental concepts of analysis, such as convergence, continuity, and compactness, have simple formulations, and the arguments involving them become more transparent. As a concrete example, consider the open interval $]-1, 1[$ and the real axis $\mathbb{R}$. These sets are topologically indistinguishable [the map $x \to \tan(\frac{1}{2}\pi x)$ furnishes a bijective correspondence between the open sets in the two spaces]. This explains why every property of $\mathbb{R}$ that only depends on the topology also is found in $]-1, 1[$. Metrically, however, the spaces are quite different. ($\mathbb{R}$ is unbounded and complete; $]-1, 1[$ enjoys the opposite properties.) A metric on a topological space may thus emphasize certain characteristics that are topologically irrelevant.

1.2.4. A subset Y in a topological space (X, τ) is a *neighborhood* of a point x in X if there is an open set A such that $x \in A \subset Y$. The system of neighborhoods of x is called the *neighborhood filter* and is denoted by $\mathcal{O}(x)$. The concept of neighborhood is fundamental in the theory, as the name topology indicates (*topos* = place; *logos* = knowledge). A rival name (now obsolete) for the theory was analysis situs, which again stresses the importance of neighborhoods (*situs* = site).

A point x is an *inner point* in a subset Y of X if there is an open set A such that $x \in A \subset Y$. The set of inner points in Y is denoted by Y^o. Note that Y^o is the set of points for which Y is a neighborhood, and Y^o is the largest open set contained in Y.

1.2.5. A subset F of a topological space (X, τ) is *closed* if $X \setminus F \in \tau$. The definition implies that $\emptyset$ and X are closed sets and that an arbitrary intersection and a finite union of closed sets is again closed.

For each subset Y of X we now define the *closure* of Y as the intersection Y^- of all closed sets containing Y. The elements in Y^- are called *limit points* for Y. Note the formulas

$$X \setminus Y^- = (X \setminus Y)^o, \qquad X \setminus Y^o = (X \setminus Y)^-.$$

We say that a set Y is *dense* in a (usually larger) set Z, if $Z \subset Y^-$.

1.2.6. Proposition. *If $Y \subset X$ and $x \in X$, then $x \in Y^-$ iff $Y \cap A \neq \emptyset$ for each A in $\mathcal{O}(x)$.*

PROOF. If $Y \cap A = \emptyset$ for some A in $\mathcal{O}(x)$, then without loss of generality we may assume that $A \in \tau$. Thus $X \setminus A$ is a closed set containing Y, so that $Y^- \subset X \setminus A$ and $x \notin Y^-$.

Conversely, if $x \notin Y^-$, then $X \setminus Y^-$ is an open neighborhood of x disjoint from Y. $\qquad \square$

1.2.7. For $Y \subset X$ the set $Y^- \setminus Y^o$ is called the *boundary* of Y and is denoted by ∂Y. We see from 1.2.6 that $x \in \partial Y$ iff every neighborhood of x meets both Y and $X \setminus Y$. In particular, $\partial Y = \partial(X \setminus Y)$. Note that a closed set contains its boundary, whereas an open set is disjoint from its boundary.

1.2.8. If (X, τ) is a topological space we define the *relative topology* on any subset Y of X to be the collection of sets of the form $A \cap Y$, $A \in \tau$. It follows that a subset of Y is closed in the relative topology iff it has the form $Y \cap F$ for some closed set F in X. To avoid ambiguity we shall refer to the relevant subsets of Y as being *relatively open* and *relatively closed*.

1.2.9. If σ and τ are two topologies on a set X, we say that σ is *weaker* than τ or that τ is *stronger* than σ, provided that $\sigma \subset \tau$. This defines an order on the set of topologies on X. There is a first element in this ordering, namely the *trivial topology*, that consists only of the two sets $\emptyset$ and X. There is also a last element, the *discrete topology*, containing every subset of X. As the next result shows, the order is a lattice in a very complete sense.

1.2.10. Proposition. *Given a system $\{\tau_j | j \in J\}$ of topologies on a set X, there is a weakest topology stronger than every τ_j, and there is a strongest topology weaker than every τ_j. These topologies are denoted $\vee \tau_j$ and $\wedge \tau_j$, respectively.*

PROOF. Define $\wedge \tau_j$ as the collection of subsets A in X such that $A \in \tau_j$ for all j in J. This is a topology, and it is weaker than every τ_j, but only minimally so.

Now let T denote the set of topologies on X that are stronger than every τ_j. The discrete topology belongs to T, so $T \neq \emptyset$. Setting $\vee \tau_j = \wedge \tau, \tau \in T$, we obtain a topology with the required property.   □

1.2.11. Given any system ρ of subsets of X there is a weakest topology $\tau(\rho)$ that contains ρ, namely, $\tau(\rho) = \wedge \tau$, where τ ranges over all topologies on X that contain ρ. We say that ρ is a *subbasis* for $\tau(\rho)$. If each set in $\tau(\rho)$ is a union of sets from ρ, we say that ρ is a *basis* for $\tau(\rho)$. It follows from 1.2.12 that this will happen iff X and each finite intersection of sets from ρ is the union of sets from ρ. In particular, ρ is a basis for $\tau(\rho)$ if it is stable under finite intersections and covers X.

Given a topological space (X, τ), we say that a system ρ of subsets of X is a *neighborhood basis* for a point x in X, if $\rho \subset \mathcal{O}(x)$ and for each A in $\mathcal{O}(x)$ there is a B in ρ, such that $B \subset A$. The reason for this terminology becomes clear from the next result.

1.2.12. Proposition. *For a system ρ of subsets of X, the topology $\tau(\rho)$ consists of exactly those sets that are unions of sets, each of which is a finite intersection of sets from ρ, together with $\emptyset$ and X.*

Conversely, a system ρ of open sets in a topological space (X, τ) is a basis (respectively, a subbasis) for τ, if ρ (respectively, the system of finite intersections of sets from ρ) contains a neighborhood basis for every point in X.

PROOF. The system of sets described in the first half of the proposition is stable under finite intersections and arbitrary unions, and it contains $\emptyset$ and X (per fiat). It therefore is a topology, and clearly the weakest one that contains ρ.

Conversely, if a system $\rho \subset \tau$ contains (or after taking finite intersections contains) a neighborhood basis for every point, let $\tau(\rho)$ be the topology it generates and note that $\tau(\rho) \subset \tau$. If $A \in \tau$, there is for each x in A a $B(x)$ in ρ [respectively, in $\tau(\rho)$] such that $x \in B(x) \subset A$. Since $A = \bigcup B(x)$, we see that ρ is a basis for τ [respectively, $\tau = \tau(\rho)$].   □

1.2.13. A topological space (X, τ) is *separable* if some sequence of points is dense in X.

A topological space (X, τ) satisfies the *first axiom of countability* if for each x in X there is a sequence $(A_n(x))$ in $\mathcal{O}(x)$, such that every A in $\mathcal{O}(x)$ contains some $A_n(x)$ (i.e. if every neighborhood filter has a countable basis).

A topological space (X, τ) satisfies the *second axiom of countability* if τ has a countable basis. According to 1.2.12 it suffices for τ to have a countable subbasis, because finite intersections of subbasis sets will then be a countable basis.

The three conditions mentioned above all say something about the "size" of τ, and the second countability axiom (which implies the two previous conditions) is satisfied for the spaces that usually occur in the applications. Thus $\mathbb{R}^n$ is second countable, because n-cubes with rational coordinates for all corners form a basis for the usual topology. Note also that any subset of a space that is first or second countable will itself be first or second countable in the relative topology.

EXERCISES

E 1.2.1. (Topology according to Hausdorff.) Suppose that to every point x in a set X we have assigned a nonempty family $\mathcal{U}(x)$ of subsets of X satisfying the following conditions:

(i) $x \in A$ for every A in $\mathcal{U}(x)$.
(ii) If $A \in \mathcal{U}(x)$ and $B \in \mathcal{U}(x)$, then there is a C in $\mathcal{U}(x)$ with $C \subset A \cap B$.
(iii) If $A \in \mathcal{U}(x)$, then for each y in A there is a B in $\mathcal{U}(y)$ with $B \subset A$.

Show that if τ is the weakest topology containing all $\mathcal{U}(x)$, $x \in X$, then $\mathcal{U}(x)$ is a neighborhood basis for x in τ for every x in X.

E 1.2.2. (Topology according to Kuratowski.) Let $\mathcal{S}(X)$ denote the system of subsets of a set X, and consider a function $Y \to \mathrm{cl}(Y)$ of $\mathcal{S}(X)$ into itself that satisfies the four *closure axioms*:

(i) $\mathrm{cl}(\emptyset) = \emptyset$.
(ii) $Y \subset \mathrm{cl}(Y)$ for every Y in $\mathcal{S}(X)$.
(iii) $\mathrm{cl}(\mathrm{cl}(Y)) = \mathrm{cl}(Y)$ for every Y in $\mathcal{S}(X)$.
(iv) $\mathrm{cl}(Y \cup Z) = \mathrm{cl}(Y) \cup \mathrm{cl}(Z)$ for all Y and Z in $\mathcal{S}(X)$.

Show that the system of sets F such that $\mathrm{cl}(F) = F$ form the closed sets in a topology on X, and that $Y^- = \mathrm{cl}(Y)$, $Y \in \mathcal{S}(X)$.

E 1.2.3. Let Y be a dense subset of a topological space (X, τ). Show that $(Y \cap A)^- = A^-$ for every open subset A of X.

E 1.2.4. Show that $\partial(Y \cup Z) \subset \partial Y \cup \partial Z$ for any two subsets Y and Z of a topological space (X, τ).

E 1.2.5. Show that the sets $]t, \infty[$, $t \in \mathbb{R}$, together with $\emptyset$ and $\mathbb{R}$ is a topology on $\mathbb{R}$. Describe the closure of a point in $\mathbb{R}$.

E 1.2.6. Let $(X, \leq)$ be a totally ordered set. The sets $\{x \in X \mid x < y\}$ and $\{x \in X \mid y < x\}$, where y ranges over X, are taken as a subbasis for a topology on X, the *order topology*. A familiar example is the order topology on $(\mathbb{R}, \leq)$. A less familiar example arises by taking X as the well-ordered set defined in E 1.1.8. Show that the order topology on this set satisfies the first but not the second axiom of countability.
    Hint: A countable union of countable sets is countable.

E 1.2.7. (The *Sorgenfrey line*.) Give the set $\mathbb{R}$ the topology τ for which a basis consists of the half-open intervals $[y, z[$, where y and z range over $\mathbb{R}$. Show that every basis set is closed in τ. Show that $(\mathbb{R}, \tau)$ is a separable space that satisfies the first but not the second axiom of countability.

  Hint: If ρ is some basis for τ and $x \in \mathbb{R}$, then ρ must contain a set A such that $x = \text{Inf}\{y \in A\}$.

E 1.2.8. (The *Sorgenfrey plane*.) Give the set $\mathbb{R}^2$ the topology τ^2, for which a basis consist of products of half-open intervals $[y_1, z_1[\times [y_2, z_2[$, where y_1, y_2, z_1, and z_2 range over $\mathbb{R}$. Show that $(\mathbb{R}^2, \tau^2)$ is a separable space. Show that the subset $\{(x, y) \in \mathbb{R}^2 | x + y = 0\}$ is discrete in the relative topology (and thus nonseparable), but closed in $\mathbb{R}^2$.

E 1.2.9. Let (X, τ) be a topological space such that τ is induced by a metric d on X. Show that τ satisfies the first axiom of countability. Show that τ satisfies the second axiom of countability iff (X, τ) is separable. Deduce from this that the Sorgenfrey line (E 1.2.7) is a nonmetrizable topological space.

E 1.2.10. A topological space (X, τ) is a *Lindelöf space* if each family σ in τ that covers X (i.e. $X = \bigcup A$, $A \in \sigma$) contains a countable subset $\{A_n | n \in \mathbb{N}\} \subset \sigma$ that covers X. Show that (X, τ) is a Lindelöf space if τ satisfies the second axiom of countability.

  Hint: If σ is an open covering of X and $\{B_n | n \in \mathbb{N}\}$ is a basis for τ, consider the countable subset $\{B_{n_k} | k \in \mathbb{N}\}$ of basis sets such that each B_{n_k} is contained in some A_k from σ. This subset must cover X.

E 1.2.11. Let σ be a family of half-open intervals in $\mathbb{R}$ (of the form $[x, y[$). Show that there is a countable subset σ' of σ such that

$$\bigcup_{A \in \sigma} A = \bigcup_{A \in \sigma'} A.$$

Deduce from this that the Sorgenfrey line (E 1.2.7) is a Lindelöf space (E 1.2.10).

  Hint: If $\sigma = \{[x_j, y_j[| j \in J\}$, then the set σ' of j's such that $x_j \notin \bigcup]x_i, y_i[$, $i \in J$, gives a family of mutually disjoint half-open intervals. Show that σ' must be countable. For the elements in $\sigma \setminus \sigma'$ we may replace half-open intervals by open intervals, and appeal to the Lindelöf property of the usual topology on $\mathbb{R}$ (which is second countable, cf. E 1.2.10).

E 1.2.12. Show that every closed subset of a Lindelöf space is a Lindelöf space in the relative topology (cf. E 1.2.10). Deduce from this that the Sorgenfrey plane (E 1.2.8) is not a Lindelöf space.

1.3. Convergence

Synopsis. Nets and subnets. Convergence of nets. Accumulation points. Universal nets. Exercises.

1.3.1. A *net* in a space X is a pair (Λ, i), where Λ is a non-empty upward-filtering ordered set (cf. 1.1.1) and i is a map from Λ into X. The standard notation for a net will, however, be $(x_\lambda)_{\lambda \in \Lambda}$, where we put $x_\lambda = i(\lambda)$ and indicate the domain of i. Since we do not ask i to be injective, we have no use for the antisymmetry in the order on Λ. Often a net is therefore defined with only a preordered index set Λ.

The most important example of a net arises when $\Lambda = \mathbb{N}$, i.e. when we have a *sequence* in X. It is, of course, this example that also motivates the notation $(x_\lambda)_{\lambda \in \Lambda}$, which is standard for sequences (although there the index set $\mathbb{N}$ is omitted).

Sequences suffice to handle all convergence problems in spaces that satisfy the first axiom of countability, in particular, all metric spaces. Certain spaces (e.g. Hilbert space in the weak topology, 3.1.10) require the more general notion of nets, and certain complicated convergence arguments (refinement of sequences by Cantor's diagonal principle) are effectively trivialized by the use of universal nets; cf. 1.3.7. Nets are also called *generalized sequences* in the literature, and as such they should be regarded.

1.3.2. A *subnet* of a net (Λ, i) in X is a net (M, j) in X together with a map $h: M \to \Lambda$, such that $j = i \circ h$, and such that for each λ in Λ there is a $\mu(\lambda)$ in M with $\lambda \leq h(\mu)$ for every $\mu \geq \mu(\lambda)$. In most cases we may choose h to be monotone [i.e. $\nu \leq \mu$ in M implies $h(\nu) \leq h(\mu)$ in Λ], and then, in order to have a subnet, it suffices to check that for each λ in Λ there is a μ in M with $\lambda \leq h(\mu)$.

The definition of subnet may sound a bit intricate, but try to formulate a strictly correct definition of the concept of subsequence!

1.3.3. We say that a net $(x_\lambda)_{\lambda \in \Lambda}$ in a set X is *eventually* in a subset Y of X, if there is a $\lambda(Y)$ such that $x_\lambda \in Y$ for every $\lambda \geq \lambda(Y)$. We say that the net is *frequently* in Y if for each λ in Λ, there is a $\mu \geq \lambda$ with x_μ in Y. Note that (x_λ) is not frequently in Y iff (x_λ) is eventually in $X \setminus Y$.

A net $(x_\lambda)_{\lambda \in \Lambda}$ in a topological space (X, τ) *converges* to a point x, if it is eventually in each A in $\mathcal{O}(x)$. We write this as $x = \lim x_\lambda$, or just $x_\lambda \to x$.

A point x in X is an *accumulation point* for a net $(x_\lambda)_{\lambda \in \Lambda}$ if the net is frequently in every A in $\mathcal{O}(x)$. Note that with these definitions x is an accumulation point of a net if some subnet of it converges to x; and, as we shall see, all accumulation points arise in this manner.

1.3.4. Fundamental Lemma. *Let $\mathcal{B}$ be a system of subsets of X that is upward-filtering under reverse inclusion. If a net $(x_\lambda)_{\lambda \in \Lambda}$ is frequently in every set B in $\mathcal{B}$, there is a subnet $(x_{h(\mu)})_{\mu \in M}$ that is eventually in every B in $\mathcal{B}$.*

PROOF. Consider the set

$$M = \{(\lambda, B) \in \Lambda \times \mathscr{B} | x_\lambda \in B\}$$

equipped with the product order as a subset of $\Lambda \times \mathscr{B}$. This order is upward filtering. Because if (λ, B) and (μ, C) belong to M, there is a D in $\mathscr{B}$ with $D \subset B \cap C$. Since the net is frequently in D, there is a ν in Λ with $\nu \geq \lambda, \nu \geq \mu$, and $x_\nu \in D$. This means that (ν, D) is a majorant for (λ, B) and (μ, C).

The map $h: M \to \Lambda$ given by $h(\lambda, B) = \lambda$ is monotone, and for each λ in Λ there is a B in $\mathscr{B}$ and a $\nu \geq \lambda$ with $x_\nu \in B$, whence $h(\nu, B) \geq \lambda$. Thus $(x_{h(\mu)})_{\mu \in M}$ is a subnet of $(x_\lambda)_{\lambda \in \Lambda}$; and for every B in $\mathscr{B}$ the subnet is eventually in B, namely when $\mu \geq (\lambda, C)$ for some (λ, C) in M with $C \subset B$. $\square$

1.3.5. Proposition. *For every accumulation point x of a net in a topological space (X, τ) there is a subnet that converges to x.*

PROOF. Apply 1.3.4 with $\mathscr{B} = \mathcal{O}(x)$. $\square$

1.3.6. Proposition. *A point x in a topological space (X, τ) belongs to the closure of a set Y iff there is a net in Y converging to x.*

PROOF. If $x \in Y^-$, then $A \cap Y \neq \emptyset$ for every A in $\mathcal{O}(x)$ by 1.2.6. Applying the axiom of choice (1.1.3) we can choose x_A in $A \cap Y$ for every A. The net $(x_A)_{A \in \mathcal{O}(x)}$ belongs to Y and obviously converges to x.

Conversely, if a net $(x_\lambda)_{\lambda \in \Lambda}$ in Y converges to x, then each A in $\mathcal{O}(x)$ contains points from Y (indeed, the whole net, eventually), whence $x \in Y^-$ by 1.2.6.
$\square$

1.3.7. A net $(x_\lambda)_{\lambda \in \Lambda}$ in X is *universal* if for every subset Y of X the net is either eventually in Y or eventually in $X \setminus Y$. In a topological space (X, τ) a universal net will therefore converge to every one of its accumulation points. Thinking of subnets as "refinements" of the original net, we see that a universal net is maximally refined. The existence of nontrivial universal nets (i.e. those that are not eventually constant) requires the axiom of choice. Note that if $f: X \to Y$ is a map and $(x_\lambda)_{\lambda \in \Lambda}$ is a net in X, then $(f(x_\lambda))_{\lambda \in \Lambda}$ is a net in Y. And if the net is universal in X, then the image net is universal in Y.

1.3.8. Theorem. *Every net $(x_\lambda)_{\lambda \in \Lambda}$ in X has a universal subnet.*

PROOF. We define a *filter* for the net to be a system $\mathscr{F}$ of nonvoid subsets of X, stable under finite intersections, containing with a set F any larger set $G \supset F$, and such that the net is frequently in every F in $\mathscr{F}$.

The set of filters for our net is nonvoid (set $\mathscr{F}_0 = \{X\}$) and ordered under inclusion. This order is inductive: If $\{\mathscr{F}_j | j \in J\}$ is a totally ordered set of filters, then $\mathscr{F} = \bigcup \mathscr{F}_j$ will be a filter for the net, majorizing every $\mathscr{F}_j$. Applying Zorn's lemma (1.1.3) we can therefore find a maximal filter $\mathscr{F}$ for $(x_\lambda)_{\lambda \in \Lambda}$.

Fix a subset Y of X. If for some λ in Λ and E, F in $\mathscr{F}$ we had both $x_\mu \notin E \cap Y$

and $x_\mu \notin F \setminus Y$ for every $\mu \geq \lambda$, then the same would hold with E and F replaced by the smaller set $E \cap F$ in $\mathscr{F}$. But

$$E \cap F = (E \cap F \cap Y) \cup (E \cap F \setminus Y),$$

and the net is frequently in $E \cap F$, a contradiction. Thus, the net is either frequently in $E \cap Y$ for every E in $\mathscr{F}$ or frequently in $F \setminus Y$ for every F in $\mathscr{F}$. In the first case we conclude that the system

$$\mathscr{F}' = \{F | F \supset E \cap Y, E \in \mathscr{F}\}$$

is a filter for the net; and since $\mathscr{F} \subset \mathscr{F}'$ and $\mathscr{F}$ is maximal, this means that $\mathscr{F} = \mathscr{F}'$, i.e. $Y \in \mathscr{F}$. In the second case we conclude analogously that $X \setminus Y \in \mathscr{F}$.

Applying 1.3.4. with $\mathscr{B}$ replaced by the maximal filter $\mathscr{F}$, we obtain a subnet of $(x_\lambda)_{\lambda \in \Lambda}$, which is universal, since for every $Y \subset X$ we have either $Y \in \mathscr{F}$ or $X \setminus Y \in \mathscr{F}$. $\qquad\square$

1.3.9. In a topological space (X, τ) satisfying the first axiom of countability (1.2.13), sequences may replace nets in almost all cases (the exception being 1.3.8). Thus, for every accumulation point x of a sequence $(x_n)_{n \in \mathbb{N}}$, there is a subsequence converging to x. Indeed, if $\{A_n | n \in \mathbb{N}\}$ is a basis for $\mathcal{O}(x)$, we may assume that $A_n \supset A_{n+1}$ for all n. By induction we can then find a subsequence $(x_{n(k)})_{k \in \mathbb{N}}$, such that $x_{n(k)} \in A_k$ for every k. But then $x_{n(k)} \to x$, as desired. Similarly, the statements in 1.3.6, 1.4.3, 1.5.2, and 1.6.2 have equivalent formulations with sequences instead of nets, under the assumption that the topological spaces mentioned are first countable.

1.3.10. Remark. It follows from 1.3.6 that a topology is determined by the family of convergent nets on the space. In principle, convergence is therefore an alternative way to describe topological phenomena (cf. your high school curriculum or freshman calculus course). One may say that a description in terms of open sets gives a static view of the problem, whereas convergence arguments yield a more dynamic description. Which one to choose often depends on the nature of the problem, so keep both in mind.

EXERCISES

E 1.3.1. Let $\mathfrak{F}$ denote the set of real-valued functions on some fixed set X. For each finite subset $E = \{x_1, \ldots, x_n\}$ of X, each $\varepsilon > 0$ and f in $\mathfrak{F}$ set

$$A(f, E, \varepsilon) = \{g \in \mathfrak{F} | |g(x_k) - f(x_k)| < \varepsilon, x_k \in E\}.$$

Supply the details for the fact that these sets are a neighborhood basis for a topology τ on $\mathfrak{F}$, called the *topology of pointwise convergence*. Consider a net $(f_\lambda)_{\lambda \in \Lambda}$ in $\mathfrak{F}$ that converges to f and convince yourself that the topology is well named.

E 1.3.2. With the notation as in E 1.1.8, consider the set $X_\Omega = X \cup \{\Omega\}$ equipped with the order topology defined in E 1.2.6. Show that Ω is in the closure of X, but that no sequence in X converges to Ω.
Hint: A countable union of countable sets is countable.

E 1.3.3. Take a bounded interval $[a, b]$ in $\mathbb{R}$ and consider the net Λ of finite sub-sets $\lambda = \{x_0, x_1, \ldots, x_n\}$ of $\mathbb{R}$ such that $a = x_0 < x_1 < \cdots < x_n = b$, ordered by inclusion. For each bounded function f on $[a, b]$ and λ in Λ we define the four numbers

$$(Sf)_k = \sup\{f(x) | x_{k-1} \le x \le x_k\}; \quad (If)_k = \inf\{f(x) | x_{k-1} \le x \le x_k\};$$

$$\sum{}^*_\lambda f = \sum_{k=1}^n (Sf)_k(x_k - x_{k-1}); \quad \sum{}_{*\lambda} f = \sum_{k=1}^n (If)_k(x_k - x_{k-1}).$$

Show that the two nets $(\sum^*_\lambda f)_{\lambda \in \Lambda}$ and $(\sum_{*\lambda} f)_{\lambda \in \Lambda}$ both converges in $\mathbb{R}$. Recall from your calculus course what it means that the nets have the same limit. Now realize that you have been using net convergence for a long time without you (and your teacher?) noticing!

E 1.3.4. A *filter* in a set X is a system $\mathscr{F}$ of nonempty subsets of X satisfying the conditions:

(i) $A \cap B \in \mathscr{F}$ for all A and B in $\mathscr{F}$.
(ii) If $A \subset B$ and $A \in \mathscr{F}$, then $B \in \mathscr{F}$.

If τ is a topology on X we say that the filter *converges* to a point x in X if $\mathcal{O}(x) \subset \mathscr{F}$. Show that a subset Y of X is open iff $Y \in \mathscr{F}$ for every filter $\mathscr{F}$ that converges to a point in Y. Show that if $\mathscr{F}$ and $\mathscr{G}$ are filters and $\mathscr{F}$ is a *subfilter* of $\mathscr{G}$ (i.e. $\mathscr{F} \subset \mathscr{G}$), then $\mathscr{G}$ converges to every convergence point for $\mathscr{F}$.

E 1.3.5. An *ultrafilter* in a set X is a filter (cf. E 1.3.4) that is not properly contained in any other filter. Show that this is the case iff for every subset Y of X we have either $Y \in \mathscr{F}$ or $X \backslash Y \in \mathscr{F}$. Show that every filter is contained in an ultrafilter.
Hint: Zornication.

E 1.3.6. Let $(x_\lambda)_{\lambda \in \Lambda}$ be a net in a set X. Show that the system $\mathscr{F}$ of subsets A of X, such that the net is in A eventually, is a filter (E 1.3.4). Show that the net converges to a point x in X iff the filter converges to x.

E 1.3.7. Let $\mathscr{F}$ be a filter in a set X (E 1.3.4) and let Λ be the set of pairs (x, A) in $X \times \mathscr{F}$ such that $x \in A$. Show that the definition $(x, A) \le (y, B)$ if $B \subset A$ gives an upward filtering preorder on Λ. Thus the map $i: \Lambda \to X$ given by $i(x, A) = x$ gives a net (Λ, i), alias $(x_\lambda)_{\lambda \in \Lambda}$. Show that the filter $\mathscr{F}$ converges to a point x in X iff the net converges to x.

E 1.3.8. Net-men and filter-fans often discuss the merits of the two means of expressing convergence. Having solved E 1.3.4–7 you are entitled to join the discussion.

1.4. Continuity

Synopsis. Continuous functions. Open maps and homeomorphisms. Initial topology. Product topology. Final topology. Quotient topology. Exercises.

1.4.1. Let (X, τ) and (Y, σ) be topological spaces. A function $f\colon X \to Y$ is said to be *continuous* if $f^{-1}(A) \in \tau$ for every A in σ. It is said to be *continuous at a point* x in X if $f^{-1}(A) \in \mathcal{O}(x)$ for every A in $\mathcal{O}(f(x))$.

1.4.2. Proposition. *A function is continuous iff it is continuous at every point.*

PROOF. If $f\colon X \to Y$ is continuous and $A \in \mathcal{O}(f(x))$ for some x in X, choose $B \subset A$ in $\mathcal{O}(f(x)) \cap \sigma$. Then $f^{-1}(B) \in \tau$ and $f^{-1}(B) \subset f^{-1}(A)$, whence $f^{-1}(A) \in \mathcal{O}(x)$.

If, conversely, f is continuous at every point and $A \in \sigma$, take x in $f^{-1}(A)$. Thus $A \in \mathcal{O}(f(x))$, whence $f^{-1}(A) \in \mathcal{O}(x)$; so that $f^{-1}(A)$ is a neighborhood of every point it contains and, consequently, is open. $\qquad\square$

1.4.3. Proposition. *For a function f between topological spaces (X, τ) and (Y, σ), and x in X, the following conditions are equivalent:*

(i) *f is continuous at x.*
(ii) *For each A in $\mathcal{O}(f(x))$ there is a B in $\mathcal{O}(x)$ such that $f(B) \subset A$.*
(iii) *For each net $(x_\lambda)_{\lambda \in \Lambda}$ such that $x_\lambda \to x$ we have $f(x_\lambda) \to f(x)$.*

PROOF. (i) $\Rightarrow$ (ii). If $A \in \mathcal{O}(f(x))$, set $B = f^{-1}(A)$. Then $B \in \mathcal{O}(x)$ by assumption and $f(B) \subset A$.

(ii) $\Rightarrow$ (iii). If $x_\lambda \to x$ and $A \in \mathcal{O}(f(x))$, choose by assumption a B in $\mathcal{O}(x)$ such that $f(B) \subset A$. Since the net $(x_\lambda)_{\lambda \in \Lambda}$ is eventually in B, it follows that the net $(f(x_\lambda))_{\lambda \in \Lambda}$ is eventually in A. This shows that $f(x_\lambda) \to f(x)$.

(iii) $\Rightarrow$ (i). If $A \in \mathcal{O}(f(x))$ and $f^{-1}(A) \notin \mathcal{O}(x)$, then $x \notin (f^{-1}(A))^\circ$, i.e.

$$x \in X \backslash (f^{-1}(A))^\circ = (X \backslash f^{-1}(A))^-.$$

By 1.3.6 there is then a net $(x_\lambda)_{\lambda \in \Lambda}$ in $X \backslash f^{-1}(A)$ converging to x. But since $f(x_\lambda) \notin A$ for all λ, we cannot have $f(x_\lambda) \to f(x)$. $\qquad\square$

1.4.4. A function $f\colon X \to Y$ between topological spaces is *open* if $f(A)$ is open in Y for every open subset A of X. In contrast to counter images for continuous functions, an open map will not necessarily take closed sets to closed sets.

A *homeomorphism* is a bijective function $f\colon X \to Y$ that is both open and continuous. Equivalently, both f and f^{-1} are continuous functions. Spaces that are homeomorphic are topologically indistinguishable. (A *topologist* is a person who cannot tell the difference between a doughnut and a coffee cup [Kelley]).

It is clear from the definitions that compositions of two continuous or open functions again produce a function of the same type.

1.4.5. Let X be a set and $\mathscr{F}$ be a family of functions $f: X \to Y_f$. If each Y_f has a topology τ_f, there is a weakest topology on X that makes all the functions in $\mathscr{F}$ continuous. A subbasis for this topology is evidently given by the system

$$\{f^{-1}(A) | A \in \tau_f, f \in \mathscr{F}\}.$$

We call it the *initial topology* induced by $\mathscr{F}$.

Note that when $\mathscr{F}$ consists of a single function $f: X \to Y$, the initial topology is simply the sets $f^{-1}(A)$, $A \in \tau_f$.

1.4.6. Proposition. *Let X have the initial topology induced by a family $\mathscr{F}$ of functions. A net $(x_\lambda)_{\lambda \in \Lambda}$ is then convergent to a point x in X iff $(f(x_\lambda))_{\lambda \in \Lambda}$ converges to $f(x)$ for every f in $\mathscr{F}$.*

PROOF. If $A \in \mathcal{O}(x)$, there is by 1.2.12 a finite number of open sets $A_k \subset Y_{f_k}$, $1 \leq k \leq n$, such that

$$x \in \bigcap f_k^{-1}(A_k) \subset A.$$

In particular, $A_k \in \mathcal{O}(f_k(x))$ for every k, and by assumption there is a $\lambda(k)$ such that $f_k(x_\lambda) \in A_k$ for all $\lambda \geq \lambda(k)$. Choosing λ_0 as a majorant for all $\lambda(k)$, $1 \leq k \leq n$, we see that $x_\lambda \in \bigcap f_k^{-1}(A_k)$ for all $\lambda \geq \lambda_0$. This proves that $x_\lambda \to x$. The converse statement is immediate from 1.4.3 (iii). □

1.4.7. Corollary. *A map $g: Z \to X$ from a topological space Z to X with initial topology induced by $\mathscr{F}$, is continuous iff all functions $f \circ g: Z \to Y_f$, for f in $\mathscr{F}$, are continuous.*

PROOF. Immediate from 1.4.2, 1.4.3, and 1.4.6. □

1.4.8. Let $\{(X_j, \tau_j) | j \in J\}$ be a family of topological spaces, and consider the cartesian product $X = \prod X_j$. For each j in J we then have the projection $\pi_j: X \to X_j$ of the product space onto its jth factor. The initial topology on X induced by the projections $\{\pi_j | j \in J\}$ is called the *product topology*. A basis for this topology is given by finite intersections $\bigcap \pi_j^{-1}(A_j)$, where $A_j \in \tau_j$, cf. 1.4.5. However, these sets are product sets of the form $\prod A_j$, where $A_j \in \tau_j$ and $A_j = X_j$ for all but finitely many j in J. Evidently, the projection of each such set is open, and since any map preserves unions, it follows that the projections $\pi_j, j \in J$, are all open maps on X.

If J is a finite set, the product topology is an easy analogy from euclidean spaces $\mathbb{R}^n$: the open boxes form a basis for the topology. When J is infinite, the analogy breaks down. Intuitively, the set of all products $\prod A_j$, where $A_j \in \tau_j$, should be the basis, giving us a much stronger topology on X than the product topology. Intuition is wrong, as the ensuing theory of product spaces (notably the Tychonoff theorem, 1.6.10) will show.

1.4.9. Let Y be a set and $\mathscr{F}$ be a family of functions $f: X_f \to Y$. If each X_f has a topology τ_f, there is a strongest topology on Y that makes all the functions in $\mathscr{F}$ continuous. It is called the *final topology* induced by $\mathscr{F}$ and consists of the sets

$$\{A \subset Y | \forall f \in \mathscr{F} : f^{-1}(A) \in \tau_f\}.$$

1.4.10. Proposition. *Let Y have the final topology induced by a family $\mathscr{F}$ of functions. A function $g: Y \to Z$ of Y into a topological space Z is then continuous iff all functions $g \circ f$, $f \in \mathscr{F}$, are continuous.*

PROOF. If all functions $g \circ f$ are continuous and A is open in Z, then $f^{-1}(g^{-1}(A))$ is open in X_f. But this means that $f^{-1}(g^{-1}(A)) \in \tau_f$ for all f, whence $g^{-1}(A)$ is open in Y. $\square$

1.4.11. Let (X, τ) be a topological space and let $\sim$ be an equivalence relation on X. With $\tilde{X}$ as the space of equivalence classes and $Q: X \to \tilde{X}$ as the quotient map, we give $\tilde{X}$ the final topology induced by Q. This is called the *quotient topology*. From the definition (1.4.9) we see that points in $\tilde{X}$ are closed sets iff each equivalence class is closed in X and that Q is an open map iff the saturation of every open set A in X under equivalence, i.e. the set

$$\tilde{A} = \{x \in X | x \sim y, y \in A\},$$

is also open in X.

EXERCISES

E 1.4.1. Find a continuous function $f: \mathbb{R} \to \mathbb{R}$ that is not open.

E 1.4.2. Find an open function $f: \mathbb{R} \to \mathbb{R}$ that is not continuous.
    Hint: If $x-\text{int}(x) = 0, \alpha_1, \alpha_2, \ldots$ is the binary expansion of the fractional part of x, define $g(x) = \lim \sup(n^{-1} \sum \alpha_j | j \leq n)$. Show that $g(I) = [0, 1]$ for every interval I in $\mathbb{R}$. Take $f = h \circ g$, where h is an arbitrary surjective function from $[0, 1]$ to $\mathbb{R}$.

E 1.4.3. Let X be the subset of points (x, y) in $\mathbb{R}^2$ such that either $x = y = 0$ or $xy = 1$, and give X the relative topology. Let $f: X \to \mathbb{R}$ be the restriction to X of the projection of $\mathbb{R}^2$ to the x-axis. Is f a continuous map? Is f an open map?

E 1.4.4. Let (X, τ) be a topological space and denote by $C(X)$ the set of continuous functions from X to $\mathbb{R}$. Show that the following combinations of elements f and g in $C(X)$ again produce elements in $C(X)$: αf [if $\alpha \in \mathbb{R}$]; $|f|$; $1/f$ [if $0 \notin f(X)$]; $f + g$; fg; $f \vee g$; $f \wedge g$.

E 1.4.5. Let $\mathbb{T}$ (torus) denote the unit circle in $\mathbb{C}$, with the relative topology. Show that the product space $\mathbb{T}^2$ (the 2-torus) is homeomorphic to a closed subset of $\mathbb{R}^3$. Consider now the map $f_\theta: \mathbb{R} \to \mathbb{T}^2$ given by $f_\theta(x) = (\exp ix, \exp i\theta x)$ for some θ in $\mathbb{R}$. Prove that f_θ is continuous, and find a condition on θ that makes f_θ injective.

E 1.4.6. Let X, Y, and Z be topological spaces and give $X \times Y$ the product topology. Show that if a function $f: X \times Y \to Z$ is continuous, then it is separately continuous in each variable [i.e., for each x in X the function $y \to f(x, y)$ is continuous from Y, to Z, and similarly for each y in Y]. Show by an example that the converse does not hold. *Hint*: Try $f(x, y) = xy(x^2 + y^2)^{-1}$ if $(x, y) \neq (0, 0)$ and $f(0, 0) = 0$.

E 1.4.7. Given a set X, show that the product space $\mathbb{R}^X$, endowed with the product topology, is homeomorphic to the space $\mathfrak{F}$ described in E 1.3.1.

E 1.4.8. Let (X, τ) and (Y, σ) be topological spaces and consider the product space $(X \times Y, \tau \times \sigma)$. Show that if $A \subset X$ and $B \subset Y$, then

$$(A \times B)^- = A^- \times B^- \quad \text{and} \quad (A \times B)^o = A^o \times B^o.$$

E 1.4.9. Let $I = [0, 1]$ be regarded both as a topological space and as an index set, and consider the product space $X = I^I$ with the product topology. Note from E 1.3.1 and E 1.4.7 that the elements in X can be "visualized" as functions $f: I \to I$. Show that the elements f in X for which the function $f: I \to I$ is continuous is a dense subset of X. Deduce from this that X is a separable space. Now consider the subset Y of X consisting of functions f for which $f(y) = 0$ for all y in I except a single point x (depending on f) for which $f(x) = 1$. Show that Y is a discrete set in the relative topology and is non-separable. Show that $Y^- \setminus Y$ consists of a single point ω in X, and that every neighborhood of ω must contain all but finitely many points from Y.

E 1.4.10. On $\mathbb{R}$ we consider the equivalence relation $\sim$ given by $x \sim y$ if $x - y \in \mathbb{Z}$. Describe the quotient space and the quotient topology. See what happens if $\mathbb{Z}$ is replaced by $\mathbb{Q}$.

E 1.4.11. (Topological direct sum.) Let (X_1, τ_1) and (X_2, τ_2) be topological spaces and let X denote the disjoint union of X_1 and X_2. Find the topology τ on X that contains X_1 and X_2 and for which the relative topology on X_j is τ_j for $j = 1, 2$. Show that τ is the final topology corresponding to the embedding maps $\iota_j: X_j \to X$ for $j = 1, 2$.

E 1.4.12. (Inductive limits.) Let (X_n, τ_n) be a sequence of topological spaces and assume that there is a continuous injective map $f_n: X_n \to X_{n+1}$ for every n. Identifying every X_n with a subset of X_{n+1} (equipped with the relative topology if f_n is a homeomorphism on its image), we form $X = \bigcup X_n$, and give it the final topology induced by the maps $f_n: X_n \to X$.

Take $X_n = \mathbb{R}^n$ with the natural embeddings $f_n: \mathbb{R}^n \to \mathbb{R}^n \times \mathbb{R}$, so that X (as a set) can be identified with the subset of points $X = (x_n)$ in $\mathbb{R}^{\mathbb{N}}$, for which $x_n = 0$ for all but finitely many n. Show that the inductive limit topology on X is stronger than the restriction of the product topology on $\mathbb{R}^{\mathbb{N}}$ to X.

Hint: Show that the intersection of X with the cube $(]0, 1[)^{\mathbb{N}}$ is open in the inductive limit topology on X.

E 1.4.13. (Connected spaces.) A topological space (X, τ) is *connected* if it cannot be decomposed as a union of two nonempty disjoint open sets. A subset of X is *clopen* if it is both open and closed. Show that X is connected iff $\emptyset$ and X are the only clopen subsets. Let $f: X \to Y$ be a surjective continuous map between topological spaces. Show that Y is connected if X is. Show that every interval in $\mathbb{R}$ is connected in the relative topology.

E 1.4.14. (Arcwise connected spaces.) A topological space (X, τ) is *arcwise connected* if for every pair x, y in X there is a continuous function $f: [0, 1] \to X$ such that $f(0) = x$ and $f(1) = y$. Geometrically speaking, $f([0, 1])$ is the curve or arc that joins x to y. Show that an arcwise connected space is connected (E 1.4.13). Show that Y is arcwise connected if it is the continuous image of an arcwise connected space (cf. E 1.4.13).

E 1.4.15. Show that $\mathbb{R}$ is not homeomorphic to $\mathbb{R}^2$.
   Hint: $\mathbb{R}^2 \backslash \{x_1, x_2\}$ is a connected space, but $\mathbb{R} \backslash \{x\}$ is disconnected.

E 1.4.16. Let E and F be closed sets in a topological space X, such that $E \cup F = X$. Show that if both X and $E \cap F$ are (arcwise) connected, then E and F are also (arcwise) connected.

E 1.4.17. Let E be a connected subset of $\mathbb{R}$. Show that E is an interval (possibly unbounded).

E 1.4.18. Show that the unit circle in $\mathbb{R}^2$ is not homeomorphic to any subset of $\mathbb{R}$.
   Hint: Use E 1.4.17.

E 1.4.19. (Homotopy.) Two continuous maps $f: X \to Y$ and $g: X \to Y$ between topological spaces X and Y are *homotopic* if there is a continuous function $F: [0, 1] \times X \to Y$ (where $[0, 1] \times X$ has the product topology), such that $F(0, x) = f(x)$ and $F(1, x) = g(x)$ for every x in X. Intuitively speaking, the homotopy F represents a continuous deformation of f into g. Show that any continuous function $f: \mathbb{R}^n \to Y$ is homotopic to a constant function, and that the same is true for any continuous function $g: X \to \mathbb{R}^n$. Show that the identity function $\iota: S^1 \to S^1$ [where $S^1 = \{(x, y) \in \mathbb{R}^2 | x^2 + y^2 = 1\}$] is not homotopic to a constant function.

E 1.4.20. (Homotopic spaces.) Two topological spaces X and Y are *homotopic* if there are continuous functions $f: X \to Y$ and $g: Y \to X$ such that $g \circ f$ is homotopic to the identity function ι_X on X and $f \circ g$ is homotopic to the identity function ι_Y on Y. Show that homotopy is an equivalence relation among topological spaces. Show that homeomorphic spaces are homotopic.

E 1.4.21. (Contractible spaces.) A topological space (X, τ) is *contractible* if it is homotopic to a point (E 1.4.20). Show that X is contractible iff the identity map ι_X is homotopic to a constant map. Show that every convex subset of $\mathbb{R}^n$ is contractible.

E 1.4.22. (The fundamental group.) Let (X, τ) be a nonempty arcwise connected (cf. E 1.4.14) topological space, and choose a base point x_0 in X. A *loop* in X is a continuous function (curve) $f: [0, 1] \to X$ such that $f(0) = f(1) = x_0$. On the space $L(X)$ of loops we define a composition fg (product) by

$$fg(t) = g(2t), \quad 0 \le t \le \tfrac{1}{2}; \qquad fg(t) = f(2t - 1), \quad \tfrac{1}{2} \le t \le 1,$$

for f and g in $L(X)$. We define homotopy of loops, written $f \sim g$, if there is a continuous function $F: [0, 1] \times [0, 1] \to X$ such that $F(s, 0) = F(s, 1) = x_0$ for every s and $F(0, t) = f(t)$, $F(1, t) = g(t)$ for every t. Show that the set $\pi(X)$ of equivalence classes (under homotopy) of loops is a group under the product $\pi(f)\pi(g) = \pi(fg)$, where $\pi: L(X) \to \pi(X)$ is the quotient map.

Hint: If F is a homotopy between the loops f_1 and f_2, and G is a homotopy between the loops g_1 and g_2, set

$$H(s, t) = F(s, 2t) \qquad \text{for} \quad 0 \le s \le 1, \quad 0 \le t \le \tfrac{1}{2};$$

$$H(s, t) = G(s, 2t - 1) \qquad \text{for} \quad 0 \le s \le 1, \quad \tfrac{1}{2} \le t \le 1;$$

and check that H is a homotopy between $f_1 g_1$ and $f_2 g_2$. The product in $\pi(X)$ is therefore well-defined. If $f \in L(X)$, define f^{-1} in $L(X)$ by $f^{-1}(t) = f(1 - t)$ and check that $f^{-1} f \sim e$, where $e(t) = x_0$ for all t. The relevant homotopy is

$$F(s, t) = f(2st) \qquad \text{for} \quad 0 \le s \le 1, \quad 0 \le t \le \tfrac{1}{2};$$

$$F(s, t) = f(2s(1 - t)) \qquad \text{for} \quad 0 \le s \le 1, \quad \tfrac{1}{2} \le t \le 1.$$

Similarly $ff^{-1} \sim e$, $fe \sim ef \sim f$, so that $\pi(e)$ is the identity in $\pi(X)$. Given f, g, h in $L(X)$ we have

$$f(gh)(t) = \begin{cases} h(4t) & \text{for} \quad 0 \le t \le \tfrac{1}{4} \\ g(4t - 1) & \text{for} \quad \tfrac{1}{4} \le t \le \tfrac{1}{2} \\ f(2t - 1) & \text{for} \quad \tfrac{1}{2} \le t \le 1 \end{cases}$$

$$(fg)h(t) = \begin{cases} h(2t) & \text{for} \quad 0 \le t \le \tfrac{1}{2} \\ g(4t - 2) & \text{for} \quad \tfrac{1}{2} \le t \le \tfrac{3}{4} \\ f(4t - 3) & \text{for} \quad \tfrac{3}{4} \le t \le 1. \end{cases}$$

To show that $f(gh) \sim (fg)h$, use the homotopy

$$F(s, t) = \begin{cases} h(4t(1 + s)^{-1}) & \text{for} \quad 4t - 1 \le s \\ g(4t - s - 1) & \text{for} \quad 4t - 2 \le s \le 4t - 1 \\ f((4t - s - 2)(2 - s)^{-1}) & \text{for} \quad s \le 4t - 2. \end{cases}$$

E 1.4.23. Show that S^2—the unit sphere in $\mathbb{R}^3$—is *simply connected* [i.e. $\pi(S^2)$ $= \{0\}$, cf. E 1.4.22] but not contractible.

E 1.4.24. Show that homotopic spaces (cf. E 1.4.20) have isomorphic fundamental groups (cf. E 1.4.22).

E 1.4.25. Show that if $f: X \to Y$ is a continuous map between arcwise connected topological spaces, and $f(x_0) = y_0$ (where x_0 and y_0 are basis points for the loop spaces on X and Y, respectively), then there is a natural group homomorphism $f_*: \pi(X) \to \pi(Y)$ between the fundamental groups (cf. E 1.4.22). Show that if $g: Y \to Z$ is a similar map, then $(g \circ f)_* = g_* \circ f_*$.

1.5. Separation

Synopsis. Hausdorff spaces. Normal spaces. Urysohn's lemma. Tietze's extension theorem. Semicontinuity. Exercises.

1.5.1. Convergence of a net in a topological space (X, τ) is not a very restrictive notion if τ has only a few sets. In a sufficiently weak topology, e.g. the trivial topology (1.2.9), a net may even converge to several points. This is undesirable (because it does not really happen very often in our applications of general topology), so one usually expects a topology to satisfy Hausdorff's separation axiom (from 1914): If $x \neq y$, there are disjoint sets A in $\mathcal{O}(x)$ and B in $\mathcal{O}(y)$. In this case we say that (X, τ) is a *Hausdorff space*.

Note that points are closed sets in a Hausdorff space (but the condition that points are closed does not imply that the space is Hausdorff). Note further that any subset of a Hausdorff space is itself a Hausdorff space in the relative topology (1.2.8).

1.5.2. Proposition. *A topological space (X, τ) is a Hausdorff space iff each net converges to at most one point.*

PROOF. If $(x_\lambda)_{\lambda \in \Lambda}$ converges to x in a Hausdorff space (X, τ), and $y \neq x$, choose disjoint neighborhoods A and B of x and y, respectively. Then (x_λ) is eventually in A, thus not in B, so (x_λ) does not converge to y.

Conversely, if each net has at most one convergence point, and $x \neq y$, consider the index set $\Lambda = \mathcal{O}(x) \times \mathcal{O}(y)$ with the product order. If for any A, B in $\mathcal{O}(x) \times \mathcal{O}(y)$ we could find $x_{A,B}$ in $A \cap B$, we would have a net $(x_{A,B})_{A,B \in \Lambda}$ converging to both x and y. This being prohibited, there must be pairs A, B with $A \cap B = \emptyset$. $\square$

1.5.3. Proposition. *Let X have the initial topology induced by a family $\mathscr{F}$ of functions that separates points in X [i.e. if $x \neq y$ there is an f in $\mathscr{F}$ with $f(x) \neq f(y)$]. If all spaces Y_f, $f \in \mathscr{F}$, are Hausdorff spaces, then X is a Hausdorff space.*

PROOF. If $x \neq y$ in X, choose f in $\mathscr{F}$ with $f(x) \neq f(y)$. Since Y_f is a Hausdorff space there are disjoint open neighborhoods A and B of $f(x)$ and $f(y)$, respectively. But then $f^{-1}(A)$ and $f^{-1}(B)$ are disjoint open neighborhoods of x and y, so that X is a Hausdorff space.   □

1.5.4. Corollary. *The product topology on a product of Hausdorff spaces is a Hausdorff topology.*

1.5.5. The Hausdorff separation axiom turns out to be a minimal demand of a decent topological space. To ascertain the existence of an ample supply of continuous real functions on the space, we need a more severe separation condition. We say that the Hausdorff space (X, τ) is *normal*, if for each disjoint pair E, F of closed sets in X there are disjoint open sets A, B such that $E \subset A$ and $F \subset B$ (Warning: some authors do not include the Hausdorff axiom in the definition of normality.) Note that normality is equivalent with the demand that for each closed set F and each open set B, such that $F \subset B$, there is an open set A with $F \subset A$ and $A^- \subset B$ (set $B = X \backslash E$).

We can now prove *Urysohn's lemma*:

1.5.6. Theorem. *In a normal topological space (X, τ) there is for each pair E, F of disjoint closed sets a continuous function $f: X \to [0, 1]$ that is 0 on E and 1 on F.*

PROOF. Set $A_1 = X \backslash F$. Then use normality to find an open set $A_{1/2}$, with $E \subset A_{1/2}$ and $A_{1/2}^- \subset A_1$. The normality condition applied to E and $A_{1/2}$ gives an open set $A_{1/4}$ with $E \subset A_{1/4}$ and $A_{1/4}^- \subset A_{1/2}$. Similarly, we obtain an open set $A_{3/4}$ with $A_{1/2}^- \subset A_{3/4}$ and $A_{3/4}^- \subset A_1$. Continuing this process by induction we obtain for each binary fraction $r = m2^{-n}$, $1 \leq m \leq 2^n$, an open set A_r containing E, such that $A_r^- \subset A_s$ whenever $r < s$.

We now define $f: X \to [0, 1]$ by letting $f(x) = 1$ if $x \notin \bigcup A_r (= A_1)$ and otherwise

$$f(x) = \inf\{r \,|\, x \in A_r\}.$$

Clearly f is 1 on F and 0 on E, so it remains to show that f is continuous.

If $0 < t \leq 1$, then $f(x) < t$ iff $x \in A_r$ for some $r < t$; whence

$$f^{-1}([0, t[) = \bigcup_{r < t} A_r,$$

which is an open set in X. If $0 \leq s < 1$, then $f(x) \leq s$ iff for each $r > s$ there is a binary fraction $p < r$ such that $x \in A_p$. Thus

$$f^{-1}([0, s]) = \bigcap_{r > s} \bigcup_{p < r} A_p = \bigcap_{q > s} A_q^-.$$

where the last equation (rather: the nontrivial inclusion $\supset$) follows from the fact that if $x \in A_q^-$ for every $q > s$, and if $r > s$ is given, we can find p and q with $r > p > q > s$, whence $x \in A_q^- \subset A_p$. Consequently, $f^{-1}(]s, 1])$ is an open set in X (since its complement was closed). The system of intervals $[0, t[$ and $]s, 1]$, where $0 \le s < t \le 1$, is a subbasis for the usual topology on $[0, 1]$ (cf. 1.2.11). It follows from the above that $f^{-1}(A)$ is open in X for every open set A in $[0, 1]$, so that f is continuous. $\square$

1.5.7. Remark. A topology τ on X induced by a metric d (cf. 1.2.2) is normal. In this case it is even quite simple to prove Urysohn's lemma directly (from which normality follows, since $E \subset f^{-1}([0, \frac{1}{3}[)$ and $F \subset f^{-1}(]\frac{2}{3}, 1]))$. Define $d(E, x) = \inf\{d(y, x) | y \in E\}$, and check that this is a continuous function of x, whose null set is precisely E^-. If now E and F are disjoint, closed sets in X, the function

$$f(x) = d(E, x)(d(E, x) + d(F, x))^{-1}$$

satisfies the requirements of the lemma.

In contrast to this, the next result, the *Tietze extension theorem*, is interesting also for metric spaces. Note, though, that in the setting of normal spaces Urysohn's result is The lemma that leads to Tietze's theorem. (However, Urysohn proved it as a step toward the metrization theorem, 1.6.14.)

1.5.8. Proposition. *In a normal topological space (X, τ), each bounded, continuous function $f: F \to \mathbb{R}$ on a closed subset F on X has an extension to a bounded, continuous function on all of X.*

PROOF. Without loss of generality we may assume that $f(F) \subset [-1, 1]$. By Urysohn's lemma (1.5.6) there is a continuous function $f_1: X \to [-\frac{1}{3}, \frac{1}{3}]$ that is $-\frac{1}{3}$ on $f^{-1}([-1, -\frac{1}{3}])$ and $\frac{1}{3}$ on $f^{-1}([\frac{1}{3}, 1])$. In the remaining part of F the values of both f and f_1 lie in $[-\frac{1}{3}, \frac{1}{3}]$, so $|f(x) - f_1(x)| \le \frac{2}{3}$ for every x in F. Repeating the argument with $f - f_1 | F$ in place of f yields a continuous function $f_2: X \to [-\frac{1}{3} \cdot \frac{2}{3}, \frac{1}{3} \cdot \frac{2}{3}]$, such that

$$|f(x) - f_1(x) - f_2(x)| \le (\tfrac{2}{3})^2$$

for every x in F. Continuing by induction we find a sequence (f_n) of continuous functions on X, with $|f_n(x)| \le (\frac{1}{3})(\frac{2}{3})^{n-1}$ for every x in X and

$$\left| f(x) - \sum_{k=1}^{n} f_k(x) \right| \le (\tfrac{2}{3})^n$$

for every x in F.

The function $\tilde{f} = \sum f_n$ satisfies

$$|\tilde{f}(x)| \le \sum_{1}^{\infty} (\tfrac{1}{3})(\tfrac{2}{3})^{n-1} = 1$$

for every x in X, and $\tilde{f}|F = f$. That $\tilde{f}$ is continuous follows from the next, familiar looking, result. $\square$

1.5.9. Proposition. *Let $C(X, Y)$ denote the space of continuous functions from a topological space (X, τ) to a complete, metric space (Y, d). Equipped with* *the uniform metric*

$$d_\infty(f,g) = \sup\{d(f(x), g(x)) \wedge 1 \,|\, x \in X\},$$

$C(X, Y)$ *is a complete, metric space.*

PROOF. Let (f_n) be a Cauchy sequence in the metric space $(C(X, Y), d_\infty)$. Then $(f_n(x))$ is a Cauchy sequence in the complete space (Y, d) for each x in X, so that we have a function $f: X \to Y$ defined by $f(x) = \lim f_n(x)$. Furthermore, we have, by definition of d_∞, for each $\varepsilon > 0$ an N such that $d_\infty(f, f_n) < \varepsilon$ whenever $n > N$. If therefore $(x_\lambda)_{\lambda \in \Lambda}$ is a net in X converging to some x, then

$$d(f(x_\lambda), f(x)) \le 2d_\infty(f_n, f) + d(f_n(x_\lambda), f_n(x))$$

$$\le 2\varepsilon + d(f_n(x_\lambda), f_n(x))$$

for $n > N$. Since f_n is continuous, we have $d(f(x_\lambda), f(x)) < \varepsilon$, eventually, which shows that f is continuous at x (cf. 1.4.3), whence $f \in C(X, Y)$ by 1.4.2. $\qquad\square$

1.5.10. A real function f on a topological space (X, τ) is *lower semicontinuous* if $f^{-1}(]t, \infty[) \in \tau$ for every real t. Analogously f is *upper semicontinuous* if $f^{-1}(]-\infty, t[) \in \tau$ for every t. Note that if f is both lower and upper semicontinuous, then

$$f^{-1}(]s, t[) = f^{-1}(]s, \infty[) \cap f^{-1}(]-\infty, t[) \in \tau$$

for all s, t; and since the open intervals form a basis for the topology on $\mathbb{R}$ (cf.1.2.11), we see that f is continuous. Note further that a function f is upper semicontinuous iff $-f$ is lower semicontinuous, so that we may restrict our attention to the class $C^{1/2}(X)$ of lower semicontinuous functions on X. This class is intimately connected with τ. If namely $A \subset X$, and if $[A]$ denotes the *characteristic function* for A {i.e. $[A](x) = 1$ if $x \in A$ and $[A](x) = 0$ if $x \notin A\}$, then $[A]^{-1}(]t, \infty[)$ will either be X (if $t < 0$), A (if $0 \le t < 1$), or $\emptyset$ (if $t \ge 1$). We conclude that $[A] \in C^{1/2}(X)$ iff $A \in \tau$.

1.5.11. Proposition. *A real function f on a topological space (X, τ) is lower semicontinuous iff for each convergent net $(x_\lambda)_{\lambda \in \Lambda}$ in X we have*

$$f(\lim x_\lambda) \le \liminf f(x_\lambda).$$

PROOF. If $f \in C^{1/2}(X)$ and $x_\lambda \to x$, then the net belongs eventually to the open neighborhood $f^{-1}(]t, \infty[)$ of x for every $t < f(x)$. Consequently, $\liminf f(x_\lambda) \ge t$, whence $\liminf f(x_\lambda) \ge f(x)$.

To prove the converse, consider the set $F = f^{-1}(]-\infty, t])$ for some real t. If $x \in F^-$, there is by 1.3.6 a net $(x_\lambda)_{\lambda \in \Lambda}$ in F that converges to x. By assumption we have

$$f(x) \le \liminf f(x_\lambda) \le t,$$

whence $x \in F$. Thus F is closed and the complement $f^{-1}(]t, \infty[)$, consequently, is open, so that $f \in C^{1/2}(X)$. $\qquad\square$

1.5.12. Proposition. *A* (*pointwise*) *supremum of any number of elements in* $C^{1/2}(X)$ *and an infimum of finitely many elements will again define an element in* $C^{1/2}(X)$. *Furthermore,* $C^{1/2}(X)$ *is stable under addition and under multiplication with positive real numbers. Finally,* $C^{1/2}(X)$ *is closed under uniform convergence.*

PROOF. If $\{f_j | j \in J\} \subset C^{1/2}(X)$ and if $\bigvee f_j$ defines a real function f (a restriction we could avoid by considering also functions with values in $[-\infty, \infty]$), then for each real t

$$f^{-1}(]t, \infty[) = \bigcup f_j^{-1}(]t, \infty[) \in \tau,$$

whence $f \in C^{1/2}(X)$. Similarly, we have (when J is finite) for $g = \bigwedge f_j$ that

$$g^{-1}(]t, \infty[) = \bigcap f_j^{-1}(]t, \infty[) \in \tau,$$

so that also $g \in C^{1/2}(X)$.

If f and g belong to $C^{1/2}(X)$, then $f(x) + g(x) > t$ iff $f(x) > s$ and $g(x) > t - s$ for some real s. Consequently,

$$(f + g)^{-1}(]t, \infty[) = \bigcup_s f^{-1}(]s, \infty[) \cap g^{-1}(]t - s, \infty[) \in \tau,$$

whence $f + g \in C^{1/2}(X)$. For $s > 0$ we have

$$(sf)^{-1}(]t, \infty[) = f^{-1}(]s^{-1}t, \infty[) \in \tau,$$

so that also $sf \in C^{1/2}(X)$.

Consider, finally, a sequence (f_n) in $C^{1/2}(X)$, which is a Cauchy sequence in the metric d_∞ defined in 1.5.9. Since $\mathbb{R}$ is complete, we find as in the proof of 1.5.9 a function f such that

$$\delta_n = \sup_x |f_n(x) - f(x)| \to 0.$$

It follows that for every convergent net $(x_\lambda)_{\lambda \in \Lambda}$ in X we have

$$f(\lim x_\lambda) \le \delta_n + f_n(\lim x_\lambda)$$

$$\le \delta_n + \liminf f_n(x_\lambda) \le 2\delta_n + \liminf f(x_\lambda).$$

Since $\delta_n \to 0$, we see from 1.5.11 that $f \in C^{1/2}(X)$. $\qquad\square$

1.5.13. Proposition. *In a topological space* (X, τ) *for which the continuous functions from* X *to* $[0, 1]$ *separate points and closed sets, in particular, in any normal space, each lower semicontinuous function* $f: X \to \mathbb{R}_+$ *is the* (*pointwise*) *supremum of continuous real functions.*

PROOF. If $f \in C^{1/2}(X)$ and $f \ge 0$, we let $M(f)$ denote the set of continuous functions g on X dominated by f. For each x in X and $\varepsilon > 0$ the set

$$F = f^{-1}(]-\infty, f(x) - \varepsilon])$$

is closed and disjoint from $\{x\}$. By assumption there is then a continuous function $g: X \to [0,1]$ such that $g|F = 0$ and $g(x) = 1$. It follows that $(f(x) - \varepsilon)g \in M(f)$. Consequently, with

$$\tilde{f} = \bigvee g, \qquad g \in M(f),$$

we see that $\tilde{f}(x) \geq f(x) - \varepsilon$. Since x and ε were arbitrary, $\tilde{f} = f$, as desired.

$\square$

EXERCISES

E 1.5.1. Let (X, τ) be a topological space and consider X^2 with the product topology. Show that X is a Hausdorff space iff the diagonal

$$\Delta = \{(x, y) \in X^2 | x = y\}$$

is a closed subset of X^2.

E 1.5.2. Let $f: X \to Y$ be a continuous map between topological spaces X and Y, and consider the *graph* of f:

$$\mathfrak{G}(f) = \{(x, y) \in X \times Y | y = f(x)\}.$$

Show that $\mathfrak{G}(f)$ is closed in $X \times Y$ if Y is a Hausdorff space.

E 1.5.3. Let $f: X \to \mathbb{R}$ be a function on a topological space X and consider the *hypergraph* of f

$$\mathfrak{G}^+(f) = \{(x, t) \in X \times \mathbb{R} | f(x) \leq t\}.$$

Show that f is lower semicontinuous on X (cf. 1.5.10) iff $\mathfrak{G}^+(f)$ is a closed subset of $X \times \mathbb{R}$.

E 1.5.4. Show that a real function f on a topological space X is lower semicontinuous iff $f: X \to \mathbb{R}$ is continuous, when $\mathbb{R}$ is given the topology from E 1.2.5.

E 1.5.5. Let f and g be continuous functions between topological spaces X and Y, where Y is a Hausdorff space. Show that the set

$$\{x \in X | f(x) = g(x)\}$$

is closed in X. Show that if f and g are equal on a dense subset of X, then $f = g$.

E 1.5.6. A *topological group* is a group G equipped with a topology τ such that the map $(x, y) \to x^{-1}y$ from $G \times G$ (with the product topology) into G is continuous. Show in this case that the maps $x \to x^{-1}$, $x \to xy$, and $x \to yx$ are homeomorphisms of G onto itself for every fixed y in G. Show that $\mathcal{O}(x) = x\mathcal{O}(e) = \mathcal{O}(e)x$ for every x in G, where e denotes the unit of G. Show that $\mathcal{O}(e)$ has a basis consisting of symmetric sets ($A^{-1} = A$). Show that a (group) homomorphism $\pi: G \to H$ between topological groups G and H is continuous iff π is continuous at e. Show that a topological group G is a Hausdorff space if for every x in G with $x \neq e$ there is either an A in $\mathcal{O}(x)$ such that $e \notin A$ or a B in $\mathcal{O}(e)$ such that $x \notin B$. Show that if $A \in \tau$, then $AB \in \tau$ and $BA \in \tau$ for every subset B of G. Show that every open subgroup of G is closed.

E 1.5.7. (*Bing's irrational slope space.*) Let $\mathbb{Q}$ denote the rational numbers and set $X = \{(x, y) \in \mathbb{Q}^2 | y \geq 0\}$. Choose an irrational number θ and for every $\varepsilon > 0$ set

$$A_\varepsilon^+(x, y) = \{z \in \mathbb{Q} | |z - (x - \theta^{-1}y)| < \varepsilon\};$$
$$A_\varepsilon^-(x, y) = \{z \in \mathbb{Q} | |z - (x + \theta^{-1}y)| < \varepsilon\}.$$

Put $B_\varepsilon(x, y) = \{(x, y)\} \cup A_\varepsilon^+(x, y) \cup A_\varepsilon^-(x, y)$, where the A-sets are regarded as subsets of X lying on the x-axis. Geometrically speaking, the sets $A_\varepsilon^\pm$ are intervals whose midpoints, joined to (x, y), give lines with slopes $\pm\theta$. Let $B_\varepsilon(x, y)$, where $(x, y) \in X$, be the basis for a topology τ on X. Show that (X, τ) is a Hausdorff space. Show that the closure of a neighborhood $B_\varepsilon(x, y)$ has the form of a W [the strokes of the letter having width $2(1 + \theta^2)^{-1/2}\theta\varepsilon$]. Use this to show that X is connected and that X is not normal (hence not metrizable). Show that every continuous real function on X is constant.

E 1.5.8. A topological Hausdorff space (X, τ) is called *regular* if for each closed set F in X and every x in $X \backslash F$ there are disjoint open sets A and B such that $x \in A$ and $F \subset B$. Show that a regular Lindelöf space (cf. E 1.2.10) is normal.
  Hint: Use the proof of 1.7.8.

E 1.5.9. A subset Y of a topological space (X, τ) is an F_σ-set, if $Y = \bigcup F_n$, where (F_n) is a sequence of closed subsets of X. Show that every F_σ-set in a normal space is normal (in the relative topology).
  Hint: Assume first that G is an open F_σ-set of the form $G = \bigcup G_n$, where each G_n is open and $G_n^- \subset G_{n+1}$. Let $f: C \to [0, 1]$ be a continuous function on a relatively closed subset C of G. Construct inductively continuous functions f_n on $(G_n^- \cap G) \cup G_{n-1}^-$ such that $f_n | G_{n-1}^- = f_{n-1} | G_{n-1}^-$ and $f_n | G_n^- \cap C = f | G_n^- \cap C$. Then (f_n) defines a unique, continuous extension of f, which proves that G is normal.
  In the general case $Y = \bigcup F_n$, let E and F be closed subsets of X such that $(E \cap Y) \cap (F \cap Y) = \emptyset$. Thus, $H = X \backslash (E \cap F)$ is open and $Y \subset H$. Construct inductively open subsets G_n such that $F_n \bigcup G_n^- \subset G_{n+1} \subset H$. Then $G = \bigcup G_n$ is open and normal, so $E \cap G$ and $F \cap G$ can be separated by open, disjoints subsets A and B in G. Consequently, $A \cap Y$ and $B \cap Y$ are relatively open subsets of Y that separate $E \cap Y$ and $F \cap Y$.

E 1.5.10. (*Completely normal spaces.*) Show that the following conditions on a topological space (X, τ) are equivalent:

(i) Every subset of X (equipped with the relative topology) is a normal space.
(ii) Every open subset of X is a normal space.
(iii) If Y and Z are subsets of X such that $Y^- \cap Z = Y \cap Z^- = \emptyset$, they can be separated by open, disjoint sets in X.

Hints: For (ii) $\Rightarrow$ (iii), consider the open, hence normal, subspace $G = X \setminus (Y^- \cap Z^-)$ and separate $Y^- \cap G$ and $Z^- \cap G$ with open, disjoint sets A and B. Show that, in fact, $Y \subset A$ and $Z \subset B$.

For (iii) $\Rightarrow$ (i) consider a pair Y, Z of relatively closed, disjoint subsets of an arbitrary subset H of X. Show that $Y^- \cap Z = Z \cap Y^- = \emptyset$.

1.6. Compactness

Synopsis. Equivalent conditions for compactness. Normality of compact Hausdorff spaces. Images of compact sets. Tychonoff's theorem. Compact subsets of $\mathbb{R}^n$. The Tychonoff cube and metrization. Exercises.

1.6.1. An *open covering* of a subset Y of a topological space (X, τ) is a subset σ of τ such that $Y \subset \bigcup A$, $A \in \sigma$. A *subcovering* of σ is a covering σ_0 that is contained in σ.

1.6.2. Theorem. *The following conditions on a topological space (X, τ) are equivalent*:

(i) *Every open covering of X has a finite subcovering.*
(ii) *If Δ is a system of closed subsets of X, such that no intersection of finitely many elements from Δ is empty, then the intersection of all elements in Δ is nonempty.*
(iii) *Every net in X has an accumulation point.*
(iv) *Every universal net in X is convergent.*
(v) *Every net in X has a convergent subnet.*

PROOF. (i) $\Rightarrow$ (ii). If $\bigcap F = \emptyset$, when F ranges over Δ, the system $\{X \setminus F | F \in \Delta\}$ is an open covering of X. By (i) there is then a subcovering $\{X \setminus F_j | 1 \leq j \leq n\}$, which implies that $\bigcap F_j = \emptyset$, in contradiction with (ii).

(ii) $\Rightarrow$ (iii). If $(x_\lambda)_{\lambda \in \Lambda}$ is a net in X, put

$$F_\lambda = \{x_\mu | \lambda \leq \mu\}^-.$$

Given $\lambda_1, \ldots, \lambda_n$ in Λ there is a common majorant λ, from which we conclude that $F_\lambda \subset F_{\lambda_k}$ for all k. In particular, $\bigcap F_{\lambda_k} \neq \emptyset$. It follows from (ii) that there is an x in $\bigcap F_\lambda$, $\lambda \in \Lambda$. By 1.2.6 this implies that for every A in $\mathcal{O}(x)$ and every λ in Λ there are elements x_μ in A with $\mu \geq \lambda$. Thus the net is frequently in A (1.3.3), which means that x is an accumulation point.

(iii) $\Rightarrow$ (iv) because a universal net converges to each of its accumulation points. Furthermore, (iv) $\Rightarrow$ (v) by 1.3.8 and (v) $\Rightarrow$ (iii) is evident from 1.3.3.

(iii) $\Rightarrow$ (i). Let σ be an open covering of X. Ordered by inclusion, the finite subsets λ of σ have an upward filtering ordering. If no λ covers X, we can, using the axiom of choice (1.1.3), find a net (x_λ) such that

$$x_\lambda \in X \backslash \bigcup_{A \in \lambda} A = \bigcap_{A \in \lambda} X \backslash A$$

for every λ. By (iii) this net has an accumulation point x in X. For any given A in σ and B in $\mathcal{O}(x)$ there is therefore a λ such that $\{A\} \leq \lambda$ (i.e. $A \in \lambda$) and $x_\lambda \in B$. In particular, $(X \backslash A) \cap B \neq \emptyset$. Since $X \backslash A$ is closed and B is arbitrary, we conclude that $x \in X \backslash A$. Since this holds for every A in σ, and σ is a covering, we have reached a contradiction. Thus, some finite subset λ of σ covers X, as desired. $\qquad\square$

1.6.3. The equivalent conditions in 1.6.2 constitute the definition of a *compact* topological space. Some authors (notably a deceased French general) include the Hausdorff separation axiom in the term compactness. They call our compact spaces quasicompact (The words bicompact and full compact are also used.) Their plausible explanation is that the shorter form should be reserved for the spaces that appear most frequently. Unfortunately, there *are* compact spaces that fail to be Hausdorff spaces (the primitive ideal space for a unital ring, equipped with the Jacobson topology). Moreover, it may facilitate the learning process to keep the axioms apart. We shall therefore stick to our definition of compactness.

A subset C of a topological space X is *compact* if it is a compact topological space in the relative topology (1.2.8). This means precisely that every open covering of C (1.6.1) has a finite subcovering. The term *relatively compact* is reserved for subsets Y of X that are contained in some compact subset of X. The covering theorem of E. Borel states that every bounded subset of $\mathbb{R}^n$ is relatively compact. For good measure we include a proof of this wellknown result in 1.6.12.

Note from 1.6.2(ii) that every closed subset of a compact set is compact. By contrast, a compact subset of a topological space need not be closed, but, as we shall see, the Hausdorff property remedies this defect.

1.6.4. Lemma. *If C is a compact subset of a Hausdorff topological space (X, τ), there are for each $x \notin C$ disjoint open subsets A and B of X, such that $C \subset A$ and $x \in B$.*

PROOF. Since X is a Hausdorff space, there are for each y in C disjoint open subsets $A(y)$ and $B(y)$, such that $y \in A(y)$ and $x \in B(y)$. The family $\{A(y) \,|\, y \in C\}$ is an open covering of C and has therefore a finite subcovering $A(y_1), A(y_2), \ldots, A(y_n)$. Set $A = \bigcup A(y_k)$ and $B = \bigcap B(y_k)$, and check that A and B are disjoint, open sets with $C \subset A$ and $x \in B$. $\qquad\square$

1.6.5. Proposition. *Each compact subset C of a Hausdorff topological space (X, τ) is closed.*

PROOF. If $x \notin C$, there is by 1.6.4 a B in $\mathcal{O}(x)$ disjoint from C. Thus $x \notin C^-$ by 1.2.6, whence $C = C^-$. ☐

1.6.6. Theorem. *Every compact Hausdorff space is normal.*

PROOF. If E and F are disjoint closed subset of the compact space X, they are themselves compact [cf. 1.6.2(ii)]. By 1.6.4 we can therefore for each x in F find disjoint, open sets $A(x)$, $B(x)$, such that $E \subset A(x)$ and $x \in B(x)$. The family $\{B(x) | x \in F\}$ is an open covering of F and therefore has a finite subcovering $B(x_1), B(x_2), \ldots, B(x_n)$. Set $B = \bigcup B(x_k)$ and $A = \bigcap A(x_k)$ and check that A and B are disjoint, open sets with $E \subset A$ and $F \subset B$. ☐

1.6.7. Proposition. *If $f: X \to Y$ is a continuous function between topological spaces X and Y, and if X is compact, then $f(X)$ is compact.*

PROOF. If σ is an open covering of $f(X)$, the system $\{f^{-1}(A) | A \in \sigma\}$ will be an open covering of X. Since X is compact, there are sets $A_1, \ldots, A_n$ in σ such that $X = \bigcup f^{-1}(A_k)$, whence $f(X) = \bigcup f \circ f^{-1}(A_k) \subset \bigcup A_k$. ☐

1.6.8. Proposition. *Let $f: X \to Y$ be a continuous, injective map from a compact space X into a Hausdorff space Y. Then f is a homeomorphism of X onto $f(X)$ in its relative topology.*

PROOF. Every closed subset F of X is compact, whence $f(F)$ is compact in Y by 1.6.7. Since Y is a Hausdorff space, this implies that $f(F)$ is closed by 1.6.5. Since f is injective, we conclude from this that $f(A)$ is relatively open in $f(X)$ for every open subset A of X, so that $f: X \to f(X)$ is a homeomorphism. ☐

1.6.9. A compact Hausdorff topology τ on X is rigid: There is no weaker topology σ on X that still makes it a Hausdorff space [apply 1.6.8 to the identical map of (X, τ) onto (X, σ)], and there is no stronger topology σ on X in which it is still compact [apply 1.6.8 to the identical map of (X, σ) onto (X, τ)].

1.6.10. Theorem. *If $\{X_j | j \in J\}$ is a family of compact topological spaces, then the product space $\prod X_j$, equipped with the product topology, is compact.*

PROOF. Let $(x_\lambda)_{\lambda \in \Lambda}$ be a universal net in $\prod X_j$. If π_j denotes the projection on X_j, the image net $(\pi_j(x_\lambda))_{\lambda \in \Lambda}$ is universal in X_j and consequently convergent to a point x_j by 1.6.2(iv). Let x denote the point in $\prod X_j$ for which $\pi_j(x) = x_j$ for all j in J. Since $\pi_j(x_\lambda) \to \pi_j(x)$ for every j, we conclude from 1.4.6 that $x_\lambda \to x$. Thus each universal net is convergent, whence $\prod X_j$ is compact by 1.6.2(iv). ☐

1.6.11. Remark. The fundamental result above is *Tychonoff's theorem*. The deceptively short proof we have presented is based on the solid preparations in 1.4.6 and 1.6.2. The proof of 1.6.2 and thus also the proof of Tychonoff's theorem uses the axiom of choice in a crucial manner. Indeed, Kelley has shown that the axiom of choice can be derived from Tychonoff's theorem (when the latter is assumed to hold for arbitrarily large index sets and all compact topological spaces).

1.6.12. Proposition. *Every closed, bounded subset C of $\mathbb{R}^n$ is compact.*

PROOF. Since C is bounded, we can find closed intervals $I_1, I_2, \ldots, I_n$ such that C is contained in the n-cube $\prod I_k$ as a closed subset. By Tychonoff's theorem it therefore suffices to show that each closed interval $I = [a, b]$ in $\mathbb{R}$ is compact.

Let σ be an open covering of I. Consider the supremum s of all those numbers t, with $a \le t \le b + 1$, such that the interval $[a, t]$ is covered by some finite subcovering of σ. If $s \le b$, there is some A_0 in σ such that $s \in A_0$. Since A_0 is open, it follows that $[s - \varepsilon, s + \varepsilon] \subset A_0$ for some $\varepsilon > 0$. By definition of s there is a finite subset $\{A_1, A_2, \ldots, A_n\}$ of σ that covers the interval $[a, s - \varepsilon]$. But then the interval $[a, s + \varepsilon]$ is covered by the sets $A_0, A_1, \ldots, A_n$ contradicting the definition of s. Consequently, $s > b$, which means that $[a, b]$ has a finite subcovering of σ, as desired. $\quad\square$

1.6.13. Our last result in this section provides a relatively concrete model for a large class of nice topological spaces. At the same time it answers the rather intriguing question: Which separable topological spaces are metrizable?

The *Tychonoff cube* is the space $\mathsf{T} = [0, 1]^{\mathbb{N}}$ obtained as a product of countably many copies of the unit interval $[0, 1]$. Equipped with the product topology, T is a compact Hausdorff space by 1.5.4, 1.6.10, and 1.6.12. For $x = \{x_n\}$ and $y = \{y_n\}$ in T define

$$d(x, y) = \sum 2^{-n}|x_n - y_n|.$$

It is easily verified that d is a metric on T. Given x in T and $\varepsilon > 0$, set $A = \{y \in \mathsf{T} \mid d(x, y) < \varepsilon\}$. Choosing m such that $2^{-m} < \frac{1}{2}\varepsilon$ we see that the set

$$B = \bigcap_{n=1}^{m} \{y \in \mathsf{T} \mid |y_n - x_n| < \tfrac{1}{2}\varepsilon\}$$

is contained in A. Since the A-sets form a basis for the topology $\tau(d)$ induced by the metric d (1.2.2), whereas the B-sets are a basis for the product topology τ (1.4.8), we see that $\tau(d) \subset \tau$. Since $\tau(d)$ is a Hausdorff topology, it follows from 1.6.9 that $\tau(d) = \tau$. Consequently, the Tychonoff cube and all its subsets are metrizable topological spaces.

1.6.14. Proposition. *Each second countable, normal space is homeomorphic to a subset of the Tychonoff cube, and, consequently, is metrizable. If the space is compact, the subset is closed.*

PROOF. Let $\{A_n | n \in \mathbb{N}\}$ be a basis for the topology on the space X. If $A_n^- \subset A_m$, then by Urysohn's lemma there is a continuous function $f: X \to [0, 1]$ that is 0 on A_n and 1 on $X \backslash A_m$. The collection of pairs (A_n, A_m) such that $A_n^- \subset A_m$ is clearly countable, and gives rise, therefore, to a sequence (f_k) of functions as described above.

We now define a map $\iota: X \to \mathsf{T}$ (where $\mathsf{T} = [0, 1]^{\mathbb{N}}$) by $\iota(x) = \{f_k(x)\}_{k \in \mathbb{N}}$. If F is a closed subset in X and $x \notin F$, there is a set A_m such that $x \in A_m \subset X \backslash F$. Thus, $\{x\} \subset A_m$ and from the normality of X we see that there is some A_n such that $x \in A_n$ and $A_n^- \subset A_m$, cf. 1.5.5. Consequently, one of the functions f_k will be 0 at x and 1 on F. If $y \in X$ with $y \neq x$, we set $F = \{y\}$ and have $f_k(y) \neq f_k(x)$. This proves that the map ι is injective. If A is open in X and $x \in A$, we set $F = X \backslash A$ and find an f_k that is 0 at x and 1 on F. Then

$$\iota(x) \in \{y \in \mathsf{T} | y_k < \tfrac{1}{2}\} \cap \iota(X) \subset \iota(A).$$

Since x is arbitrary, this proves that $\iota(A)$ is relatively open in $\iota(X)$. Clearly $\iota: X \to \mathsf{T}$ is continuous (1.4.7), and we just showed that it was a relatively open map and therefore a homeomorphism between X and $\iota(X)$.

If X is compact, $\iota(X)$ is closed in T by 1.6.7 and 1.6.5. $\square$

EXERCISES

E 1.6.1. Let $f: X \to Y$ be a function between compact Hausdorff spaces X and Y, such that the graph of f is closed in $X \times Y$ (see E 1.5.2). Show that f is continuous.

E 1.6.2. Let $f: X \to \mathbb{R}$ be a continuous function on a compact space X. Show that f is bounded. Show that f attains its maximum and its minimum on X. Assume only that f is semicontinuous on X and prove "half" of the two previous assertions.

E 1.6.3. Let C and D be compact subsets of topological spaces X and Y, respectively. Show that if G is an open subset of $X \times Y$ containing $C \times D$, there are open sets A and B in X and Y, respectively, such that $C \subset A, D \subset B$, and $A \times B \subset G$.

E 1.6.4. Let C and F be closed subsets of a metric space (X, d). Show that if C is compact, there is an x_0 in C such that

$$\inf\{d(x, y) | x \in C, y \in F\} = \inf\{d(x_0, y) | y \in F\}.$$

E 1.6.5. Let C be a compact subset of a metric space (X, d). Show that for every open covering σ of C there is an $\varepsilon > 0$, such that for each x in C the ball $\{y \in X | d(x, y) < \varepsilon\}$ is contained in some element from σ.

E 1.6.6. (Dini's lemma.) Let $(f_\lambda)_{\lambda \in \Lambda}$ be a net of real continuous functions on a compact space X. Assume that $\lambda \leq \mu$ implies $f_\lambda(x) \leq f_\mu(x)$ for every x in X and that there is a continuous function f on X such that $\lim f_\lambda(x) = f(x)$ for every x in X. Prove that (f_λ) converges uniformly to f, i.e. $\|f_\lambda - f\|_\infty \to 0$.

E 1.6.7. Let X be an infinite set and let τ denote the system of subsets A of X such that $X \setminus A$ is finite, together with the set $\emptyset$. Show that τ is a topology on X, and that (X, τ) is compact, but not a Hausdorff space. Show that points are closed sets in X. Show that every infinite subset of X is dense and deduce that (X, τ) is separable. Assume that X is uncountable and show that (X, τ) satisfies neither the first nor the second axiom of countability.

E 1.6.8. (*The Cantor set.*) Let C denote the set of real numbers x of the form

$$x = \sum_{n=1}^{\infty} \alpha_n 3^{-n}, \quad \text{where } \alpha_n = 0 \text{ or } \alpha_n = 2.$$

Show that $C = \bigcap C_n$, where $C_1 = [0, 1]$ and where C_{n+1} is obtained from C_n by deleting the (open) middle third of each of the intervals that belong to C_n. Deduce from this that C (in the relative topology induced from $\mathbb{R}$) is a compact Hausdorff space. Show that C as a subset of $\mathbb{R}$ has empty interior, but no isolated points. Show that every x in C has a unique expression $x = \sum \alpha_n 3^{-n}$ and that C is homeomorphic with the product space $\{0, 1\}^{\mathbb{N}}$ (and thus uncountable). Show that the map $f: C \to [0, 1]$ given by $f(x) = \sum \alpha_n 2^{-n-1}$, where $x = \sum \alpha_n 3^{-n}$, is continuous and surjective.

E 1.6.9. Consider the Cantor set C defined in E 1.6.8. Show that if x, y, and z are points in C with $y < x < z$ and $x = \frac{1}{2}(y + z)$, then, if $x = \sum \alpha_n 3^{-n}$, $y = \sum \beta_n 3^{-n}$, $z = \sum \gamma_n 3^{-n}$, there is an m such that $\alpha_n = \beta_n = \gamma_n$ for all $n < m$; then either $\alpha_m = \beta_m = 0$, $\gamma_m = 2$ whence $\alpha_n = 2$ and $\beta_n = \gamma_n = 0$ for $n > m$, or $\alpha_m = \gamma_m = 2$, $\beta_m = 0$ and $\beta_n = \gamma_n = 2$ and $\alpha_n = 0$ for $n > m$. Deduce that $\mathbb{R} \setminus C$ contains an interval, one of whose endpoints is x.

E 1.6.10. Show that for each closed subset F of the Cantor set C (cf. E 1.6.8) there is a continuous function $f: C \to F$ such that $f | F$ is the identity map.

Hint: Define $f(x)$ to be the nearest point to x in F. In case y and z in F have the same minimal distance to x we have $x = \frac{1}{2}(y + z)$ with $y < x < z$, and may define $f(x) = y$ if $]x, x + \varepsilon[\subset \mathbb{R} \setminus C$, but $f(x) = z$ if $]x - \varepsilon, x[\subset \mathbb{R} \setminus C$, using E 1.6.9.

E 1.6.11. Show that every compact metric space X is the continuous, surjective image of the Cantor set C (cf. E 1.6.8).

Hint: Identify C with $\{0, 1\}^{\mathbb{N}}$ and for $\alpha = (\alpha_n)$ in C let $1(\alpha) = \{n \in \mathbb{N} | \alpha_n = 1\}$. Let D be the open set of points α in C for which $1(\alpha)$ is finite, and for each α in $C \setminus D$ define $f(\alpha) = \bigcap A_n$, $n \in 1(\alpha)$, where (A_n) is a basis for the topology on X consisting of closed balls with radii tending to 0 as $n \to \infty$. Show that the set E of points α in $C \setminus D$ for which $f(\alpha) = \emptyset$ is open in $C \setminus D$. Take $F = C \setminus (D \cup E)$ and show that we have a surjective function $f: F \to X$. Show that

$f^{-1}(A_n) = \{\alpha \in F \,|\, \alpha_n = 1\}$ and hence that f is continuous. Now apply E 1.6.10.

E 1.6.12. Let $(X, \leq)$ be a totally ordered set equipped with the order topology from E 1.2.6. A subset Y of X is called order bounded if for some x and z in X we have $x \leq y \leq z$ for all y in Y. If every order bounded subset of X has an infimum and a supremum, we say that X is order complete. Show that X is order complete iff every closed, order-bounded subset of X is compact. Show that X is connected iff it is order complete and contains no holes [i.e. no pairs (x, z), such that $x < z$ but $\{y \,|\, x < y < z\} = \emptyset$].
  Hint: Imitate the proof of 1.6.12.

E 1.6.13. Let $X = [0, 1] \times [0, 1]$ equipped with the lexicographic order $[(x_1, y_1) \leq (x_2, y_2)$ if either $x_1 < x_2$ or if $x_1 = x_2$ and $y_1 \leq y_2]$. Show that X equipped with the order topology is a connected, compact Hausdorff space. Show that X satisfies the first axiom of countability, but that X is not separable and thus not metrizable.
  Hint: Use E 1.6.12.

E 1.6.14. Let $\mathbb{T}$ (torus) denote the unit circle in $\mathbb{C}$ regarded as a multiplicative group. Show that $\mathbb{T}$ is a compact topological group (cf. E 1.5.6). Show that $\mathbb{T}^X$ in the product topology is a compact topological group for every index set X.

E 1.6.15. Given a discrete group G, take $\mathbb{T}$ as in E 1.6.14 and let $\hom(G, \mathbb{T})$ denote the set of maps $\alpha: G \to \mathbb{T}$ such that $\alpha(g_1 g_2) = \alpha(g_1)\alpha(g_2)$ for all g_1, g_2 in G. Show that $\hom(G, \mathbb{T})$ is a compact topological group.
  Hint: Find a group isomorphism of $\hom(G, \mathbb{T})$ to a closed subgroup of $\mathbb{T}^G$.

1.7. Local Compactness

Synopsis. One-point compactification. Continuous functions vanishing at infinity. Normality of locally compact, σ-compact spaces. Paracompactness. Partition of unity. Exercises.

1.7.1. A topological space (X, τ) is *locally compact* if every point in X has a compact neighborhood. Equivalently, τ has a basis of relatively compact sets. The class of locally compact spaces is very important in analysis and geometry and contains by 1.6.12 all euclidean (and all locally euclidean) spaces. The class also contains all discrete spaces, which gives us a large but not very serious collection of examples.

  Some problems about locally compact spaces can be solved by adjoining a "point at infinity," and then using compactness arguments on the enlarged space. We first discuss this process.

1.7.2. A *compactification* of a topological space (X, τ) is a triple $(\tilde{X}, \tilde{\tau}, \iota)$, where $(\tilde{X}, \tilde{\tau})$ is a compact topological space and ι is a homeomorphism of X on a dense subset of $\tilde{X}$. Usually the embedding is understood, and we just write $X \subset \tilde{X}$. It may happen that a compactification of a Hausdorff space is not itself Hausdorff, and therefore worthless. Fortunately, the class of topological spaces that admit a Hausdorff compactification is quite large. These are the *Tychonoff spaces* or *completely regular spaces* (defined in 4.3.17), a class which contains all normal spaces. For our purposes here, the simplest compactification—the *one-point compactification*—will suffice, because it gives a Hausdorff compactification of every locally compact Hausdorff space.

1.7.3. Proposition. *Every noncompact topological space (X, τ) admits a compactification $(\tilde{X}, \tilde{\tau})$ with $\tilde{X} = X \cup \{\infty\}$. The space $\tilde{X}$ is Hausdorff iff X is a locally compact Hausdorff space.*

PROOF. Let $\tilde{X}$ denote the union of X with a single extra point, which we suggestively denote by ∞. Consider now the system τ' of subsets A of $\tilde{X}$ such that $\infty \in A$ and $\tilde{X} \backslash A$ is a closed, compact subset of X. Since every closed subset of a compact set is compact, we see that τ' is stable under formation of arbitrary unions and finite intersections. Furthermore, if $A \in \tau'$ and $B \in \tau$, then $A \cup B \in \tau'$ and $A \cap B \in \tau$, since

$$\tilde{X} \backslash (A \cup B) = (\tilde{X} \backslash A) \cap (X \backslash B); \qquad A \cap B = X \backslash ((\tilde{X} \backslash A) \cup (X \backslash B)).$$

It follows that the system $\tilde{\tau} = \tau \cup \tau'$ is a topology on $\tilde{X}$, such that the relative topology on X is τ (because $X \cap A \in \tau$ if $A \in \tau'$). Thus, the embedding of X in $\tilde{X}$ is a homeomorphism. Every neighborhood of ∞ meets X (because X is noncompact), so X is dense in $\tilde{X}$. Every open covering of $\tilde{X}$ must contain at least one set A_0 from τ'. The remaining sets give an open covering of the compact set $\tilde{X} \backslash A_0$, and therefore have a finite number $A_1, \ldots, A_n$ covering $\tilde{X} \backslash A_0$. Taken together the sets $A_0, A_1, \ldots, A_n$ form a finite subcovering of $\tilde{X}$. This proves that $\tilde{X}$ is a compactification of X.

If X is a locally compact Hausdorff space and $x \in X$, we can choose a compact neighborhood C of x. By 1.6.5 C is closed, so $\tilde{X} \backslash C \in \tau'$. It follows that C^o and $\tilde{X} \backslash C$ are disjoint, open neighborhoods of x and ∞ in $\tilde{X}$. All other pairs of points can be separated by sets in τ, and we conclude that $\tilde{X}$ is a Hausdorff space. Conversely, if $\tilde{X}$ is a Hausdorff space, then X must be locally compact (Hausdorff), because it is an open subset of $\tilde{X}$ so that the next proposition applies. □

1.7.4. Proposition. *Every open and every closed subset of a locally compact Hausdorff space is a locally compact Hausdorff space in the relative topology.*

PROOF. If A is an open subset of X and $x \in A$, we choose a compact neighborhood C of x in X. Applying 1.6.4 to x and $C \backslash A$ we obtain disjoint open sets B and D in X, such that $x \in B$ and $C \backslash A \subset D$. Then $C \backslash D$ is a compact neighborhood of x contained in A.

If F is a closed subset of X and $x \in F$, we take, as before, C to be a compact neighborhood of x in X. Then $C \cap F$ is compact and is a neighborhood of x in the relative topology on F. $\square$

1.7.5. Proposition. *Let (X, τ) be a locally compact Hausdorff space. To each compact set C and every open set A containing C there is a continuous function $f: X \to [0, 1]$, that is 1 on C and for which the set $\{x \in X \mid f(x) > 0\}^-$ is compact and contained in A.*

PROOF. Consider the compact Hausdorff space $X \cup \{\infty\}$ (1.7.3). Since C is closed in $X \cup \{\infty\}$, which is normal by 1.6.6, we can find a closed neighborhood D of ∞ disjoint from C; and we may assume that $D \supset X \cup \{\infty\} \backslash A$. Applying Urysohn's lemma (1.5.6) to the sets C and D we obtain a continuous function g on $X \cup \{\infty\}$ that is 1 on C and 0 on D. We define another continuous function f by setting $f(x) = 2(g(x) - \frac{1}{2}) \vee 0$. Then

$$\{x \in X \mid f(x) > 0\}^- = \{x \in X \mid g(x) > \tfrac{1}{2}\}^-$$
$$\subset \{x \in X \mid g(x) \geq \tfrac{1}{2}\} \subset \{x \in X \mid g(x) > 0\}$$
$$\subset X \cup \{\infty\} \backslash D \subset A.$$

Moreover, the first three of these sets are compact because g is continuous on $X \cup \{\infty\}$ and $g(\infty) = 0$. $\square$

1.7.6. We say that a function f on X has *compact support* if the set $\{x \in X \mid f(x) \neq 0\}^-$ is compact, and we denote by $C_c(X)$ the set of continuous scalar-valued functions on X with compact support. It follows from 1.7.5 that the class $C_c(X)$ is large enough to separate any compact set from any disjoint closed set.

A slightly larger class of functions will be important to us later (see 2.1.13 and 4.3.14). This is the set $C_0(X)$ of continuous functions f on X, such that for every $\varepsilon > 0$ the set

$$\{x \in X \mid |f(x)| \geq \varepsilon\}$$

is compact. Note that $f \in C_0(X)$ iff it has an extension $\tilde{f}$ in $C(X \cup \{\infty\})$ such that $\tilde{f}(\infty) = 0$. We say that the function f *vanishes at infinity*.

1.7.7. A locally compact Hausdorff space X need not be normal. This unpleasant phenomenon will not arise, however, if in addition we demand that X is σ-compact, i.e. a countable union of compact subsets. For future use we note that in a locally compact, σ-compact Hausdorff space X there is an increasing sequence (E_n) of open, relatively compact subsets, such that $E_n^- \subset E_{n+1}$ for every n, and $\bigcup E_n = X$. To see this we use 1.7.5 to observe that every compact subset of X is contained in a relatively compact, open set. Since X by definition is a union of a sequence (C_n) of compact sets, we may construct a sequence (E_n) inductively by taking E_{n+1} as a relatively compact open set containing $C_n \cup E_n^-$.

1.7.8. Proposition. *Every locally compact, σ-compact Hausdorff space is normal.*

PROOF. If E and F are disjoint, closed sets in the space X, we choose for each x in E an open neighborhood $A(x)$ in $\mathcal{O}(x)$ such that $A(x)^- \cap F = \emptyset$. For this, 1.7.5 is more than enough. Since E is σ-compact, the covering $\{A(x)|x \in E\}$ has a countable subcovering (A_n). Similarly, we can find a sequence (B_n) of open sets that covers F, such that $B_n^- \cap E = \emptyset$ for every n. Now set

$$A'_n = A_n \Big\backslash \bigcup_{k \leq n} B_k^-, \qquad B'_n = B_n \Big\backslash \bigcup_{k \leq n} A_k^-.$$

If $m \leq n$, then $A'_n \subset X \backslash B_m$ and $B'_m \subset B_m$, so that $A'_n \cap B'_m = \emptyset$. By symmetry this also holds if $n \leq m$, so that $A = \bigcup A'_n$ and $B = \bigcup B'_n$ are disjoint, open subsets of X. Since (A_n) covers E, whereas $E \cap B_k^- = \emptyset$, it follows that $E \subset A$. Similarly, $F \subset B$, and the proof is complete. $\qquad\square$

1.7.9. Corollary. *Every second countable, locally compact Hausdorff space is metrizable.*

PROOF. If X is second countable and locally compact, choose a basis (A_n) for the topology. The subsequence of relatively compact A_n's is still a basis, because X is locally compact; in particular, X is σ-compact. By 1.7.8 we see that X is normal and thus metrizable by 1.6.14. $\qquad\square$

1.7.10. A *refinement* of an open covering $\{A_j|j \in J\}$ of a topological space (X, τ) is an open covering $\{B_i|i \in I\}$, such that each B_i is contained in some A_j. A covering $\{A_j|j \in J\}$ is called *locally finite* if each x in X has a neighborhood A such that $A \cap A_j = \emptyset$ for all but finitely many j in J. A topological space is *paracompact* if every open covering has a locally finite refinement.

Paracompactness, like local compactness, is clearly a definition that attempts to localize the useful properties of compactness. The class of paracompact spaces is pleasantly large and includes, in fact, all metrizable spaces. There are therefore spaces that are paracompact without being locally compact (and vice versa). In view of 1.7.11 it is worth mentioning that every locally compact, paracompact Hausdorff space is the topological disjoint union of a family of σ-compact spaces.

1.7.11. Proposition. *Every locally compact, σ-compact Hausdorff space is paracompact.*

PROOF. Given an open covering $\{A_j|j \in J\}$ of the space X, we choose a sequence (E_n) of open, relatively compact subsets covering X, such that $E_n^- \subset E_{n+1}$ for every n, cf. 1.7.7. We write $C_n = E_n^- \backslash E_{n-1}$ (with $E_n = \emptyset$ for $n \leq 0$) and let

$$B_j = A_j \cap (E_{n+1} \backslash E_{n-2}^-), \qquad j \in J.$$

The sets B_j are open and cover C_n. Since C_n is compact, there is therefore a finite subset J_n of J such that $C_n \subset \bigcup B_j, j \in J_n$. Put $I = \bigcup J_n$ and note that the family $\{B_j | j \in I\}$ is an open covering of X and a refinement of the first covering. Given an x in X we have $x \in E_n \setminus E_{n-1}$ for (exactly) one n, so that $E_n \setminus E_{n-2}^-$ is an open neighborhood of x. Now $B_j \cap (E_n \setminus E_{n-2}^-) = \emptyset$ unless $j \in J_m$, where

$$n - 3 \leq m \leq n + 2;$$

and this holds only for a finite number of j's. $\square$

1.7.12. Proposition. *Let* (A_n) *be a locally finite, open covering of a normal space* X. *There is then a sequence* (f_n) *of continuous functions from* X *to* $[0,1]$, *such that* $f_n(x) = 0$ *for* $x \notin A_n$ *and* $\sum f_n(x) = 1$ *for every* x *in* X.

PROOF. Assume that for each natural number $k < n$ we have chosen open sets B_k such that $B_k^- \subset A_k$ and

$$\left(\bigcup_{k<n} B_k \right) \cup \left(\bigcup_{j \geq n} A_j \right) = X \qquad (*)$$

(so for $n = 1$ no choices have been made). Put

$$F_n = X \Big\backslash \left(\left(\bigcup_{k<n} B_k \right) \cup \left(\bigcup_{j>n} A_j \right) \right).$$

It follows from $(*)$ that $F_n \subset A_n$, and since F_n is closed and X is normal, there is an open set B_n such that

$$F_n \subset B_n \subset B_n^- \subset A_n,$$

cf. the remark in 1.5.5. Consequently,

$$\left(\bigcup_{k<n+1} B_k \right) \cup \left(\bigcup_{j \geq n+1} A_j \right) \supset \left(\bigcup_{k<n} B_k \right) \cup \left(\bigcup_{j>n} A_j \right) \cup F_n = X,$$

which is the condition $(*)$ for $n + 1$. By induction, we can therefore find a sequence (B_n) of open sets such that $B_n^- \subset A_n$ and $(*)$ is satisfied for every n. Since the covering (A_n) is locally finite, the "tails" in $\bigcup A_j, j \geq n$, have empty intersection, which by $(*)$ implies that (B_n) is a covering of X.

By Urysohn's lemma (1.5.6) there is for every n a continuous function $g_n: X \rightarrow [0,1]$ that is 1 on B_n and 0 on $X \setminus A_n$. Set $f = \sum g_n$ and note that f is a finite and continuous real function because each sum $\sum g_n(x)$ only contains finitely many nonzero summands (of which at least one is 1). Defining $f_n = g_n / f$ we obtain the desired sequence (f_n). $\square$

1.7.13. Remark. The construction in 1.7.12 is called *partition of unity*. Appealing to Zorn's lemma it can be performed for an arbitrary locally finite covering. Note, however, that the proof of 1.7.11 showed that in a locally compact, σ-compact Hausdorff space any open covering has a *countable* locally finite refinement.

EXERCISES

E 1.7.1. Show that the unit circle S^1 in $\mathbb{R}^2$ and the unit interval $[0, 1]$ both are (Hausdorff) compactifications of $\mathbb{R}$.

Hint: Use the fact that $\mathbb{R}$ is homeomorphic to the open interval $]0, 1[$ and (therefore also) homeomorphic to $S^1 \setminus \{(1, 0)\}$.

E 1.7.2. Show for every n in $\mathbb{N}$ that $\mathbb{R}^n$ has the closed unit ball $B_n = \{x \in \mathbb{R}^n | x_1^2 + \cdots + x_n^2 \leq 1\}$ in $\mathbb{R}^n$ as a compactification.

E 1.7.3. Show for every n in $\mathbb{N}$ that $\mathbb{R}^n$ has the unit sphere $S^n = \{x \in \mathbb{R}^{n+1} | x_1^2 + \cdots + x_{n+1}^2 = 1\}$ in $\mathbb{R}^{n+1}$ as a compactification.

E 1.7.4. (The *Bohr compactification*.) Show that there is a compact, abelian topological group $\check{\mathbb{R}}$ and a continuous, injective group homomorphism $\iota\colon \mathbb{R} \to \check{\mathbb{R}}$, such that $\iota(\mathbb{R})$ is dense in $\check{\mathbb{R}}$.

Hint: Let $\check{\mathbb{R}}$ denote $\mathbb{R}$ as a discrete group and put $\check{\mathbb{R}} = \text{hom}(\check{\mathbb{R}}, \mathbb{T})$, cf. E 1.6.15. Define $\iota(x)$ to be the homomorphism $y \to \exp(ixy)$, $y \in \check{\mathbb{R}}$ for each x in $\mathbb{R}$.

E 1.7.5. Let X be a Hausdorff topological space and Y be a dense subset. Assume that A is an open set and that $A \subset B$ for some subset B such that $B \cap Y$ is compact. Show that $A \subset Y$.

Hint: If $x \in A \setminus Y$, there is an open neighborhood A_0 of x disjoint from $B \cap Y$. But $A \cap A_0 \cap Y \neq \emptyset$, a contradiction.

E 1.7.6. Let X be a Hausdorff topological space and Y be a subset that is locally compact in the relative topology. Assume first that Y is dense in X and show that Y is open. Show in general that $Y = A \cap F$, where A is open and F is closed in X. Compare this result with 1.7.4.

Hint: Use E 1.7.5.

E 1.7.7. Show that it is impossible to find a compact Hausdorff topological group G and an injective group homomorphism $\iota\colon \mathbb{R} \to G$ such that (G, ι) is a compactification of $\mathbb{R}$. Compare this with E 1.7.4 and conclude that the word compactification there is used in a weaker sense than ours.

Hint: Use E 1.7.6 to see that $\iota(\mathbb{R})$ would be an open subgroup of G, and kill this possibility with E 1.5.6 (last line).

E 1.7.8. Fix a prime number p, and take the sets

$$A(n, \alpha) = \{m \in \mathbb{Z} | m = n + qp^\alpha, q \in \mathbb{Z}\},$$

where $n \in \mathbb{Z}$ and $\alpha \in \mathbb{N} \cup \{0\}$, to be the basis for a topology τ on $\mathbb{Z}$. Show that τ is induced by the metric d given by $d(n, m) = p^{-\alpha}$, where α is the largest number (in $\mathbb{N} \cup \{0\}$) such that p^α divides $|n - m|$. Show in particular that $A(n, \alpha) = \{m \in \mathbb{Z} | d(n, m) \leq p^{-\alpha}\}$. Show that $(\mathbb{Z}, \tau)$ has no isolated points and that the space is not locally compact.

E 1.7.9. Let X be the well-ordered set of countable ordinal numbers (see

E 1.1.8) and let Ω denote the first uncountable ordinal number. Show that X and $X \cup \{\Omega\}$, equipped with the order topology (cf. E 1.2.6), are, respectively, a locally compact and a compact Hausdorff space.

Hint: Use E 1.6.12.

E 1.7.10. Let X be as in E 1.7.9. Show that every increasing sequence in X is convergent. Show that if (x_n) and (y_n) are increasing sequences in X such that $x_n \leq y_n \leq x_{n+1}$ for all n, then the sequences have the same limit. Conclude from this that if E and F are disjoint closed subsets of X, then there is an x in X such that either $E \subset [1, x]$ or $F \subset [1, x]$. Show finally that X is a normal space.

Hint: If $E \subset [1, x]$, then E is compact, so 1.7.5 applies to E and $A = X \backslash F$.

E 1.7.11. Take X and $X \cup \{\Omega\}$ as in E 1.7.9, and show that the product space $Z = X \times (X \cup \{\Omega\})$ is a nonnormal, locally compact Hausdorff space.

Hint: Take $E = \{(x, y) \in Z \,|\, x = y\}$ and $F = X \times \{\Omega\}$. If A was an open set containing E, such that $A^- \cap F = \emptyset$, define the function $f: X \to X$ by letting $f(x) = y$, where y is the first element such that $y \geq x$ and $(x, y) \notin A$. Use E 1.7.10 to show that $Z \backslash A$ contains a sequence $(x_n, f(x_n))$ that converges to a point in E.

E 1.7.12. Show that a paracompact Hausdorff space X is *regular* in the sense of E 1.5.8.

E 1.7.13. Show that a paracompact Hausdorff space is normal.

Hint: If E and F are disjoint, closed subsets of X, use regularity to cover E with a family $\{A_j \,|\, j \in J\}$ of open sets such that $A_j^- \cap F = \emptyset$. Use paracompactness to conclude that the covering may be taken to be locally finite. Set

$$A = \bigcup A_j, \qquad B = X \backslash \bigcup A_j^-, \qquad B^o = X \backslash (\bigcup A_j^-)^-.$$

Show that $E \subset A$, $F \subset B$ and $A \cap B = \emptyset$. Use the local finiteness to conclude that $B = B^o$.

CHAPTER 2

Banach Spaces

Assuming a basic knowledge of linear algebra we now infuse extra topological concepts and arrive at the theory of topological vector spaces. Keeping to the essentials we only develop the theory of locally convex spaces, and we make these appear as seminormed spaces (with initial topology).

The central theme in this chapter is the interplay between normed spaces and their dual spaces (equipped with the w^*-topology). We first prove the basic results about normed spaces and their operators, and give some examples. We then use the category theorem to establish the open mapping theorem, the closed graph theorem, and the uniform boundedness principle for Banach spaces. Then dual spaces and weak topologies are introduced. Following modern usage we divide the Hahn–Banach theorem in two: an extension theorem (used mainly in normed spaces) and a separation theorem (used mainly in dual spaces). We conclude with applications of compactness in dual spaces, notably the theorems associated with the names Alaoglu, Krein, Milman, and Smulian.

2.1. Normed Spaces

Synopsis. Normed spaces. Bounded operators. Quotient norm. Finite-dimensional spaces. Completion. Examples. Sum and product of normed spaces. Exercises.

2.1.1. We consider a vector space $\mathscr{X}$ over the number field $\mathbb{F}$, where $\mathbb{F} = \mathbb{R}$ or $\mathbb{F} = \mathbb{C}$. A *norm* on $\mathscr{X}$ is (as you well know) a function $\|\cdot\|: \mathscr{X} \to \mathbb{R}_+$ satisfying the conditions:

(i) $\|x + y\| \le \|x\| + \|y\|$ (subadditivity).
(ii) $\|ax\| = |\alpha| \|x\|$, $\alpha \in \mathbb{F}$ (homogeneity).
(iii) $\|x\| = 0 \Rightarrow x = 0$ (faithfulness).

If only (i) and (ii) are satisfied, we say that $\|\cdot\|$ is a *seminorm*. Every normed (vector) space is a metric space under the metric $d(x, y) = \|x - y\|$. If $(\mathfrak{X}, d)$ is complete (i.e. if every Cauchy sequence is convergent), we say that the normed space $\mathfrak{X}$ is a *Banach space*.

In a normed space $\mathfrak{X}$ we denote by $\mathfrak{B}(x, r)$ the closed ball in $\mathfrak{X}$ with center x and radius r. These balls evidently form a basis for the neighborhood filter $\mathcal{O}(x)$. Since $\mathfrak{B}(x, r) = \mathfrak{B}(0, r) + x$ and, more generally, $\mathcal{O}(x) = \mathcal{O}(0) + x$, we see that the topology on $\mathfrak{X}$ has a very uniform character.

Finally, note that the norm function is continuous on $\mathfrak{X}$ since $|\|x\| - \|y\|| \le \|x - y\|$.

2.1.2. Proposition. *Let* $T: \mathfrak{X} \to \mathfrak{Y}$ *be an operator (i.e. a linear function) between normed vector spaces* $\mathfrak{X}$ *and* $\mathfrak{Y}$. *Then the following conditions are equivalent:*

(i) T *is continuous.*
(ii) T *is continuous at some point in* $\mathfrak{X}$.
(iii) T *is bounded (i.e. there is a constant* $\alpha \ge 0$
    such that $\|Tx\| \le \alpha\|x\|$ *for every* x *in* $\mathfrak{X}$).

PROOF. (i) $\Rightarrow$ (ii) is trivial; we prove (ii) $\Rightarrow$ (iii). If T is continuous at x, there is a $\delta > 0$ such that $\|x - y\| \le \delta$ implies $\|Tx - Ty\| \le 1$ for all y in $\mathfrak{X}$ [cf. 1.4.3(ii)]. But for every $z \ne 0$ in $\mathfrak{X}$ we have

$$\|(\delta\|z\|^{-1}z + x) - x\| \le \delta,$$

whence $\|T(\delta\|z\|^{-1}z)\| \le 1$ or $\|Tz\| \le \delta^{-1}\|z\|$.
    (iii) $\Rightarrow$ (i). The inequality

$$\|Ty - Tx\| = \|T(y - x)\| \le \alpha\|y - x\|$$

shows that T is continuous at x for every x in $\mathfrak{X}$, whence T is continuous by 1.4.2. $\square$

2.1.3. The set of bounded (i.e. continuous) operators between $\mathfrak{X}$ and $\mathfrak{Y}$ is denoted by $\mathbf{B}(\mathfrak{X}, \mathfrak{Y})$. It is clearly a vector space. Setting

$$\|T\| = \sup\{\|Tx\| \,|\, x \in \mathfrak{X}, \|x\| \le 1\}$$

we obtain a norm on $\mathbf{B}(\mathfrak{X}, \mathfrak{Y})$ called the *operator norm*.

If $\mathfrak{Z}$ is also a normed space and $S \in \mathbf{B}(\mathfrak{Y}, \mathfrak{Z})$, $T \in \mathbf{B}(\mathfrak{X}, \mathfrak{Y})$, then $S \circ T$ (from now on just written ST) belongs to $\mathbf{B}(\mathfrak{X}, \mathfrak{Z})$, and $\|ST\| \le \|S\| \|T\|$. (The operator norm is *submultiplicative*.) In particular, the set $\mathbf{B}(\mathfrak{X}, \mathfrak{X})$ [from now on written $\mathbf{B}(\mathfrak{X})$] is a normed algebra, i.e. an algebra equipped with a submultiplicative norm.

2.1.4. Proposition. *If $\mathfrak{X}$ and $\mathfrak{Y}$ are normed spaces and $\mathfrak{Y}$ is complete (i.e. a Banach space), then $\mathbf{B}(\mathfrak{X}, \mathfrak{Y})$ is a Banach space. In particular, $\mathbf{B}(\mathfrak{X})$ is a Banach algebra for every Banach space $\mathfrak{X}$.*

PROOF. Let (T_n) be a Cauchy sequence in $\mathbf{B}(\mathfrak{X}, \mathfrak{Y})$. For each x in $\mathfrak{X}$ we then have the Cauchy sequence $(T_n x)$ in $\mathfrak{Y}$, and thus a limit vector denoted by Tx. This defines a function $T: \mathfrak{X} \to \mathfrak{Y}$, which is clearly linear. The inequality

$$\|Tx - T_n x\| = \lim_m \|T_m x - T_n x\|$$

$$\leq \limsup \|T_m - T_n\| \, \|x\|$$

shows that T is bounded and that $T_n \to T$. □

2.1.5. Proposition. *Let $\mathfrak{X}$ be a normed vector space and $\mathfrak{Y}$ be a subspace of $\mathfrak{X}$. Denote by Q the quotient map of $\mathfrak{X}$ onto the linear space $\mathfrak{X}/\mathfrak{Y}$ of cosets $x + \mathfrak{Y}$, $x \in \mathfrak{X}$. The definition*

$$\|Qx\| = \inf\{\|x - y\| \, | \, y \in \mathfrak{Y}\}$$

gives a seminorm on $\mathfrak{X}/\mathfrak{Y}$. If $\mathfrak{Y}$ is closed in $\mathfrak{X}$, we actually have a norm; and, furthermore, if $\mathfrak{X}$ is a Banach space, $\mathfrak{X}/\mathfrak{Y}$ is a Banach space.

PROOF. Given x_1 and x_2 in $\mathfrak{X}$, there is for every $\varepsilon > 0$ elements y_1 and y_2 in $\mathfrak{Y}$ such that

$$\|Qx_1\| + \|Qx_2\| + \varepsilon \geq \|x_1 - y_1\| + \|x_2 - y_2\|$$

$$\geq \|x_1 + x_2 - (y_1 + y_2)\| \geq \|Q(x_1 + x_2)\|.$$

Since ε is arbitrary, we have shown the subadditivity on $\mathfrak{X}/\mathfrak{Y}$. The homogeneity is evident since Q is linear, so $\mathfrak{X}/\mathfrak{Y}$ is a seminormed space.

If $\|Qx\| = 0$, there is a sequence (y_n) in $\mathfrak{Y}$ such that $\|x - y_n\| \to 0$. It follows that $\mathfrak{X}/\mathfrak{Y}$ is a normed space iff $\mathfrak{Y}$ is closed in $\mathfrak{X}$.

If (z_n) is a Cauchy sequence in $\mathfrak{X}/\mathfrak{Y}$, we can find a subsequence (z_n') such that $\|z_{n+1}' - z_n'\| < 2^{-n}$ for all n. By induction we now choose x_n in $\mathfrak{X}$ such that $Qx_n = z_n'$ and $\|x_{n+1} - x_n\| < 2^{-n}$. If $\mathfrak{X}$ is a Banach space, (x_n) converges to an element x in $\mathfrak{X}$; and since Q is norm decreasing, (z_n') converges to Qx. Then (z_n) converges to Qx also, so that $\mathfrak{X}/\mathfrak{Y}$ is complete, as desired. □

2.1.6. Note that the quotient map Q takes the open unit ball of $\mathfrak{X}$ *onto* the open unit ball of $\mathfrak{X}/\mathfrak{Y}$. A similar statement about the closed unit balls holds only in special cases.

Note further that the (metric) topology on $\mathfrak{X}/\mathfrak{Y}$ is the quotient topology corresponding to the equivalence relation $\sim$, where $x_1 \sim x_2$ iff $x_1 - x_2 \in \mathfrak{Y}$; see 1.4.11.

2.1.7. Proposition. *If* $T \in \mathbf{B}(\mathfrak{X}, \mathfrak{Y})$, *where* $\mathfrak{X}$ *and* $\mathfrak{Y}$ *are normed spaces, and if* $\mathfrak{Z}$ *is a closed subspace of* $\mathfrak{X}$ *contained in* $\ker T$, *then the equation* $\tilde{T}Qx = Tx$ *defines an operator* $\tilde{T}$ *in* $\mathbf{B}(\mathfrak{X}/\mathfrak{Z}, \mathfrak{Y})$ *with* $\|\tilde{T}\| = \|T\|$.

PROOF. It follows from linear algebra that $\tilde{T}$ is an operator on $\mathfrak{X}/\mathfrak{Z}$, which is a normed space by 2.1.5. Since

$$\|\tilde{T}Qx\| = \|Tx\| = \|T(x - z)\| \le \|T\| \|x - z\|$$

for every z in $\mathfrak{Z}$, it follows from the definition of the quotient norm (2.1.5) that $\|\tilde{T}Qx\| \le \|T\| \|Qx\|$, whence $\|\tilde{T}\| \le \|T\|$. The reverse inequality $\|\tilde{T}\| \ge \|T\|$ is clear, since Q is norm decreasing. $\qquad\square$

2.1.8. Proposition. *If* $\mathfrak{X}$ *is a normed space and* $\mathfrak{Y}$ *is a closed subspace, such that both* $\mathfrak{Y}$ *and* $\mathfrak{X}/\mathfrak{Y}$ *are Banach spaces, then* $\mathfrak{X}$ *is a Banach space.*

PROOF. If (x_n) is a Cauchy sequence in $\mathfrak{X}$, then (Qx_n) is a Cauchy sequence in $\mathfrak{X}/\mathfrak{Y}$, since Q is norm decreasing. By assumption, there is therefore an x in $\mathfrak{X}$ such that $Qx_n \to Qx$. By the definition of the quotient norm, see 2.1.5, we can find y_n in $\mathfrak{Y}$ such that $\|x_n - x - y_n\| < 1/n + \|Q(x_n - x)\|$ for every n. But then

$$\|y_n - y_m\| = \|(y_n + x - x_n) - (y_m + x - x_m) + x_n - x_m\|$$
$$\le \frac{1}{n} + \|Q(x_n - x)\| + \frac{1}{m} + \|Q(x_m - x)\| + \|x_n - x_m\|,$$

from which we conclude that (y_n) is a Cauchy sequence and consequently convergent to some y in $\mathfrak{Y}$ by assumption. Finally,

$$\|x_n - (x + y)\| \le \|x_n - x - y_n\| + \|y_n - y\|,$$

which shows that $x_n \to x + y$, as desired. $\qquad\square$

2.1.9. Proposition. *Every finite-dimensional subspace* $\mathfrak{Y}$ *of a normed space* $\mathfrak{X}$ *is a Banach space and, consequently, closed in* $\mathfrak{X}$. *Moreover, if* $\dim(\mathfrak{Y}) = n$, *every linear isomorphism of* $\mathbb{F}^n$ *onto* $\mathfrak{Y}$ *is a homeomorphism.*

PROOF. Given a basis $x_1, \ldots, x_n$ for $\mathfrak{Y}$, consider the linear isomorphism $\Phi \colon \mathbb{F}^n \to \mathfrak{Y}$ given by $\Phi(\alpha) = \sum \alpha_k x_k$, where $\alpha = (\alpha_1, \ldots, \alpha_n)$. Clearly Φ is continuous, so if we set

$$S^{n-1} = \{\alpha \in \mathbb{F}^n \mid \|\alpha\| = 1\}$$

(using, say, the euclidean norm on $\mathbb{F}^n$), then $\Phi(S^{n-1})$ is compact, hence closed in $\mathfrak{Y}$ (1.6.7 & 1.6.5), because the sphere S^{n-1} is compact in $\mathbb{F}^n$. By assumption $0 \notin \Phi(S^{n-1})$, so there is an open, convex neighborhood C of 0 (e.g. an open ball of radius > 0), such that $C \cap \Phi(S^{n-1}) = \varnothing$. But then $\Phi^{-1}(C)$ is an open, convex neighborhood of 0 in $\mathbb{F}^n$, and by convexity it is contained in the open unit ball B of $\mathbb{F}^n$, because it contains 0 and is disjoint from the boundary S^{n-1} of B. It follows that $\Phi^{-1}(\varepsilon C) \subset \varepsilon B$, which shows that Φ^{-1} is continuous, so that Φ is a homeomorphism. $\square$

2.1.10. Remark. We see from 2.1.9 that operations in a normed space, which only involve a finite number of vectors, can be performed as if we were dealing with ordinary finite-dimensional vector spaces, with their ordinary topologies and norms. Note also that the proof of 2.1.9 has been phrased so that it will be valid also for finite-dimensional subspaces of locally convex topological vector spaces, cf. 2.4.

2.1.11. Proposition. *If $\mathfrak{X}$ and $\mathfrak{Y}$ are Banach spaces and $\mathfrak{X}_0$ is a dense subspace of $\mathfrak{X}$, then every operator T_0 in $\mathbf{B}(\mathfrak{X}_0, \mathfrak{Y})$ has a unique extension to an operator T in $\mathbf{B}(\mathfrak{X}, \mathfrak{Y})$, and $\|T\| = \|T_0\|$.*

PROOF. To each x in $\mathfrak{X}$ we can find a sequence (x_n) in $\mathfrak{X}_0$ converging to x. Since T_0 is bounded and $\mathfrak{Y}$ is a Banach space, the sequence $(T_0 x_n)$ converges to an element Tx in $\mathfrak{Y}$. It is easy to check that Tx is independent of the sequence (x_n) and depends linearly on x. Thus the map $x \to Tx$ is an operator that extends T_0, and, clearly, $\|T\| = \|T_0\|$. $\square$

2.1.12. Proposition. *To each normed space $\mathfrak{X}$ there is a Banach space $\tilde{\mathfrak{X}}$, uniquely determined up to isometric isomorphism, such that $\tilde{\mathfrak{X}}$ contains $\mathfrak{X}$ as a dense subspace.*

PROOF. If $\tilde{\mathfrak{X}}_1$ and $\tilde{\mathfrak{X}}_2$ are two Banach spaces for which we have isometric embeddings $T_j: \mathfrak{X} \to \tilde{\mathfrak{X}}_j$, $j = 1, 2$, of $\mathfrak{X}$ as a dense subspace, then $T_0 = T_2 T_1^{-1}$ is an isometry of $T_1(\mathfrak{X})$ onto $T_2(\mathfrak{X})$. Extending T_0 by continuity, as in 2.1.11, we obtain an isometry T of $\tilde{\mathfrak{X}}_1$ onto $\tilde{\mathfrak{X}}_2$ [because $T(\tilde{\mathfrak{X}}_1)$ is both closed and dense in $\tilde{\mathfrak{X}}_2$]. Since $TT_1 = T_2$, it follows that the two completions $(\tilde{\mathfrak{X}}_1, T_1)$ and $(\tilde{\mathfrak{X}}_2, T_2)$ are isometrically isomorphic.

The existence of a completion $\tilde{\mathfrak{X}}$ of $\mathfrak{X}$ can be established in several ways. The most pedestrian is to imitate the construction of the real numbers from the rationals, and define $\tilde{\mathfrak{X}}$ to be the space of Cauchy sequences in $\mathfrak{X}$ modulo the space of null sequences. We choose instead to let $\tilde{\mathfrak{X}}$ be the closure of $\mathfrak{X}$ embedded in its bidual (Banach) space $\mathfrak{X}^{**}$. The details are given in 2.3.7. $\square$

2.1.13. Finally, we mention a series of examples of classical Banach spaces and constructions that generate new Banach spaces from old ones.

The euclidean spaces $\mathbb{F}^n$ can be normed in many ways. Of major interest are the p-norms (for $1 \le p \le \infty$), where

$$\|x\|_p = (\sum |x_k|^p)^{1/p}, \quad \text{for } p < \infty;$$

$$\|x\|_\infty = \max |x_k|.$$

Of these, the 2-norm (the euclidean norm) is the most important; although we see from 2.1.9 and 2.1.2 that all norms on $\mathbb{F}^n$ are *equivalent*, i.e. there are constants α and β such that $\|\cdot\| \leq \alpha \||\cdot\|\|$ and $\||\cdot\|\| \leq \beta \|\cdot\|$ for any pair $\|\cdot\|$ and $\||\cdot\|\|$. This phenomenon is atypical, however. General experience confirms that the Banach spaces that occur in applications have only one "natural" norm. The result in 2.2.6 offers perhaps a partial explanation of this fact. Recall that for metric spaces (without any linear structure) we have no such uniqueness; and very often different metrics give the same topology.

2.1.14. For a locally compact Hausdorff space X we let $C_c(X)$ denote the vector space of continuous functions $f: X \to \mathbb{F}$ such that the support of f (i.e. the closure of the co-null set of f) is compact; cf. 1.7.6. On $C_c(X)$ we define the ∞-norm (the uniform norm) by

$$\|f\|_\infty = \sup\{|f(x)| \,|\, x \in X\}.$$

In general, $C_c(X)$ is not complete in the ∞-norm. It is easy to verify, however, that the vector space $C_b(X)$ of bounded, continuous functions on X is a Banach space in the ∞-norm (for every topological space X, cf. 1.5.9). To construct the completion of $C_c(X)$ it therefore suffices by 2.1.12 to find the closure of $C_c(X)$ in $C_b(X)$. Simple computations show this to be the Banach space $C_0(X)$ of continuous functions vanishing at infinity; see 1.7.6.

If X is an open or a closed subset of $\mathbb{R}^n$ (or maybe an intersection of such subsets, cf. 1.7.4), we define the p-norms on $C_c(X)$ for $1 \leq p < \infty$ with the aid of the Riemann integral:

$$\|f\|_p = \left(\int |f(x)|^p \, dx \right)^{1/p}.$$

It requires the Hölder and Minkowski inequalities (6.4.6 and 6.4.7) to see that these definitions actually define norms on $C_c(X)$ (except for the case $p = 1$, which is evident). The completions of these spaces are the Lebesgue spaces $L^p(X)$.

2.1.15. Let $\int$ denote a Radon integral on a locally compact Hausdorff space X; cf. 6.1.2. As in 6.4.5, let $\mathscr{L}^p(X)$ denote the space of $\int$-measurable functions $f: X \to \mathbb{F}$, such that

$$\int |f|^p < \infty \quad \text{for } 1 \leq p < \infty,$$

$$\text{ess sup} |f| = \inf\left\{ s \,\Big|\, \int (|f| - |f| \wedge s) = 0 \right\} < \infty,$$

for $p = \infty$. Then the definitions

$$\|f\|_p = \left(\int |f|^p \right)^{1/p} \quad \text{for } 1 \leq p < \infty;$$

$$\|f\|_\infty = \text{ess sup} |f| \quad \text{for } p = \infty$$

give seminorms on each $\mathscr{L}^p(X)$ by the Hölder and Minkowski inequalities. If $\mathscr{N}(X)$ denotes the space of null functions (i.e. functions f such that $\int |f| = 0$), the quotient spaces $L^p(X) = \mathscr{L}^p(X)/\mathscr{N}(X)$ are normed spaces in the respective p-norms; and they are actually Banach spaces. This result, the Riesz–Fischer theorem, is proved in 6.4.10. For $X \subset \mathbb{R}^n$ and $\int$ the Lebesgue integral, we recover the classical Lebesgue spaces from 2.1.14.

2.1.16. Given a family $\{\mathfrak{X}_j | j \in J\}$ of normed spaces, we consider those elements x in the cartesian product space $\prod \mathfrak{X}_j$ for which

$$\|x\|_\infty = \sup \|P_j(x)\| < \infty$$

(where P_j denotes the projection onto $\mathfrak{X}_j$). These elements evidently form a normed vector space with $\|\cdot\|_\infty$ as norm. We call it the *direct product* of the normed spaces (and note that it is not the full cartesian product, unless the set J is finite). If all the $\mathfrak{X}_j$'s are Banach spaces, their direct product is also a Banach space.

It is often more convenient to consider norms on the *algebraic direct sum* $\sum \mathfrak{X}_j$, consisting of those elements x in $\prod \mathfrak{X}_j$ for which $P_j(x) = 0$ for all but finitely many j in J. On this vector space we define p-norms for $1 \leq p \leq \infty$ by

$$\|x\|_p = (\sum \|P_j(x)\|^p)^{1/p} \quad \text{for } 1 \leq p < \infty;$$

$$\|x\|_\infty = \sup \|P_j(x)\| \quad \text{for } p = \infty.$$

That these expressions define norms is proved (for $1 < p < \infty$) by applications of the Minkowski inequality for sequences of numbers.

2.1.17. Proposition. *If $\{\mathfrak{X}_j | j \in J\}$ is a family of Banach spaces, the completion of $\sum \mathfrak{X}_j$ in p-norm ($1 \leq p < \infty$) consists of those elements x in $\prod \mathfrak{X}_j$ for which $\sum \|P_j(x)\|^p < \infty$. The completion of $\sum \mathfrak{X}_j$ in ∞-norm consists of those x in $\prod \mathfrak{X}_j$ for which the norm function $j \to \|P_j(x)\|$ belongs to $C_0(J)$, where J is considered as a discrete, topological space.*

PROOF. Let $\mathfrak{X}_p$ denote the space of elements in $\prod \mathfrak{X}_j$ that satisfies the requirements of the proposition. As in the case of $\sum \mathfrak{X}_j$ we show that $\|\cdot\|_p$ is a norm on $\mathfrak{X}_p$. Moreover, we see that $\mathfrak{X}_p$ contains $\sum \mathfrak{X}_j$ as a dense subspace. To prove that $\mathfrak{X}_p$ is a Banach space it therefore suffices to show that each Cauchy sequence (x_n) in $\sum \mathfrak{X}_j$ has a limit in $\mathfrak{X}_p$. Since each projection P_j is norm decreasing, and every $\mathfrak{X}_j$ is a Banach space, we can define $\tilde{x}_j = \lim P_j(x_n)$ in $\mathfrak{X}_j$. Moreover, for $p < \infty$ and for every finite subset J_0 of J we have

$$\sum_{J_0} \|\tilde{x}_j\|^p = \lim \sum_{J_0} \|P_j(x_n)\|^p \leq \lim \|x_n\|_p^p < \infty.$$

We conclude that $\sum \|\tilde{x}_j\|^p < \infty$, so that we can define an element $\tilde{x}$ in $\mathfrak{X}_p$ by $P_j(\tilde{x}) = \tilde{x}_j$, $j \in J$. Finally, since

$$\sum_{J_0} \|P_j(\tilde{x}) - P_j(x_n)\|^p = \lim_m \sum \|P_j(x_m - x_n)\|^p$$

$$\leq \lim_m \|x_m - x_n\|_p^p$$

for every finite subset J_0, we see that

$$\|\tilde{x} - x_n\|_p \leq \lim_m \|x_m - x_n\|_p,$$

so that $x_n \to \tilde{x}$ as desired.

For $p = \infty$, the element $\tilde{x}$ in $\prod \mathfrak{X}_j$ defined by $P_j(\tilde{x}) = \tilde{x}_j$ satisfies

$$\|P_j(\tilde{x}) - P_j(x_n)\| = \lim_m \|P_j(x_m - x_n)\|$$

$$\leq \lim_m \|x_m - x_n\|_\infty;$$

and it follows immediately that $\tilde{x} \in \mathfrak{X}_\infty$ and that $x_n \to \tilde{x}$.   □

2.1.18. The completion of $\sum \mathfrak{X}_j$ in ∞-norm is usually called the *direct sum* of the $\mathfrak{X}_j$'s. The completion in 2-norm will play a prominent role later (in 3.1.5), when we define the orthogonal sum of Hilbert spaces.

An important case of the construction in 2.1.16 arises by taking $\mathfrak{X}_j = \mathbb{F}$ for all j in J. The direct product, denoted by $\ell^\infty(J)$, is then the space of all bounded functions from J to $\mathbb{F}$. For $p < \infty$, we obtain the spaces $\ell^p(J)$, which, via the counting measure on J, can be identified with the measure-theoretic spaces $L^p(J)$ defined in 2.1.15. The direct sum is the space $c_0(J)$ of functions from J to $\mathbb{F}$ that vanish at infinity.

The most common index set is $J = \mathbb{N}$. In this case we talk about the *sequence spaces* denoted by ℓ^∞, ℓ^p $(1 \leq p < \infty)$, and c_0.

If the index set J is finite, the direct sum and product of the Banach spaces $\mathfrak{X}_j$ will be isometrically isomorphic, and all p-norms will be equivalent.

EXERCISES

E 2.1.1. Show that a normed vector space $\mathfrak{X}$ is a Banach space iff for every sequence (x_n) in $\mathfrak{X}$, such that $\sum \|x_n\| < \infty$, the series $\sum x_n$ converges in $\mathfrak{X}$.

E 2.1.2. Given normed spaces $\mathfrak{X}$ and $\mathfrak{Y}$, where $\mathfrak{Y}$ is finite-dimensional. Show that an operator $T \colon \mathfrak{X} \to \mathfrak{Y}$ is continuous iff $\ker T$ is a closed subspace of $\mathfrak{X}$.

E 2.1.3. Show that the closed unit ball in a normed space $\mathfrak{X}$ is compact iff $\mathfrak{X}$ is finite-dimensional.

E 2.1.4. Let $\mathfrak{Y}$ and $\mathfrak{Z}$ be closed subspaces of a Banach space $\mathfrak{X}$. Show that $\mathfrak{Y} + \mathfrak{Z}$ is closed if $\mathfrak{Z}$ is finite-dimensional.

   Hint: Consider the quotient map $Q \colon \mathfrak{X} \to \mathfrak{X}/\mathfrak{Y}$ and apply 2.1.9 to $Q(\mathfrak{Z})$. Note that $\mathfrak{Z} + \mathfrak{Y} = Q^{-1}(Q(\mathfrak{Z}))$.

E 2.1.5. Let ℓ^∞ denote the space of bounded sequences equipped with the ∞-norm, cf. 2.1.18. Show that ℓ^∞ is a Banach space and is nonseparable.

E 2.1.6. Consider the subspaces c_0 and c of ℓ^∞ (see E 2.1.5) defined by $(x_n) \in c_0$ if $x_n \to 0$ and $(x_n) \in c$ if $x_n \to x_\infty$ for some x_∞ in $\mathbb{F}$. Show that both c_0 and c are closed subspaces of ℓ^∞ and that both are separable. Define e in ℓ^∞ by $e_n = 1$ for all n, and show that $c = \mathbb{F}e + c_0$.

E 2.1.7. On the space ℓ^∞ (see E 2.1.5 and E 2.1.6) define

$$\|x\|_Q = \lim \sup |x_n|, \qquad x = (x_n) \in \ell^\infty.$$

Show that $\|\cdot\|_Q$ is a seminorm on ℓ^∞ and that its null space is precisely c_0. Show that $\|x\|_Q = \|Qx\|$, for every x in ℓ^∞, where Q denotes the quotient map onto ℓ^∞/c_0, and this space is equipped with the quotient norm.

E 2.1.8. Let Ω be an open subset of $\mathbb{C}$ and let $H_b(\Omega)$ denote the bounded complex functions that are holomorphic in Ω. Show that $H_b(\Omega)$ is a Banach space under the ∞-norm.

E 2.1.9. Let $C^n(I)$ denote the space of n times continuously differentiable functions on the interval $I = [a, b]$. Show that $C^n(I)$ is a Banach space under the norm

$$\|f\| = \sum_{k=0}^{n} \|f^{(k)}\|_\infty,$$

where $f^{(k)}$ as usual denotes the kth derivative of the function f in $C^n(I)$.

E 2.1.10. (Lipschitz functions of order α.) For $0 < \alpha \leq 1$ let $\mathrm{lip}^\alpha(I)$ denote the space of functions f on the interval $I = [a, b]$, such that

$$L(f) = \sup\{|f(x) - f(y)| |x - y|^{-\alpha} | (x, y) \in I^2, x \neq y\} < \infty.$$

Show that $\mathrm{lip}^\alpha(I)$ is a Banach space under the norm $\|f\| = L(f) + |f(a)|$. Show that this norm is equivalent with the norm $\||f\|| = L(f) + \|f\|_\infty$.

E 2.1.11. Let $\mathfrak{X}$ and $\mathfrak{Y}$ be normed spaces and consider families of vectors $\{x_j | j \in J\}$ in $\mathfrak{X}$ and $\{y_j | j \in J\}$ in $\mathfrak{Y}$, with the same index set J, such that the x_j's span a dense subset of $\mathfrak{X}$. Show that there is an operator T in $\mathbf{B}(\mathfrak{X}, \mathfrak{Y})$ such that $Tx_j = y_j$ for every j in J iff there is a constant α such that for every finite subset λ of J we have

$$\left\| \sum_{j \in \lambda} \alpha_j y_j \right\| \leq \alpha \left\| \sum_{j \in \lambda} \alpha_j x_j \right\|$$

for all choices of scalars α_j, $j \in \lambda$. Show in this case that T is uniquely determined.

E 2.1.12. Let $\mathfrak{X}$ be a real Banach space. Show that the space $\mathfrak{Y} = \mathfrak{X} \times \mathfrak{X}$ is a complex Banach space with the operations

$$(x_1, x_2) + (y_1, y_2) = (x_1 + y_1, x_2 + y_2),$$

$$(\alpha + i\beta)(x_1, x_2) = (\alpha x_1 - \beta x_2, \alpha x_2 + \beta x_1),$$

$$\|(x_1, x_2)\| = \sup_{\theta} \|x_1 \cos \theta + x_2 \sin \theta\|.$$

Show that the set $\mathfrak{Y}_r = \mathfrak{X} \times \{0\}$ is a real subspace of $\mathfrak{Y}$ such that $\mathfrak{Y} = \mathfrak{Y}_r + i\mathfrak{Y}_r$, and that the map $x \to (x, 0)$ is an isometry of $\mathfrak{X}$ onto $\mathfrak{Y}_r$.

E 2.1.13. Let $\mathfrak{X}$ be a real Banach algebra with unit element 1. Define the linear operations in $\mathfrak{Y} = \mathfrak{X} \times \mathfrak{X}$ as in E 2.1.12, an define the product by

$$(x_1, x_2)(y_1, y_2) = (x_1 y_1 - x_2 y_2, x_1 y_2 + x_2 y_1).$$

Show that $\mathfrak{Y}$ is a complex Banach algebra under the "operator norm"

$$\||(x_1, x_2)\|| = \sup\{\|(x_1, x_2)(y_1, y_2)\| \, \|(y_1, y_2)\|^{-1} | (y_1, y_2) \in \mathfrak{Y}\},$$

where $\|\cdot\|$ is the norm on $\mathfrak{Y}$ defined in E 2.1.12. Show that the map $x \to (x, 0)$ is an isometry of $\mathfrak{X}$ into $\mathfrak{Y}$, and a real homomorphism.

2.2. Category

Synopsis. The Baire category theorem. The open mapping theorem. The closed graph theorem. The principle of uniform boundedness. Exercises.

2.2.1. Three fundamental results on operators in Banach spaces (2.2.4, 2.2.7, and 2.2.9) are based on Baire's category theorem for complete metric spaces (2.2.2). The name category theorem derives from Baire's nomenclature that a set is of *first category* if it can be written as a countable union of closed sets with empty interior. All other sets are of the *second category*. The content of Baire's theorem is then that every first category set has empty interior. We prefer not to use the terminology in the sequel.

2.2.2. Proposition. *If (A_n) is a sequence of open, dense subsets of a complete metric space (X, d), then the intersection $\bigcap A_n$ is dense in X.*

PROOF. Let B_0 be a closed ball in X with radius $r > 0$. Since A_1 is dense in X and open, the set $A_1 \cap B_0^o$ contains a closed ball B_1 with radius $< 2^{-1}r$. Since A_2 is also dense and open, $A_2 \cap B_1^o$ contains a closed ball B_2 with radius $< 4^{-1}r$. By induction we find a sequence (B_n) of closed balls in X, such that $B_n \subset A_n \cap B_{n-1}^o$ and radius$(B_n) < 2^{-n}r$ for every n. Since X is complete, there is evidently a point x in X such that

$$\{x\} = \bigcap B_n \subset B_0 \cap (\bigcap A_n).$$

This shows that $\bigcap A_n$ intersects every nontrivial ball in X, ensuring its density.   □

2.2.3. Lemma. *If $T \in \mathbf{B}(\mathfrak{X}, \mathfrak{Y})$, where $\mathfrak{X}$ and $\mathfrak{Y}$ are Banach spaces, and the image of the unit ball $\mathfrak{B}(0,1)$ in $\mathfrak{X}$ is dense in some ball $\mathfrak{B}(0,r)$ in $\mathfrak{Y}$ with $r > 0$, then*

$$\mathfrak{B}(0,(1-\varepsilon)r) \subset T(\mathfrak{B}(0,1))$$

for every $\varepsilon > 0$.

PROOF. Put $\mathfrak{A} = T(\mathfrak{B}(0,1))$. If $y \in \mathfrak{B}(0,r)$ and $\varepsilon > 0$, there is by assumption an element y_1 in $\mathfrak{A}$ with $\|y - y_1\| < \varepsilon r$. By homogeneity there is now a y_2 in $\varepsilon \mathfrak{A}$ such that $\|y - y_1 - y_2\| < \varepsilon^2 r$. Proceeding by induction we find a sequence (y_n), such that $y_n \in \varepsilon^{n-1} \mathfrak{A}$ and

$$\left\| y - \sum_{k=1}^{n} y_k \right\| < \varepsilon^n r.$$

Choose x_n in $\mathfrak{X}$ such that $Tx_n = y_n$ and $\|x_n\| \leq \varepsilon^{n-1}$. Then $x = \sum x_n \in \mathfrak{X}$ and $Tx = y$. Moreover, $\|x\| \leq \sum \varepsilon^{n-1} = (1-\varepsilon)^{-1}$, which proves that $(1-\varepsilon)^{-1}\mathfrak{A} \supset \mathfrak{B}(0,r)$, as desired. $\square$

2.2.4. Theorem. *If $\mathfrak{X}$ and $\mathfrak{Y}$ are Banach spaces and $T \in \mathbf{B}(\mathfrak{X}, \mathfrak{Y})$ with $T(\mathfrak{X}) = \mathfrak{Y}$, then T is an open map.*

PROOF. By assumption

$$\mathfrak{Y} = T(\mathfrak{X}) = \bigcup (T(\mathfrak{B}(0,n)))^-.$$

By the category theorem (2.2.2) not all of the closed sets $T(\mathfrak{B}(0,n))^-$ have empty interior. Consequently, there is an n and a ball $\mathfrak{B}(y,\varepsilon)$ such that $T(\mathfrak{B}(0,n))^-$ contains $\mathfrak{B}(y,\varepsilon)$. This means that $T(\mathfrak{B}(0,1))$ is dense in $\mathfrak{B}(n^{-1}y, n^{-1}\varepsilon)$, and therefore also dense in $\mathfrak{B}(0, n^{-1}\varepsilon)$ [because $2\mathfrak{B}(0, n^{-1}\varepsilon) \subset \mathfrak{B}(n^{-1}y, n^{-1}\varepsilon) - \mathfrak{B}(n^{-1}y, n^{-1}\varepsilon)$]. It follows from 2.2.3 that $\mathfrak{B}(0,\delta)$ is contained in $T(\mathfrak{B}(0,1))$ for every $\delta < n^{-1}\varepsilon$. Since every open set $\mathfrak{A}$ in $\mathfrak{X}$ is a union of balls, we conclude from the linearity of T that $T(\mathfrak{A})$ contains a neighborhood (actually a closed ball) around each of its points, whence $T(\mathfrak{A})$ is open. This is the *open mapping theorem*. $\square$

2.2.5. Corollary. *Every bounded, bijective operator between two Banach spaces has a bounded inverse.*

2.2.6. Corollary. *If a vector space $\mathfrak{X}$ is a Banach space under two norms $\|\cdot\|$ and $\|\|\cdot\|\|$, and if $\|\cdot\| \leq \alpha \|\|\cdot\|\|$ for some constant $\alpha > 0$, there is a $\beta > 0$ such that $\|\|\cdot\|\| \leq \beta \|\cdot\|$.*

2.2.7. Theorem. *If $T: \mathfrak{X} \to \mathfrak{Y}$ is an operator between Banach spaces $\mathfrak{X}$ and $\mathfrak{Y}$, such that the graph*

$$\mathfrak{G}(T) = \{(x,y) \in \mathfrak{X} \times \mathfrak{Y} \mid Tx = y\}$$

is closed in $\mathfrak{X} \times \mathfrak{Y}$, then T is bounded.

PROOF. We use the ∞-norm on $\mathfrak{X} \times \mathfrak{Y}$ (cf. 2.1.16) and note that $\mathfrak{G}(T)$ by assumption is a closed subspace of $\mathfrak{X} \times \mathfrak{Y}$. The projection map $P_1\colon \mathfrak{G}(T) \to \mathfrak{X}$ given by $P_1(x, Tx) = x$ is norm decreasing and bijective and therefore has a bounded inverse P_1^{-1} by 2.2.5. The other projection $P_2\colon \mathfrak{G}(T) \to \mathfrak{Y}$, given by $P_2(x, Tx) = Tx$, is norm decreasing; and since $T = P_2 P_1^{-1}$, it follows that T is bounded. $\qquad\square$

2.2.8. Remark. The usual way to use the *closed graph theorem* (2.2.7) is to check that the graph $\mathfrak{G}(T)$ of a given operator T contains all its limit points (1.3.6). Thus, for every sequence (x_n) in $\mathfrak{X}$ such that $x_n \to x$ and $Tx_n \to y$ for some elements x and y in $\mathfrak{X}$ and $\mathfrak{Y}$, respectively, we must show that $Tx = y$. This is a tremendous advantage from proving continuity of T from scratch, because in that case we have no control over the sequence (Tx_n).

2.2.9. Theorem. *Consider a family $\{T_\lambda | \lambda \in \Lambda\}$ in $\mathbf{B}(\mathfrak{X}, \mathfrak{Y})$, where $\mathfrak{X}$ and $\mathfrak{Y}$ are Banach spaces. If each set $\{T_\lambda x | \lambda \in \Lambda\}$ is bounded in $\mathfrak{Y}$ for every x in $\mathfrak{X}$, the set $\{\|T_\lambda\| | \lambda \in \Lambda\}$ is bounded.*

PROOF. For each λ let $\mathfrak{Y}_\lambda$ denote a copy of $\mathfrak{Y}$, and form the direct product $\mathfrak{Y}_\Lambda$ of the $\mathfrak{Y}_\lambda$'s as defined in 2.1.16. Define an operator $T\colon \mathfrak{X} \to \mathfrak{Y}_\Lambda$ by

$$Tx = \{T_\lambda x | \lambda \in \Lambda\},$$

and note from the pointwise boundedness of the family $\{T_\lambda\}$ that T is well defined. To show that the graph of T is closed in $\mathfrak{X} \times \mathfrak{Y}_\Lambda$, consider a sequence (x_n) in $\mathfrak{X}$ such that $x_n \to x$ in $\mathfrak{X}$ and $Tx_n \to y$ in $\mathfrak{Y}_\Lambda$. With P_λ as the projection of $\mathfrak{Y}_\Lambda$ on $\mathfrak{Y}_\lambda$ we see that $T_\lambda x_n \to P_\lambda y$. But each T_λ is bounded, so $T_\lambda x_n \to T_\lambda x$, whence $T_\lambda x = P_\lambda y$ for all λ. But this means that $y = Tx$, and thus T is bounded by 2.2.7. Since $T_\lambda = P_\lambda T$, it follows that $\|T_\lambda\| \leq \|T\|$ for all λ in Λ, as desired. $\qquad\square$

2.2.10. Corollary. *Consider a net $(T_\lambda)_{\lambda \in \Lambda}$ in $\mathbf{B}(\mathfrak{X}, \mathfrak{Y})$, such that each net $(T_\lambda x)_{\lambda \in \Lambda}$ is bounded and convergent in $\mathfrak{Y}$ for every x in $\mathfrak{X}$. There is then a T in $\mathbf{B}(\mathfrak{X}, \mathfrak{Y})$ such that $T_\lambda x \to Tx$ for every x in $\mathfrak{X}$.*

PROOF. Clearly we can define an operator $T\colon \mathfrak{X} \to \mathfrak{Y}$ by $Tx = \lim T_\lambda x$. From 2.2.9 we see that $\|T_\lambda\| \leq \alpha$ for some α and all λ, and therefore $\|Tx\| \leq \alpha \|x\|$ for every x in $\mathfrak{X}$, i.e. T is bounded. $\qquad\square$

2.2.11. Remark. The *principle of uniform boundedness* (2.2.9), which allows us to pass from pointwise boundedness to uniform boundedness, only uses completeness of the space $\mathfrak{X}$ (because we may replace $\mathfrak{Y}$ with its completion, cf. 2.1.12). The very useful result in 2.2.10 is particularly applicable if the net Λ is a sequence, because then convergence of each sequence $(T_n x)$, $x \in \mathfrak{X}$, automatically implies boundedness. Moreover, we see that it suffices in 2.2.10 to know that each net $(T_\lambda x)_{\lambda \in \Lambda}$ is bounded and that the net is convergent for a dense set of elements x in $\mathfrak{X}$; cf. 2.1.11.

EXERCISES

E 2.2.1. (*Complementary subspaces.*) Let $\mathfrak{Y}$ and $\mathfrak{Z}$ be closed subspaces in a Banach space $\mathfrak{X}$. Show that each element x in $\mathfrak{X}$ has a unique decomposition $x = y + z$, $y \in \mathfrak{Y}$, $z \in \mathfrak{Z}$ iff $\mathfrak{Y} + \mathfrak{Z} = \mathfrak{X}$ and $\mathfrak{Y} \cap \mathfrak{Z} = \{0\}$. Show in this case that there is a constant $\alpha > 0$ such that $\|y\| + \|z\| \leq \alpha \|x\|$ for every x in $\mathfrak{X}$.

Hint: Apply 2.2.6 to $\mathfrak{Y} + \mathfrak{Z}$ with the 1-norm (2.1.17).

E 2.2.2. Let $\mathfrak{Y}$ and $\mathfrak{Z}$ be complementary subspaces of $\mathfrak{X}$ as defined in E 2.2.1. Define the operator $P: \mathfrak{X} \to \mathfrak{X}$ by $Px = y$, where $x = y + z$. Show that $P \in \mathbf{B}(\mathfrak{X})$ and that P is *idempotent*, i.e. $P^2 = P$. Show, conversely, that if P is an idempotent operator in $\mathbf{B}(\mathfrak{X})$, then $\mathfrak{Y} = P(\mathfrak{X})$ and $\mathfrak{Z} = \ker P$ are complementary subspaces. Show finally that an operator T in $\mathbf{B}(\mathfrak{X})$ commutes with P (i.e. $PT = TP$) iff $T(\mathfrak{Y}) \subset \mathfrak{Y}$ and $T(\mathfrak{Z}) \subset \mathfrak{Z}$.

E 2.2.3. Let $\mathfrak{Y}$ and $\mathfrak{Z}$ be complementary subspaces of $\mathfrak{X}$ as defined in E 2.2.1. Show that every operator T in $\mathbf{B}(\mathfrak{X})$ gives rise to four operators T_{ij}, $1 \leq i, j \leq 2$, where $T_{11} \in \mathbf{B}(\mathfrak{Y})$, $T_{22} \in \mathbf{B}(\mathfrak{Z})$, $T_{12} \in \mathbf{B}(\mathfrak{Z}, \mathfrak{Y})$, and $T_{21} \in \mathbf{B}(\mathfrak{Y}, \mathfrak{Z})$, such that T can be regarded as an operator matrix.

$$T = \begin{pmatrix} T_{11} & T_{12} \\ T_{21} & T_{22} \end{pmatrix}$$

Check that sum and product of elements in $\mathbf{B}(\mathfrak{X})$ are compatible with matrix sum and product of the corresponding matrices. Give necessary and sufficient conditions on the matrix for having $T(\mathfrak{Y}) \subset \mathfrak{Y}$ or for $T(\mathfrak{Z}) \subset \mathfrak{Z}$. Compare with the last question in E 2.2.2.

E 2.2.4. Consider the complex Banach space

$$\mathfrak{X} = \{ f \in C([0, 2\pi]) \mid f(0) = f(2\pi) \},$$

with the ∞-norm. For each n, let T_n be the operator in $\mathbf{B}(\mathfrak{X})$ that assigns to f the nth partial sum of its Fourier series, i.e.,

$$T_n f = \sum_{k=-n}^{n} \frac{1}{2\pi} \int_0^{2\pi} f(t) e_{-k}(t)\, dt\, e_k,$$

where $e_k(x) = \exp(ikx)$. Show that $\|T_n\| \to \infty$ and deduce from 2.2.10 that there are functions f whose Fourier series is not uniformly convergent.

Hint: Use the fact that $T_n f = D_n \times f$ (convolution product, $\times$; see 6.6.21), where D_n is the Dirichlet kernel,

$$D_n(x) = \sum_{k=-n}^{n} e_k(x) = \sin((n + \tfrac{1}{2})x)(\sin(\tfrac{1}{2}x))^{-1}.$$

Use the fact that

$$\|T_n\| = \|D_n\|_1 > 4\pi^{-2} \sum_{k=1}^{n} k^{-1}$$

E 2.2.5. Take $\mathfrak{X} = L^1([0, 2\pi])$; cf. 2.1.14. By the Riemann–Lebesgue lemma the Fourier coefficients of an integrable function tend to zero. We can therefore define an operator $T\colon \mathfrak{X} \to c_0(\mathbb{Z})$ by

$$Tf(n) = \frac{1}{2\pi} \int_0^{2\pi} f(t) \exp(-int)\, dt.$$

It is well known that T is injective, and easy to prove that T is bounded. Show that T cannot be surjective.

Hint: Use the Dirichlet kernel from E 2.2.4 to prove that there is no $\varepsilon > 0$ such that $\|T(D_n)\|_\infty \geq \varepsilon \|D_n\|_1$ for all n. Then apply 2.2.5.

E 2.2.6. Consider the Banach space $\mathfrak{X} = C([0, 1])$ with the ∞-norm. For each n, let $\mathfrak{F}_n$ denote the set of functions f in $\mathfrak{X}$, for which there is some x in $[0, 1]$ such that $|f(x) - f(y)| \leq n|x - y|$ for all y in $[0, 1]$. Show that $\mathfrak{F}_n$ is closed in $\mathfrak{X}$ and has empty interior. Conclude from 2.2.2 that there is a dense set of functions in $\mathfrak{X}$ that are not differentiable at any point in $[0, 1]$.

Hint: Every function f in $\mathfrak{X}$ can be uniformly approximated by a piecewise linear function whose slope is everywhere numerically larger than n.

2.3. Dual Spaces

Synopsis. The Hahn–Banach extension theorem. Spaces in duality. Adjoint operator. Exercises.

2.3.1. A *functional* on a vector space $\mathfrak{X}$ over the field $\mathbb{F}$ is a linear map $\varphi\colon \mathfrak{X} \to \mathbb{F}$. If $\mathfrak{X}$ is a normed space, we denote by $\mathfrak{X}^*$ the space $\mathbf{B}(\mathfrak{X}, \mathbb{F})$ of bounded ($=$ continuous by 2.1.2) functionals on $\mathfrak{X}$. It follows from 2.1.4 that $\mathfrak{X}^*$ is a Banach space, called the *dual space* of $\mathfrak{X}$. The fundamental result about the existence of elements in $\mathfrak{X}^*$ is the *Hahn–Banach extension theorem* (2.3.3). In the theory of topological vector spaces this theorem plays the same role as Urysohn's lemma (1.5.6) for general topology.

We define a *Minkowski functional* to be a function $m\colon \mathfrak{X} \to \mathbb{R}$ that is subadditive $[m(x + y) \leq m(x) + m(y)]$ and positive homogeneous $[m(tx) = tm(x)$ for $t \geq 0$ in $\mathbb{R}]$. A Minkowski functional (an abuse of our definition of functionals) resembles a seminorm, except that the condition (ii) in 2.1.1 is only supposed to hold for positive scalars.

2.3.2. Fundamental Lemma. *If m is a Minkowski functional on a real vector space $\mathfrak{X}$, and φ is a functional on a linear subspace $\mathfrak{Y}$ of $\mathfrak{X}$ that is dominated by m [i.e. $\varphi(y) \leq m(y)$ for every y in $\mathfrak{Y}$], there exists a functional $\tilde{\varphi}$ on $\mathfrak{X}$, dominated by m, such that $\tilde{\varphi}|\mathfrak{Y} = \varphi$.*

PROOF. If $x \in \mathfrak{X} \setminus \mathfrak{Y}$, any extension of φ from $\mathfrak{Y}$ to $\mathfrak{Y} + \mathbb{R}x$ is determined by

$$\tilde{\varphi}(y + sx) = \varphi(y) + s\alpha, \qquad s \in \mathbb{R}, \quad y \in \mathfrak{Y}.$$

We wish to choose α such that $\varphi(y) + s\alpha \leq m(y + sx)$. Owing to the positive homogeneity of m and φ, it suffices to check these inequalities for $s = 1$ and $s = -1$. Therefore, the demand is that

$$\varphi(y) - m(y - x) \leq \alpha \leq -\varphi(z) + m(z + x)$$

for all y and z in $\mathfrak{Y}$. Now by assumption

$$-\varphi(z) + m(z + x) - \varphi(y) + m(y - x)$$
$$= m(y - x) + m(z + x) - \varphi(y + z) \geq m(y + z) - \varphi(y + z) \geq 0,$$

and we may therefore as our α take any number in the nonempty, closed interval spanned by the numbers

$$\sup_{y} \{\varphi(y) - m(y - x)\}, \qquad \inf_{z} \{-\varphi(z) + m(z + x)\}.$$

Consider the family Λ of pairs $(\mathfrak{Z}, \psi)$, where $\mathfrak{Z}$ is a linear subspace of $\mathfrak{X}$ containing $\mathfrak{Y}$ and ψ is a functional on $\mathfrak{Z}$ dominated by m such that $\psi|\mathfrak{Y} = \varphi$. If we define $(\mathfrak{Z}_1, \psi_1) \leq (\mathfrak{Z}_2, \psi_2)$ to mean that $\mathfrak{Z}_1 \subset \mathfrak{Z}_2$ and $\psi_2|\mathfrak{Z}_1 = \psi_1$, then Λ is an ordered set. We claim that Λ is inductively ordered. Indeed, if $N = \{(\mathfrak{Z}_\mu, \psi_\mu) | \mu \in M\}$ is a totally ordered subset of Λ, we let $\mathfrak{Z} = \bigcup \mathfrak{Z}_\mu$ and define ψ on $\mathfrak{Z}$ by $\psi(z) = \psi_\mu(z)$ if $z \in \mathfrak{Z}_\mu$. Owing to the total ordering, we see that $\mathfrak{Z}$ is a linear subspace of $\mathfrak{X}$ containing all the $\mathfrak{Z}_\mu$'s, and ψ is a well-defined functional on $\mathfrak{Z}$ dominated by m that simultaneously extends all the ψ_μ's. Thus, $(\mathfrak{Z}, \psi) \in \Lambda$ and is a majorant for N, as claimed. By Zorn's lemma (1.1.3) there is therefore a maximal extension $(\tilde{\mathfrak{Y}}, \tilde{\varphi})$ of $(\mathfrak{Y}, \varphi)$ dominated by m. If we had $\tilde{\mathfrak{Y}} \neq \mathfrak{X}$, the first part of the proof, applied to $(\tilde{\mathfrak{Y}}, \tilde{\varphi})$, would give a dominated extension to the space $\tilde{\mathfrak{Y}} + \mathbb{R}x$, contradicting the maximality of $(\tilde{\mathfrak{Y}}, \tilde{\varphi})$. Consequently, $\tilde{\mathfrak{Y}} = \mathfrak{X}$, as desired. $\qquad\square$

2.3.3. Theorem. *If m is a seminorm on a vector space $\mathfrak{X}$, and φ is a functional on a subspace $\mathfrak{Y}$ of $\mathfrak{X}$ such that $|\varphi| \leq m$, there is a functional $\tilde{\varphi}$ on $\mathfrak{X}$ with $|\tilde{\varphi}| \leq m$ and $\tilde{\varphi}|\mathfrak{Y} = \varphi$.*

PROOF. If $\mathbb{F} = \mathbb{R}$, the result is contained in 2.3.2, because $\varphi \leq m$ is equivalent with $|\varphi| \leq m$ by the symmetry of m.

If $\mathbb{F} = \mathbb{C}$, we first regard $\mathfrak{X}$ as a real vector space and consider the real functional $\operatorname{Re} \varphi$ on $\mathfrak{Y}$. By 2.3.2 we can then find a real functional ψ on $\mathfrak{X}$ that extends $\operatorname{Re} \varphi$ and is dominated by m. Define $\tilde{\varphi} \colon \mathfrak{X} \to \mathbb{C}$ by

$$\tilde{\varphi}(x) = \psi(x) - i\psi(ix), \quad x \in \mathfrak{X}.$$

Note that $\tilde{\varphi}(ix) = i\tilde{\varphi}(x)$, so that $\tilde{\varphi}$ is a complex functional. Note further that $\tilde{\varphi}|\mathfrak{Y} = \varphi$, because

$$\operatorname{Re} \tilde{\varphi}|\mathfrak{Y} = \psi|\mathfrak{Y} = \operatorname{Re} \varphi,$$

and complex functionals with the same real part are identical. Finally, if $x \in \mathfrak{X}$, choose α in $\mathbb{C}$ with $|\alpha| = 1$ such that $\tilde{\varphi}(\alpha x) \in \mathbb{R}_+$. Then

$$|\tilde{\varphi}(x)| = \tilde{\varphi}(\alpha x) = \psi(\alpha x) \leq m(\alpha x) = m(x),$$

whence $|\tilde{\varphi}| \leq m$ as desired. $\qquad\square$

2.3.4. Corollary. *To every $x \neq 0$ in a normed space $\mathfrak{X}$ there is a φ in $\mathfrak{X}^*$ with $\|\varphi\| = 1$ and $\varphi(x) = \|x\|$.*

PROOF. Define φ on $\mathbb{F}x$ by $\varphi(\alpha x) = \alpha \|x\|$, and note that $\|\varphi\| = 1$. Then apply 2.3.3 with $\mathfrak{Y} = \mathbb{F}x$ and $m = \|\cdot\|$. $\qquad \square$

2.3.5. Corollary. *To every closed subspace $\mathfrak{Y}$ of a normed space $\mathfrak{X}$ and every x in $\mathfrak{X} \backslash \mathfrak{Y}$, there is a φ in $\mathfrak{X}^*$ with $\|\varphi\| = 1$, $\varphi|\mathfrak{Y} = 0$, and $\varphi(x) = \inf\{\|x - y\| \,|\, y \in \mathfrak{Y}\}$.*

PROOF. Apply 2.3.4 on the normed space $\mathfrak{X}/\mathfrak{Y}$, cf. 2.1.5, and note that functionals in $(\mathfrak{X}/\mathfrak{Y})^*$ may be regarded as elements in $\mathfrak{X}^*$ that annihilate $\mathfrak{Y}$. (We have more to say about this situation in 2.4.13.) $\qquad \square$

2.3.6. For a subspace $\mathfrak{Y}$ of a normed space $\mathfrak{X}$, the *annihilator* of $\mathfrak{Y}$ is defined as

$$\mathfrak{Y}^\perp = \{\varphi \in \mathfrak{X}^* \,|\, \varphi(y) = 0, \, \forall y \in \mathfrak{Y}\}.$$

Similarly, we define the annihilator of a subspace $\mathfrak{Z}$ of $\mathfrak{X}^*$ as

$$\mathfrak{Z}^\perp = \{x \in \mathfrak{X} \,|\, \varphi(x) = 0, \, \forall \varphi \in \mathfrak{Z}\}.$$

It is immediate that $\mathfrak{Y} \subset (\mathfrak{Y}^\perp)^\perp$, and it follows from 2.3.5 that actually $\mathfrak{Y} = (\mathfrak{Y}^\perp)^\perp$ if it is a closed subspace. By contrast, we cannot expect that $\mathfrak{Z} = (\mathfrak{Z}^\perp)^\perp$ for every norm closed subspace $\mathfrak{Z}$ of $\mathfrak{X}^*$. After all, our definition of annihilators for subspaces of $\mathfrak{X}^*$ give spaces in $\mathfrak{X}$, which is, in general, much smaller than the dual space of $\mathfrak{X}^*$.

2.3.7. Given a normed space $\mathfrak{X}$, we form the Banach dual space $\mathfrak{X}^*$ and, by iteration, we obtain the *bidual space* $\mathfrak{X}^{**}$. We define a map $\iota \colon \mathfrak{X} \to \mathfrak{X}^{**}$ by

$$\iota(x)(\varphi) = \varphi(x), \qquad x \in \mathfrak{X}, \quad \varphi \in \mathfrak{X}^*.$$

Clearly ι is a norm decreasing operator, and we see from 2.3.4 that ι actually is an isometry. We may therefore identify $\mathfrak{X}$ with a subspace of $\mathfrak{X}^{**}$, and, if $\mathfrak{X}$ is not a Banach space, we obtain a completion $\bar{\mathfrak{X}}$ of $\mathfrak{X}$ by letting $\bar{\mathfrak{X}}$ be the norm closure of $\iota(\mathfrak{X})$ in the Banach space $\mathfrak{X}^{**}$.

A Banach space $\mathfrak{X}$ is *reflexive* if the isometry ι defined above is surjective. Typical examples of reflexive spaces (apart from finite-dimensional spaces) are the L^p-spaces for $1 < p < \infty$. Read about this in 6.5.11.

2.3.8. We say that two vector spaces $\mathfrak{X}$ and $\mathfrak{Y}$ are in (algebraic) *duality* [or that they form a (algebraic) *dual pair*], if there is a bilinear form

$$\langle \cdot, \cdot \rangle \colon \mathfrak{X} \times \mathfrak{Y} \to \mathbb{F},$$

such that the space $\langle \cdot, \mathfrak{Y} \rangle$ of functionals on $\mathfrak{X}$ separates points in $\mathfrak{X}$ and, similarly, $\langle \mathfrak{X}, \cdot \rangle$ is a separating space of functionals on $\mathfrak{Y}$. If $\mathfrak{X}$ and $\mathfrak{Y}$ are normed spaces, we delete the word algebraic above if $\langle \cdot, \mathfrak{Y} \rangle \subset \mathfrak{X}^*$ and $\langle \mathfrak{X}, \cdot \rangle \subset \mathfrak{Y}^*$.

With this definition of duality it follows that for every normed space $\mathfrak{X}$, the pair $(\mathfrak{X}, \mathfrak{X}^*)$ are in duality; the bilinear form being given by

$$\langle x, \varphi \rangle = \varphi(x), \qquad x \in \mathfrak{X}, \quad \varphi \in \mathfrak{X}^*.$$

We obtain another dual pair $(\mathfrak{X}^{**}, \mathfrak{X}^*)$, with the bilinear form

$$\langle z, \varphi \rangle = z(\varphi), \qquad z \in \mathfrak{X}^{**}, \quad \varphi \in \mathfrak{X}^*;$$

and we may regard this form as an extension of the former, if we identify $\mathfrak{X}$ with its image in $\mathfrak{X}^{**}$ under the isometry ι described in 2.3.7.

2.3.9. If $\mathfrak{X}$ and $\mathfrak{Y}$ are normed spaces and $T \in \mathbf{B}(\mathfrak{X}, \mathfrak{Y})$, we define $T^* \colon \mathfrak{Y}^* \to \mathfrak{X}^*$ by the equation

$$\langle x, T^*\varphi \rangle = \langle Tx, \varphi \rangle, \qquad x \in \mathfrak{X}, \quad \varphi \in \mathfrak{Y}^*. \tag{$*$}$$

We say that T^* is the *adjoint* operator to T.

It is clear from this definition that if $S \in \mathbf{B}(\mathfrak{X}, \mathfrak{Y})$ as well, and $\alpha \in \mathbb{F}$, then $(\alpha T + S)^* = \alpha T^* + S^*$. Furthermore, we see that if $\mathfrak{Z}$ is a third normed space and $R \in \mathbf{B}(\mathfrak{Y}, \mathfrak{Z})$, then $(RT)^* = T^*R^*$.

2.3.10. Proposition. *If* $T \in \mathbf{B}(\mathfrak{X}, \mathfrak{Y})$, *where* $\mathfrak{X}$ *and* $\mathfrak{Y}$ *are normed spaces, then* $T^* \in \mathbf{B}(\mathfrak{Y}^*, \mathfrak{X}^*)$ *and* $\|T^*\| = \|T\|$.

Proof. From the defining relation $(*)$ in 2.3.9 we have

$$\|T^*\varphi\| = \sup\{|\langle Tx, \varphi \rangle| \,|\, x \in \mathfrak{X}, \|x\| \le 1\}$$
$$\le \sup\{\|T\| \|x\| \|\varphi\| \,|\, x \in \mathfrak{X}, \|x\| \le 1\} = \|T\| \|\varphi\|.$$

Thus $T^* \in \mathbf{B}(\mathfrak{Y}^*, \mathfrak{X}^*)$ and $\|T^*\| \le \|T\|$.

Given $\varepsilon > 0$ we can find x in $\mathfrak{X}$ with $\|x\| = 1$ such that $\|Tx\| \ge \|T\| - \varepsilon$. Choosing φ in $\mathfrak{Y}^*$ for Tx as in 2.3.4 we obtain

$$\|T\| \le \varepsilon + \|Tx\| = \varepsilon + |\langle Tx, \varphi \rangle|$$
$$= \varepsilon + |\langle x, T^*\varphi \rangle| \le \varepsilon + \|T^*\|.$$

Since ε is arbitrary, we conclude that $\|T^*\| = \|T\|$. $\qquad\qquad\square$

2.3.11. Proposition. *Given operators* $T \colon \mathfrak{X} \to \mathfrak{Y}$ *and* $S \colon \mathfrak{Y}^* \to \mathfrak{X}^*$ *between Banach spaces* $\mathfrak{X}$ *and* $\mathfrak{Y}$ *and their duals, satisfying*

$$\langle Tx, \varphi \rangle = \langle x, S\varphi \rangle,$$

for all x *in* $\mathfrak{X}$ *and* φ *in* $\mathfrak{Y}^*$, *we have that both* S *and* T *are bounded and* $S = T^*$.

Proof. Let (x, y) be a limit point on the graph of T in $\mathfrak{X} \times \mathfrak{Y}$. Thus, we have a sequence (x_n) in $\mathfrak{X}$ such that $x_n \to x$ and $Tx_n \to y$. Now for each φ in $\mathfrak{Y}^*$ we have

$$\langle y, \varphi \rangle = \lim \langle Tx_n, \varphi \rangle$$
$$= \lim \langle x_n, S\varphi \rangle = \langle x, S\varphi \rangle = \langle Tx, \varphi \rangle.$$

Since $\mathfrak{Y}^*$ separates points in $\mathfrak{Y}$ (2.3.4), it follows that $y = Tx$. This means that the graph of T is closed, whence $T \in \mathbf{B}(\mathfrak{X}, \mathfrak{Y})$ by 2.2.7. Thus $T^* \in \mathbf{B}(\mathfrak{Y}^*, \mathfrak{X}^*)$ by 2.3.10, and $T^* = S$ by (*) in 2.3.9, since certainly $\mathfrak{X}$ separates points in $\mathfrak{X}^*$. $\square$

2.3.12. Remark. Despite its formulation, the content of the preceding result is negative. There is namely no hope of explaining away the existence in the applications of important unbounded operators with well-defined adjoint operators. We learn from 2.3.11 that these operators cannot be defined on Banach spaces, but at best on dense subspaces. Especially for operators on Hilbert space (see Chapter 3), where the adjoint operator is fundamental, we see that the theory of unbounded (self-adjoint) operators necessarily must come to terms with operators that are not everywhere defined.

EXERCISES

E 2.3.1. If m_1 and m_2 are seminorms on a vector space $\mathfrak{X}$, and φ is a functional on $\mathfrak{X}$ such that $|\varphi| \leq m_1 + m_2$, then there are functionals φ_1 and φ_2 on $\mathfrak{X}$ with $\varphi_1 + \varphi_2 = \varphi$ and $|\varphi_i| \leq m_i$ for $i = 1, 2$.
   Hint: Define the seminorm m on $\mathfrak{X} \times \mathfrak{X}$ by $m(x_1, x_2) = m_1(x_1) + m_2(x_2)$, and consider the functional ψ on the subspace $\mathfrak{D}$ of $\mathfrak{X} \times \mathfrak{X}$, where $(x_1, x_2) \in \mathfrak{D}$ if $x_1 = x_2$, and $\psi(x, x) = \varphi(x)$. Then apply 2.3.3.

E 2.3.2. Consider the Banach spaces c_0, c, and ℓ^∞ defined in E 2.1.5 and E 2.1.6. Furthermore, consider the Banach space ℓ^1 of absolutely summable sequences, cf. 2.1.18. Show that the bilinear form

$$\langle x, y \rangle = \sum x_n y_n$$

gives rise to isometric isomorphisms of ℓ^1 onto $(c_0)^*$ and of ℓ^∞ onto $(\ell^1)^*$. Describe c^* and show that neither c_0 nor c are reflexive spaces.

E 2.3.3. (*Banach limits.*) On the real Banach space ℓ^∞ (cf. 2.1.18 and E 2.1.5) we define the shift operator S in $\mathbf{B}(\ell^\infty)$ given by $(Sx)_n = x_{n+1}$ for every $x = (x_n)$ in ℓ^∞. Show that there is a functional L in $(\ell^\infty)^*$ satisfying

(i) $L(Sx) = L(x)$;
(ii) $\liminf x_n \leq L(x) \leq \limsup x_n$;

for every $x = (x_n)$ in ℓ^∞. Interpret the result as a process of taking limits in a generalized sense.
   Hint: Define m_n on ℓ^∞ by $m_n(x) = n^{-1}(x_1 + \cdots + x_n)$, and let $\mathfrak{Y}$ be the subspace in ℓ^∞ of elements x for which $\lim m_n(x)$ exists. Define L on $\mathfrak{Y}$ by $L(x) = \lim m_n(x)$, and define m on ℓ^∞ by $m(x) = \limsup x_n$. Then apply 2.3.2.

E 2.3.4. Show that a normed space $\mathfrak{X}$ for which $\mathfrak{X}^*$ is separable is itself separable.
   Hint: If (φ_n) is a dense sequence in $\mathfrak{X}^*$, choose (x_n) in $\mathfrak{X}$ with

$\|x_n\| = 1$ such that $|\varphi_n(x_n)| \geq \frac{1}{2}\|\varphi_n\|$ for every n. Then use 2.3.5 to show that the subspace spanned by (x_n) is dense in $\mathfrak{X}$.

E 2.3.5. Show that a Banach space $\mathfrak{X}$ is reflexive iff $\mathfrak{X}^*$ is reflexive.

E 2.3.6. Let $\mathfrak{X}$ be a normed space and consider an element φ in $\mathfrak{X}^*$ as an operator in $\mathbf{B}(\mathfrak{X}, \mathbb{F})$. Compute φ^* in $\mathbf{B}(\mathbb{F}, \mathfrak{X}^*)$.

E 2.3.7. Consider the Banach spaces c_0, ℓ^1, and ℓ^∞ from 2.1.18 (and E 2.3.2). Define $T: \ell^1 \to c_0$ by

$$(Tx)_n = \sum_{m \geq n} x_m, \qquad x = (x_n) \in \ell^1.$$

Show that $T \in \mathbf{B}(\ell^1, c_0)$ and give an explicit formula for T^* in $\mathbf{B}(\ell^1, \ell^\infty)$.

E 2.3.8. Consider an operator T in $\mathbf{B}(\mathfrak{X}, \mathfrak{Y})$, where $\mathfrak{X}$ and $\mathfrak{Y}$ are Banach spaces, such that $T(\mathfrak{X})$ is closed in $\mathfrak{Y}$. Show that $T^*(\mathfrak{Y}^*) = (\ker T)^\perp$.

Hint: If $\varphi \in (\ker T)^\perp$ in $\mathfrak{X}^*$, the equation $\psi_0(Tx) = \varphi(x)$ determines a functional ψ_0 on $T(\mathfrak{X})$. Use the open mapping theorem to prove that ψ_0 is continuous, and extend ψ_0 to an element ψ in $\mathfrak{Y}^*$ by 2.3.3. Show that $T^*\psi = \varphi$.

E 2.3.9. Given Banach spaces $\mathfrak{X}$, $\mathfrak{Y}$, and $\mathfrak{Z}$ we say that a map $B: \mathfrak{X} \times \mathfrak{Y} \to \mathfrak{Z}$ is a bounded bilinear operator if $x \to B(x, y)$ belongs to $\mathbf{B}(\mathfrak{X}, \mathfrak{Z})$ for every fixed y and $y \to B(x, y)$ belongs to $\mathbf{B}(\mathfrak{Y}, \mathfrak{Z})$ for every fixed x. Show in this case that

$$\sup\{\|B(x, y)\| \mid \|x\| \leq 1, \|y\| \leq 1\} < \infty. \qquad (*)$$

Show that the space $\mathbf{BIL}(\mathfrak{X} \times \mathfrak{Y}, \mathfrak{Z})$ of bounded bilinear operators from $\mathfrak{X} \times \mathfrak{Y}$ to $\mathfrak{Z}$ is a Banach space under the norm determined by $(*)$ and the pointwise vector operations.

Hint: Use the principle of uniform boundedness.

E 2.3.10. Given Banach spaces $\mathfrak{X}$ and $\mathfrak{Y}$ put $\mathbf{BIL}(\mathfrak{X} \times \mathfrak{Y}) = \mathbf{BIL}(\mathfrak{X} \times \mathfrak{Y}, \mathbb{F})$; cf. E 2.3.9. Show that if $B \in \mathbf{BIL}(\mathfrak{X} \times \mathfrak{Y})$, the definitions

$$\langle x, S_B(y) \rangle = B(x, y) = \langle y, T_B(x) \rangle$$

give operators S_B in $\mathbf{B}(\mathfrak{Y}, \mathfrak{X}^*)$ and T_B in $\mathbf{B}(\mathfrak{X}, \mathfrak{Y}^*)$. Show that the maps $B \to S_B$ and $B \to T_B$ are isometric isomorphisms of $\mathbf{BIL}(\mathfrak{X} \times \mathfrak{Y})$ onto $\mathbf{B}(\mathfrak{Y}, \mathfrak{X}^*)$ and $\mathbf{B}(\mathfrak{X}, \mathfrak{Y}^*)$, respectively.

E 2.3.11. *(Projective tensor product.)* Take $\mathfrak{X}$, $\mathfrak{Y}$, and $\mathbf{BIL}(\mathfrak{X} \times \mathfrak{Y})$ as in E 2.3.10. Show that there is a bounded bilinear operator $\otimes$ from $\mathfrak{X} \times \mathfrak{Y}$ to $\mathbf{BIL}(\mathfrak{X} \times \mathfrak{Y})^*$ defined by $x, y \to x \otimes y$, where

$$\langle B, x \otimes y \rangle = B(x, y), \qquad x \in \mathfrak{X}, \quad y \in \mathfrak{Y}, \quad B \in \mathbf{BIL}(\mathfrak{X} \times \mathfrak{Y}). \quad (*)$$

Show that we also have a bounded bilinear operator $\otimes$ from $\mathfrak{X}^* \times \mathfrak{Y}^*$ to $\mathbf{BIL}(\mathfrak{X} \times \mathfrak{Y})$ defined by $\varphi, \psi \to \varphi \otimes \psi$, where

$$\varphi \otimes \psi(x, y) = \varphi(x)\psi(y), \qquad x \in \mathfrak{X}, \quad y \in \mathfrak{Y}, \quad \varphi \in \mathfrak{X}^*, \quad \psi \in \mathfrak{Y}^*.$$

Prove that $\|x \otimes y\| = \|x\| \, \|y\|$ and $\|\varphi \otimes \psi\| = \|\varphi\| \, \|\psi\|$.

Define $\mathfrak{X} \otimes \mathfrak{Y}$ to be the norm closed subspace of $(\mathbf{BIL}(\mathfrak{X} \times \mathfrak{Y}))^*$ spanned by the vectors $x \otimes y$, for x in $\mathfrak{X}$ and y in $\mathfrak{Y}$. If $\varphi \in (\mathfrak{X} \otimes \mathfrak{Y})^*$ and $B \in \mathbf{BIL}(\mathfrak{X} \times \mathfrak{Y})$, set $B_\varphi(x, y) = \varphi(x \otimes y)$ and $\varphi_B(a) = \langle a, B \rangle$, cf. (*), and show that the maps $\varphi \to B_\varphi$ and $B \to \varphi_B$ are the inverse of each other and determine an isometric isomorphism between $(\mathfrak{X} \otimes \mathfrak{Y})^*$ and $\mathbf{BIL}(\mathfrak{X} \times \mathfrak{Y})$.

Finally, show that if $\mathfrak{Z}$ is a Banach space and B is a bounded bilinear operator in $\mathbf{BIL}(\mathfrak{X} \times \mathfrak{Y}, \mathfrak{Z})$, there is a unique $\tilde{B}$ in $\mathbf{B}(\mathfrak{X} \otimes \mathfrak{Y}, \mathfrak{Z})$ such that

$$\tilde{B}(x \otimes y) = B(x, y), \qquad x \in \mathfrak{X}, \quad y \in \mathfrak{Y}.$$

Hint: If $a \in \mathfrak{X} \otimes \mathfrak{Y}$, define $\tilde{B}(a)$ in $\mathfrak{Z}^{**}$ by

$$\langle \tilde{B}(a), \varphi \rangle = \langle a, \varphi \circ B \rangle, \qquad \varphi \in \mathfrak{Z}^*.$$

Note that $\tilde{B}(x \otimes y) = B(x, y) \in \mathfrak{Z}$, when $\mathfrak{Z}$ is identified with its image in $\mathfrak{Z}^{**}$ (2.3.7), and finally use that the span of elements $x \otimes y$ is dense in $\mathfrak{X} \otimes \mathfrak{Y}$.

2.4. Seminormed Spaces

Synopsis. Topological vector spaces. Seminormed spaces. Continuous functionals. The Hahn–Banach separation theorem. Weak* topology. W^*-closed subspaces and their duality theory. Exercises.

2.4.1. A *topological vector space* is a vector space $\mathfrak{X}$ equipped with a Hausdorff topology such that the vector operations are continuous. Thus both maps

$$(x, y) \to x + y \text{ from } \mathfrak{X} \times \mathfrak{X} \text{ to } \mathfrak{X},$$

$$(\alpha, x) \to \alpha x \text{ from } \mathbb{F} \times \mathfrak{X} \text{ to } \mathfrak{X}$$

are continuous (when $\mathfrak{X} \times \mathfrak{X}$ and $\mathbb{F} \times \mathfrak{X}$ have the product topologies).

A topological vector space $\mathfrak{X}$ is said to be *locally convex* if the neighborhood filter around 0 has a basis consisting of convex sets. Since $\mathcal{O}(x) = x + \mathcal{O}(0)$ in any topological vector space, this implies that every point in $\mathfrak{X}$ has a basis of convex sets for its neighborhood filter.

Given a topological vector space $\mathfrak{X}$ we denote by $\mathfrak{X}^*$ the set of continuous functionals on $\mathfrak{X}$, cf. 2.3.1. This is clearly a vector space, called the *dual space* of $\mathfrak{X}$. In general, $\mathfrak{X}^*$ need not be very large, indeed it can be $\{0\}$. However, using the *Hahn–Banach separation theorem* (2.4.7), we shall show that if $\mathfrak{X}$ is locally convex, $\mathfrak{X}^*$ contains an abundance of elements.

The vast majority of topological vector spaces arising in analysis are locally convex, and we shall be exclusively concerned with this case. To facilitate the presentation—drawing on intuition gained from normed spaces—we shall (seemingly) be even more restrictive and only define topologies on vector spaces that arise from seminorms. It is a good exercise to show, using 2.4.6, that every locally convex space arises in this manner (E 2.4.17).

2.4.2. Consider a vector space $\mathfrak{X}$ with a separating family $\mathfrak{F}$ of seminorms. This means that if $x \neq y$ in $\mathfrak{X}$, there is an m in $\mathfrak{F}$ with $m(x - y) \neq 0$. Equivalently the set $\mathfrak{F} \times \mathfrak{X}$ of functions $x \to m(x - y)$, where $(m, y) \in \mathfrak{F} \times \mathfrak{X}$, separates points in $\mathfrak{X}$. The initial topology (cf. 1.4.5) induced by the family $\mathfrak{F} \times \mathfrak{X}$ is called the *seminorm topology* induced by $\mathfrak{F}$. By 1.5.3 it is a Hausdorff topology on $\mathfrak{X}$, but much more is true. We see from 1.4.5 that a subbasis for the neighborhood filter $\mathcal{O}(x)$ of a point x in $\mathfrak{X}$ is given by sets

$$\{y \in \mathfrak{X} \mid |m(y - z) - m(x - z)| < \varepsilon\}$$

for m in $\mathfrak{F}$, z in $\mathfrak{X}$ and $\varepsilon > 0$. Taking here $z = x$ we obtain the subfamily of sets $\{y \in \mathfrak{X} \mid m(y - x) < \varepsilon\}$. On the other hand, by the triangle inequality for m each of these sets is contained in any of the former (with the same m and ε), so that it suffices to consider the latter family. We conclude that sets of the form

$$\bigcap_k \{y \in \mathfrak{X} \mid m_k(y - x) < \varepsilon\} \qquad (*)$$

for $\varepsilon > 0$ and $\{m_1, \ldots, m_n\}$ a finite subset of $\mathfrak{F}$, constitute a basis for $\mathcal{O}(x)$ in the seminorm topology.

From the description $(*)$ above it is easy to describe continuity and convergence in a seminormed space $\mathfrak{X}$: a net (x_λ) converges to a point x in $\mathfrak{X}$ precisely if $m(x_\lambda - x) \to 0$ for each m in $\mathfrak{F}$. Also we see (cf. 1.4.7) that a map $f: Y \to \mathfrak{X}$ from a topological space Y into $\mathfrak{X}$ is continuous iff $y_\lambda \to y$ implies that $m(f(y_\lambda) - f(y)) \to 0$ for each m in $\mathfrak{F}$ and every convergent net (y_λ) in Y.

It is now easy to verify that every seminormed space $\mathfrak{X}$ is a locally convex topological vector space. Thus, if $x_\lambda \to x$ and $y_\lambda \to y$ in $\mathfrak{X}$, we must verify that $x_\lambda + y_\lambda \to x + y$. But from the above we have

$$m(x_\lambda + y_\lambda - (x + y)) \leq m(x_\lambda - x) + m(y_\lambda - y) \to 0$$

for each m in $\mathfrak{F}$, as desired. Similarly, if also $\alpha_\lambda \to \alpha$ in $\mathbb{F}$, then

$$m(\alpha_\lambda x_\lambda - \alpha x) \leq |\alpha_\lambda| m(x_\lambda - x) + |\alpha_\lambda - \alpha| m(x) \to 0.$$

Thus $\mathfrak{X}$ is a topological vector space, and from $(*)$ we get the desired neighborhood basis for $\mathcal{O}(x)$ of convex sets.

If $\mathfrak{F} = \{m\}$, then by the separation condition m must be a norm, and we are back in a known situation. The new phenomena will occur, therefore, when $\mathfrak{F}$ is infinite and each m in $\mathfrak{F}$ has a kernel. Typically, $m(x) = |\varphi(x)|$, where φ belongs to a (separating) family $\mathfrak{Y}$ of functionals on $\mathfrak{X}$. In this case we say that $\mathfrak{X}$ has the *weak topology* induced by $\mathfrak{Y}$.

2.4.3. Lemma. *For a set $\varphi, \varphi_1, \ldots, \varphi_n$ of functionals on a vector space $\mathfrak{X}$ the following conditions are equivalent:*

(i) $\varphi = \sum \alpha_k \varphi_k$, where $\{\alpha_1, \ldots, \alpha_n\} \subset \mathbb{F}$.
(ii) *For some $\alpha > 0$ we have $|\varphi(x)| \leq \alpha \max |\varphi_k(x)|$ for every x in $\mathfrak{X}$.*
(iii) $\bigcap \ker \varphi_k \subset \ker \varphi$.

PROOF. The implications (i) $\Rightarrow$ (ii) $\Rightarrow$ (iii) are evident, so it suffices to show that (iii) $\Rightarrow$ (i). Toward this end, consider the operator $T: \mathfrak{X} \to \mathbb{F}^n$ given by

$$Tx = (\varphi_1(x), \ldots, \varphi_n(x)).$$

By assumption we have $\ker T \subset \ker \varphi$, and it follows from linear algebra that $\varphi = f \circ T$, where f is a functional on $\mathbb{F}^n$. But then f corresponds to a vector $(\alpha_1, \ldots, \alpha_n)$ in $\mathbb{F}^n$ and thus

$$\varphi(x) = f(Tx) = \sum \alpha_k \varphi_k(x). \qquad \square$$

2.4.4. Proposition. *Consider vector spaces $\mathfrak{X}$ and $\mathfrak{Y}$ such that $\mathfrak{Y}$ is a separating space of functionals on $\mathfrak{X}$, and give $\mathfrak{X}$ the weak topology induced by $\mathfrak{Y}$. If φ is a weakly continuous functional on $\mathfrak{X}$, then $\varphi \in \mathfrak{Y}$.*

PROOF. Since φ is weakly continuous at 0, there is by (∗) in 2.4.1 an $\varepsilon > 0$ and $\varphi_1, \ldots, \varphi_n$ in $\mathfrak{Y}$ such that

$$\bigcap \{x \in \mathfrak{X} \,|\, |\varphi_k(x)| < \varepsilon\} \subset \{x \in \mathfrak{X} \,|\, |\varphi(x)| < 1\}.$$

By homogeneity this means that

$$|\varphi(x)| \leq \varepsilon^{-1} \max |\varphi_k(x)|$$

for every x in $\mathfrak{X}$. Thus φ is a linear combination of the φ_k's by 2.4.3, in particular, $\varphi \in \mathfrak{Y}$. $\qquad \square$

2.4.5. Remark. The result in 2.4.4 has a more symmetric version: If $\mathfrak{X}$ and $\mathfrak{Y}$ are vector spaces in algebraic duality via a bilinear form $\langle \cdot, \cdot \rangle$, as described in 2.3.8, and if we give $\mathfrak{X}$ and $\mathfrak{Y}$ the weak topologies induced by the families $\langle \cdot, \mathfrak{Y} \rangle$ and $\langle \mathfrak{X}, \cdot \rangle$, respectively, we obtain locally convex topological vector spaces $\mathfrak{X}$ and $\mathfrak{Y}$ such that $\mathfrak{X}^* = \mathfrak{Y}$ and $\mathfrak{Y}^* = \mathfrak{X}$.

2.4.6. Lemma. *Let $\mathfrak{C}$ be an open, convex neighborhood of 0 in a topological vector space $\mathfrak{X}$, and define*

$$m(x) = \inf\{s > 0 \,|\, s^{-1}x \in \mathfrak{C}\}.$$

Then $m: \mathfrak{X} \to \mathbb{R}_+$ is a Minkowski functional on $\mathfrak{X}$ and

$$\mathfrak{C} = \{x \in \mathfrak{X} \,|\, m(x) < 1\}.$$

PROOF. Since $n^{-1}x \to 0$ as $n \to \infty$, and $\mathfrak{C}$ is a neighborhood of 0, it follows that $n^{-1}x \in \mathfrak{C}$ eventually. This shows that $m(x) < \infty$ for every x in $\mathfrak{X}$.

It follows from the definition that m is positive homogeneous. To prove subadditivity, take x and y in $\mathfrak{X}$ and consider s and t in $\mathbb{R}_+$ such that $s^{-1}x \in \mathfrak{C}$ and $t^{-1}y \in \mathfrak{C}$. Since $\mathfrak{C}$ is convex we then have

$$(s + t)^{-1}(x + y) = (s + t)^{-1}s(s^{-1}x) + (s + t)^{-1}t(t^{-1}y) \in \mathfrak{C},$$

whence $m(x + y) \leq s + t$. As this holds for all choices of s and t, it follows that $m(x + y) \leq m(x) + m(y)$; so that m is indeed a Minkowski functional.

If $x \in \mathfrak{C}$, then also $(1 + \varepsilon)x \in \mathfrak{C}$ for some $\varepsilon > 0$, because $\mathfrak{C}$ is open. Thus $m(x) \leq (1 + \varepsilon)^{-1} < 1$. Conversely, if $m(x) < 1$ we can find $s < 1$ such that $s^{-1}x \in \mathfrak{C}$. Since $0 \in \mathfrak{C}$ and $\mathfrak{C}$ is convex, it follows that

$$x = (1 - s)0 + s(s^{-1}x) \in \mathfrak{C}. \qquad \square$$

2.4.7. Theorem. *Let $\mathfrak{A}$ and $\mathfrak{B}$ be disjoint, nonempty, convex subsets of a topological vector space $\mathfrak{X}$. If $\mathfrak{A}$ is open, there is a φ in $\mathfrak{X}^*$ and a t in $\mathbb{R}$ such that*

$$\operatorname{Re} \varphi(x) < t \leq \operatorname{Re} \varphi(y)$$

for every x in $\mathfrak{A}$ and y in $\mathfrak{B}$.

PROOF. First consider the case $\mathbb{F} = \mathbb{R}$. Choose x_0 in $\mathfrak{A}$ and y_0 in $\mathfrak{B}$ and put $z = y_0 - x_0$ and $\mathfrak{C} = \mathfrak{A} - \mathfrak{B} + z$. Then $\mathfrak{C}$ is an open, convex neighborhood of 0, because

$$\mathfrak{C} = \bigcup_{y \in \mathfrak{B}} \mathfrak{A} - y + z,$$

a union of open sets. Let m denote the Minkowski functional associated with $\mathfrak{C}$ as defined in 2.4.6. Since $\mathfrak{A} \cap \mathfrak{B} = \emptyset$, we know that $z \notin \mathfrak{C}$, whence $m(z) \geq 1$. Define φ_0 on $\mathbb{R}z$ by $\varphi_0(sz) = s$. For $s \geq 0$ this implies that

$$\varphi_0(sz) = s \leq sm(z) = m(sz),$$

so that $\varphi_0 \leq m$. By 2.3.2 we can therefore extend φ_0 to a functional φ on $\mathfrak{X}$ dominated by m. To see that φ is continuous, note that $\varphi(x) < 1$ if $x \in \mathfrak{C}$ by 2.4.6, which means that $|\varphi| < \varepsilon$ on the neighborhood $-\varepsilon\mathfrak{C} \cap \varepsilon\mathfrak{C}$ of 0.

If $x \in \mathfrak{A}$ and $y \in \mathfrak{B}$, then $x - y + z \in \mathfrak{C}$, whence $\varphi(x - y + z) < 1$. Since $\varphi(z) = 1$, it follows that $\varphi(x) < \varphi(y)$ for all pairs x and y. Thus, $\varphi(\mathfrak{A})$ and $\varphi(\mathfrak{B})$ are disjoint intervals in $\mathbb{R}$, and since $\mathfrak{A}$ is open, so is $\varphi(\mathfrak{A})$. Taking t as the right endpoint of $\varphi(\mathfrak{A})$ we obtain the desired result.

Next, if $\mathbb{F} = \mathbb{C}$, we regard it first as a real vector space and find as above a real functional ψ such that $\psi(\mathfrak{A}) < t \leq \psi(\mathfrak{B})$. Then, as in the proof of 2.3.3, we define the complex functional φ as $\varphi(x) = \psi(x) - i\psi(ix)$. Since $\operatorname{Re} \varphi = \psi$, we are done, because the continuity of φ follows from that of ψ. $\qquad \square$

2.4.8. If $\mathfrak{X}$ is a normed space, it may be considered as a separating family of functionals on its dual space $\mathfrak{X}^*$ via the bilinear form $\langle \cdot, \cdot \rangle$ described in 2.3.8. The ensuing weak topology on $\mathfrak{X}^*$ is known as the *w*-topology*, and turns $\mathfrak{X}^*$ into a locally convex topological vector space, having $\mathfrak{X}$ as its dual space, cf. 2.4.4. Convergence in the w*-topology is pointwise convergence: a net $\{\varphi_\lambda\}_{\lambda \in \Lambda}$ in $\mathfrak{X}^*$ is w*-convergent to an element φ iff $\varphi_\lambda(x) \to \varphi(x)$ for every x in $\mathfrak{X}$.

It is clear from the definition of the w*-topology as an initial topology that it is weaker than the norm topology on $\mathfrak{X}^*$. That the two topologies in fact only coincide when $\dim(\mathfrak{X}) < \infty$ will be seen from Alaoglu's theorem (2.5.2), because the unit ball in an infinite-dimensional normed space never is compact, cf. E 2.1.3.

Again using the duality between $\mathfrak{X}$ and $\mathfrak{X}^*$ we define the *weak topology* on $\mathfrak{X}$ as the one induced from $\mathfrak{X}^*$. Thus a net $(x_\lambda)_{\lambda \in \Lambda}$ in $\mathfrak{X}$ is weakly convergent to x iff $\varphi(x_\lambda) \to \varphi(x)$ for every φ in $\mathfrak{X}^*$. The weak topology is weaker than the norm topology, and, in general, they are different. However, we see from 2.4.4 that in both topologies $\mathfrak{X}^*$ is the dual space. This has a curious, but very useful, consequence: every norm closed convex subset $\mathfrak{C}$ of $\mathfrak{X}$ is weakly closed. Because if $x \notin \mathfrak{C}$, then $\mathfrak{B} \cap \mathfrak{C} = \emptyset$ for some small open ball $\mathfrak{B}$ around x. Applying 2.4.7 we find an element φ in $\mathfrak{X}^*$ that separates x and $\mathfrak{C}$, and thus a weakly closed neighborhood of $\mathfrak{C}$ of the form

$$\mathfrak{A} = \{y \in \mathfrak{X} | \operatorname{Re} \varphi(y) \geq t\},$$

such that $x \notin \mathfrak{A}$.

2.4.9. We see from 2.4.8 that a normed space $\mathfrak{X}$ may have other interesting topologies, so that the words and symbols for interior and closure have to be qualified. Therefore, from now on we shall use the symbol $\mathfrak{Y}^=$ for any subset $\mathfrak{Y}$ of $\mathfrak{X}$ to denote *norm closure* (as distinct from other weak closures of $\mathfrak{Y}$). This closure symbol also has the advantage that it cannot be confused with complex conjugation, an operation that will occur frequently in the following chapters.

2.4.10. Proposition. *Let $\mathfrak{X}$ be a normed space and $\mathfrak{Z}$ be a w^*-closed subspace of $\mathfrak{X}^*$. For every φ in $\mathfrak{X}^* \backslash \mathfrak{Z}$ there is an x in $\mathfrak{Z}^\perp$ such that $\langle x, \varphi \rangle \neq 0$.*

PROOF. Choose a w^*-open, convex neighborhood $\mathfrak{C}$ of φ disjoint from $\mathfrak{Z}$ (cf. 2.4.2). By 2.4.7 in conjunction with 2.4.4 there is then an x in $\mathfrak{X}$ and a real t such that

$$\operatorname{Re}\langle x, \varphi \rangle \in \operatorname{Re}\langle x, \mathfrak{C} \rangle < t \leq \operatorname{Re}\langle x, \mathfrak{Z} \rangle.$$

Since $\mathfrak{Z}$ is a subspace, this means that $t \leq 0$ and $x \in \mathfrak{Z}^\perp$. $\square$

2.4.11. Corollary. *Every w^*-closed subspace of $\mathfrak{X}^*$ has the form $\mathfrak{Y}^\perp$ for some norm closed subspace $\mathfrak{Y}$ of $\mathfrak{X}$.*

2.4.12. Proposition. *If $T \in \mathbf{B}(\mathfrak{X}, \mathfrak{Y})$, for Banach spaces $\mathfrak{X}$ and $\mathfrak{Y}$, the adjoint $T^*: \mathfrak{Y}^* \to \mathfrak{X}^*$ is w^*-continuous. Conversely, every w^*-continuous operator $S: \mathfrak{Y}^* \to \mathfrak{X}^*$ has the form $S = T^*$ for some T in $\mathbf{B}(\mathfrak{X}, \mathfrak{Y})$; in particular S is bounded.*

PROOF. By 1.4.7 it suffices to show that $x \circ T^*$ is w^*-continuous on $\mathfrak{Y}^*$ for each x in $\mathfrak{X}$, which is trivially true, because $x \circ T^* = Tx \in \mathfrak{Y}$.

Assume now that $S: \mathfrak{Y}^* \to \mathfrak{X}^*$ is a w^*-continuous operator. Then $x \circ S$ is a w^*-continuous functional on $\mathfrak{Y}^*$ for each x in $\mathfrak{X}$, whence $x \circ S \in \mathfrak{Y}$ by 2.4.4. We can therefore define an operator $T: \mathfrak{X} \to \mathfrak{Y}$ uniquely, such that $Tx = x \circ S$

for every x in $\mathfrak{X}$. But then

$$\langle Tx, \varphi \rangle = x \circ S(\varphi) = \langle x, S\varphi \rangle$$

for x in $\mathfrak{X}$ and φ in $\mathfrak{Y}^*$; whence $T \in \mathbf{B}(\mathfrak{X}, \mathfrak{Y})$, $S \in \mathbf{B}(\mathfrak{Y}^*, \mathfrak{X}^*)$, and $S = T^*$ by 2.3.11. $\qquad\square$

2.4.13. Proposition. *Consider a closed subspace* $\mathfrak{Y}$ *of a normed space* $\mathfrak{X}$. *Let* $I: \mathfrak{Y} \to \mathfrak{X}$ *denote the inclusion map and* $Q: \mathfrak{X} \to \mathfrak{X}/\mathfrak{Y}$ *denote the quotient map. Then we may identify* Q^* *with the inclusion map of* $\mathfrak{Y}^\perp$ *into* $\mathfrak{X}^*$ *and* I^* *with the quotient map of* $\mathfrak{X}^*$ *onto* $\mathfrak{X}^*/\mathfrak{Y}^\perp$.

PROOF. If $\varphi \in (\mathfrak{X}/\mathfrak{Y})^*$, then evidently $Q^*\varphi \in \mathfrak{Y}^\perp$. Since Q maps the open unit ball of $\mathfrak{X}$ onto that of $\mathfrak{X}/\mathfrak{Y}$, cf. 2.1.6, we have

$$\|Q^*\varphi\| = \sup\{|\langle x, Q^*\varphi\rangle|\,|\,x \in \mathfrak{X},\ \|x\| < 1\}$$

$$= \sup\{|\langle Qx, \varphi\rangle|\,|\,x \in \mathfrak{X},\ \|x\| < 1\} = \|\varphi\|;$$

which shows that Q^* is an isometry of $(\mathfrak{X}/\mathfrak{Y})^*$ into $\mathfrak{Y}^\perp$ in $\mathfrak{X}^*$. On the other hand, each ψ in $\mathfrak{Y}^\perp$ determines a unique φ in $(\mathfrak{X}/\mathfrak{Y})^*$ such that $\langle x, \psi \rangle = \langle Qx, \varphi \rangle$, whence $Q^*\varphi = \psi$, so that Q^* is surjective.

Since $QI = 0$ we know that $I^*Q^* = 0$, cf. 2.3.9, whence $\mathfrak{Y}^\perp \subset \ker I^*$. By definition, however, I^* is the restriction map $\varphi \to \varphi|\mathfrak{Y}$ of $\mathfrak{X}^*$, so $\ker I^* = \mathfrak{Y}^\perp$. Denoting by $\tilde{Q}$ the quotient map of $\mathfrak{X}^*$ to $\mathfrak{X}^*/\mathfrak{Y}^\perp$, we have by 2.1.7 an operator $\tilde{I}^*$ in $\mathbf{B}(\mathfrak{X}^*/\mathfrak{Y}^\perp, \mathfrak{Y}^*)$ such that $I^* = \tilde{I}^*\tilde{Q}$. By the Hahn–Banach extension theorem (2.3.3) each ψ in $\mathfrak{Y}^*$ extends to a φ in $\mathfrak{X}^*$ with $\|\varphi\| = \|\psi\|$. Thus, $\psi = I^*\varphi$, so that I^* is surjective. Moreover, $\tilde{I}^*$ is an isometry, because $\|\tilde{I}^*\| = \|I^*\| = \|I\| = 1$ and

$$\|\tilde{I}^*(\varphi + \mathfrak{Y}^\perp)\| = \|\psi\| = \|\varphi\| \geq \|\varphi + \mathfrak{Y}^\perp\|.$$

We may therefore identify I^* with $\tilde{Q}$, as desired. $\qquad\square$

EXERCISES

E 2.4.1. Let $\{\mathfrak{C}_j|\,j \in J\}$ be a family of convex subsets in a vector space $\mathfrak{X}$. Show that the smallest convex set in $\mathfrak{X}$ containing all the $\mathfrak{C}_j$'s (the *convex hull*, denoted by $\text{conv}\{\mathfrak{C}_j|\,j \in J\}$) consists of all points of the form $x = \sum \lambda_j x_j$, where $x_j \in \mathfrak{C}_j$, $\lambda_j \geq 0$ for all j and $\lambda_j = 0$ for all but finitely many j, and $\sum \lambda_j = 1$.

E 2.4.2. Let $\mathfrak{C}_1$ and $\mathfrak{C}_2$ be convex, compact subsets of a topological vector space $\mathfrak{X}$. Show that $\text{conv}\{\mathfrak{C}_1 \cup \mathfrak{C}_2\}$ (cf. E 2.4.1) is compact.
    Hint: Use the fact that the set $\{(s, t) \in \mathbb{R}_+ \times \mathbb{R}_+|\,s + t = 1\}$ is compact in $\mathbb{R}^2$.

E 2.4.3. Let $\mathfrak{C}$ be a convex subset of a topological vector space $\mathfrak{X}$. Show that if $x \in \mathfrak{C}^\circ$ and $y \in \mathfrak{C}^-$, every point of the form $z = \lambda x + (1 - \lambda)y$ for $0 < \lambda \leq 1$ belongs to $\mathfrak{C}^\circ$. Show that if $\mathfrak{C}^\circ \neq \emptyset$, then $(\mathfrak{C}^\circ)^- = \mathfrak{C}^-$ and $(\mathfrak{C}^-)^\circ = \mathfrak{C}^\circ$.

E 2.4.4. Let $\mathfrak{X}$ be a normed space and $\mathfrak{Y}$ be a subspace of $\mathfrak{X}^*$. Show that $\mathfrak{Y}$ is separating for $\mathfrak{X}$ iff $\mathfrak{Y}$ is w^*-dense in $\mathfrak{X}^*$.

E 2.4.5. Let $\mathfrak{X}$ be a normed space, identified with a subspace of $\mathfrak{X}^{**}$, cf. 2.3.9. Let $\mathfrak{B}$ and $\mathfrak{B}^{**}$ denote the closed unit balls in $\mathfrak{X}$ and $\mathfrak{X}^{**}$, respectively. Show that $\mathfrak{B}$ is dense in $\mathfrak{B}^{**}$, when $\mathfrak{X}^{**}$ is given the weak topology induced by $\mathfrak{X}^*$.

  Hint: If $a \in \mathfrak{B}^{**} \backslash \mathfrak{B}^{-w}$, it has an open, convex neighborhood disjoint from $\mathfrak{B}^{-w}$, and 2.4.7 applies.

E 2.4.6. Let (x_n) be a sequence in a normed space $\mathfrak{X}$, such that $\varphi(x_n) \to \varphi(x)$ for some x in $\mathfrak{X}$ and all φ in $\mathfrak{X}^*$. Show that for each $\varepsilon > 0$ and m there is a convex combination $y = \sum \lambda_n x_n$, with all $n \geq m$, such that $\|x - y\| < \varepsilon$.

  Hint: Let $\mathfrak{C} = (\mathrm{conv}\{x_n | n \in \mathbb{N}\})^=$ and use 2.4.7 to see that $x \in \mathfrak{C}$; cf. the last paragraph in 2.4.8.

E 2.4.7. Consider the space ℓ^1, both as a Banach space and as the locally convex topological space with the weak topology induced by its dual space ℓ^∞; cf. E 2.3.2. Show that every weakly convergent sequence in ℓ^1 is norm convergent. Explain why this fact does not imply that the two topologies coincide.

  Hint: If $x_n \to 0$ weakly, but $\|x_n\|_1 \geq \varepsilon > 0$ for infinitely many n, we can find (passing if necessary to a subsequence) an increasing sequence of numbers $a(n)$, with $a(0) = 0$, such that

$$\sum_{k=1}^{a(n-1)} |x_n(k)| \leq n^{-1}, \qquad \sum_{k=a(n-1)+1}^{a(n)} |x_n(k)| \geq \|x_n\|_1 - 2n^{-1},$$

for every n. Define x in ℓ^∞ by $x(k) = \mathrm{sign}\, \overline{x_n(k)}$ for $a(n-1) < k \leq a(n)$, and check that

$$|\langle x_n, x \rangle| \geq \|x_n\|_1 - 4n^{-1} \geq \varepsilon - 4n^{-1}.$$

E 2.4.8. Let $\mathfrak{Y}$ be a closed subspace of a reflexive Banach space $\mathfrak{X}$. Show that both $\mathfrak{Y}$ and $\mathfrak{X}/\mathfrak{Y}$ are reflexive.

  Hint: Use 2.4.13.

E 2.4.9. (*Uniformly convex spaces.*) A Banach space $\mathfrak{X}$ is uniformly convex if whenever we have sequences (x_n) and (y_n) in $\mathfrak{X}$, with $\|x_n\| = \|y_n\| = 1$ for all n and $\|\frac{1}{2}(x_n + y_n)\| \to 1$, then $\|x_n - y_n\| \to 0$. Show in this case that every closed, convex subset of $\mathfrak{X}$ has a uniquely determined element with smallest norm.

E 2.4.10. Show that if J is not a singleton, the Banach space $c_0(J)$ (cf. 2.1.18) is not uniformly convex (see E 2.4.9).

E 2.4.11. Let $\mathfrak{X}$ be a uniformly convex Banach space (cf. E 2.4.9), equipped also with the weak topology induced by a separating subspace $\mathfrak{Y}$ of $\mathfrak{X}^*$

such that the unit ball of $\mathfrak{Y}$ is w^*-dense in the unit ball of $\mathfrak{X}^*$. Show that if (x_n) converges weakly to some x in $\mathfrak{X}$, and $\|x_n\| \to \|x\|$ as well, then (x_n) converges to x in norm.

E 2.4.12. Show that every uniformly convex Banach space is reflexive.

Hint: By assumption there is to each $\varepsilon > 0$ a δ such that $\|\frac{1}{2}(x + y)\| > 1 - \delta$ implies $\|x - y\| < \varepsilon$ for all vectors x, y in $\mathfrak{B}(0, 1)$; cf. E 2.4.9. Take z in $\mathfrak{X}^{**}$ with $\|z\| = 1$ and choose φ in $\mathfrak{X}^*$ with $\|\varphi\| = 1$ and $|\langle z, \varphi \rangle - 1| < \delta$. Put

$$\mathfrak{C} = \{x \in \mathfrak{B}(0, 1) \mid |\langle x, \varphi \rangle - 1| < \delta\}$$

and use E 2.4.5 to show that z belongs to the weak closure of $\mathfrak{C}$. Note that $\|x - y\| < \varepsilon$ for all x and y in $\mathfrak{C}$ and conclude that $\|z - x\| \leq \varepsilon$ for some x in $\mathfrak{C}$.

E 2.4.13. Show that the Banach spaces $L^p(X)$, defined in 2.1.14 (or 2.1.15), are uniformly convex (E 2.4.9) for $p \geq 2$.

Hint: Use the inequality

$$|s + t|^p + |s - t|^p \leq 2^{p-1}(|s|^p + |t|^p),$$

valid for all real numbers s and t.

E 2.4.14. Given Banach spaces $\mathfrak{X}$ and $\mathfrak{Y}$, such that $\mathfrak{X}$ is reflexive and $T: \mathfrak{Y} \to \mathfrak{X}^*$ is an isometry of $\mathfrak{Y}$ on a w^*-dense subspace of $\mathfrak{X}^*$. Show that $T(\mathfrak{Y}) = \mathfrak{X}^*$, so that also $\mathfrak{Y}$ is reflexive with $\mathfrak{Y}^* = T^*(\mathfrak{X})$.

E 2.4.15. Show that the Banach spaces $L^p(X)$, defined in 2.1.14 (or 2.1.15), are reflexive, and that $L^p(X)^* = L^q(X)$ whenever $p^{-1} + q^{-1} = 1$.

Hint: Take $p \geq 2$, and use the Hölder inequality (6.4.6) to construct an isometry T of $L^q(X)$ into $L^p(X)^*$. Now use E 2.4.13 and E 2.4.12 to see that the assumptions in E 2.4.14 are satisfied.

E 2.4.16. Show that a continuous Minkowski functional m on a topological vector space $\mathfrak{X}$ is a seminorm iff the corresponding open convex set $\mathfrak{C} = \{x \in \mathfrak{X} \mid m(x) < 1\}$ is *balanced*, i.e. $\alpha \mathfrak{C} = \mathfrak{C}$, whenever $\alpha \in \mathbb{F}$ and $|\alpha| = 1$. (If $\mathbb{F} = \mathbb{R}$ the term *symmetrical* is used.)

E 2.4.17. Show that if $\mathfrak{X}$ is a locally convex topological vector space there is a basis for the neighborhood filter around 0 consisting of balanced, open, convex sets. Conclude that every locally convex topological vector space is a seminormed space.

Hint: If $\mathfrak{C}$ is an open, convex neighborhood of 0 and $x_\lambda \to 0$, then with $S = \{\alpha \in \mathbb{F} \mid |\alpha| = 1\}$ we have $Sx_\lambda \subset \mathfrak{C}$, eventually. Otherwise, passing to a subnet, we could find α_λ in S such that $\alpha_\lambda x_\lambda \notin \mathfrak{C}$ but $\alpha_\lambda \to \alpha$ in S. Since scalar multiplication is continuous as a function from $\mathbb{F} \times \mathfrak{X}$ to $\mathfrak{X}$ this leads to a contradiction. It follows that $\bigcap \alpha \mathfrak{C}$, $\alpha \in S$, is a neighborhood of 0 whose interior will provide a balanced, convex neighborhood.

E 2.4.18. Let $\mathfrak{X}$ and $\mathfrak{Y}$ be locally convex topological vector spaces in duality via a bilinear form $\langle \cdot, \cdot \rangle$, cf. 2.4.5. For each subset $\mathfrak{C}$ of $\mathfrak{X}$ define the *polar* of $\mathfrak{C}$ as

$$\mathfrak{C}^0 = \{\varphi \in \mathfrak{Y} | \forall x \in \mathfrak{C}: \operatorname{Re}\langle x, \varphi \rangle \geq -1\}$$

Show that $\mathfrak{C}^0$ is a closed, convex subset of $\mathfrak{Y}$. Define the polar of any subset of $\mathfrak{Y}$ by the corresponding formula to obtain a closed, convex subset of $\mathfrak{X}$. Show that the *bi-polar* $\mathfrak{C}^{00}$ contains $\mathfrak{C}$. Show finally— as a generalization of 2.3.6 and 2.4.11 — that $\mathfrak{C}^{00} = \mathfrak{C}$ iff $\mathfrak{C}$ is a closed, convex subset of $\mathfrak{X}$.

Hint: If $x \in \mathfrak{C}^{00} \backslash \mathfrak{C}$, apply 2.4.7 to obtain a contradiction.

2.5. w^*-Compactness

Synopsis. Alaoglu's theorem. Krein–Milman's theorem. Examples of extremal sets. Extremal probability measures. Krein–Smulian's theorem. Vector-valued integration. Exercises.

2.5.1. In this section we consider a normed space $\mathfrak{X}$ and its dual space $\mathfrak{X}^*$. We shall be particularly interested in the closed unit ball of $\mathfrak{X}^*$, which we denote by $\mathfrak{B}^*$; but also other convex (compact) sets may occur. Note that if $\mathfrak{X}^*$ is given the w^*-topology (2.4.8), then $\mathfrak{B}^*$ is a w^*-closed subset of $\mathfrak{X}^*$, but in general not a w^*-neighborhood of 0 [unless $\dim(\mathfrak{X}) < \infty$]. The following easy, but fundamental, result is known as *Alaoglu's theorem*.

2.5.2. Theorem. *For each normed space $\mathfrak{X}$, the unit ball $\mathfrak{B}^*$ of $\mathfrak{X}^*$ is w^*-compact.*

PROOF. Let $(\varphi_\lambda)_{\lambda \in \Lambda}$ be a universal net in $\mathfrak{B}^*$ (cf. 1.3.7). For every x in $\mathfrak{X}$ we have $|\varphi_\lambda(x)| \leq \|x\|$, so that the image net $(\varphi_\lambda(x))_{\lambda \in \Lambda}$ is contained in a compact subset of $\mathbb{F}$, and thus convergent to a number $\varphi(x)$ by 1.6.2(iv). For all x and y in $\mathfrak{X}$ and α in $\mathbb{F}$ we have, by straightforward computations with limits, that

$$|\varphi(x)| \leq \|x\|, \qquad \varphi(x + y) = \varphi(x) + \varphi(y), \qquad \varphi(\alpha x) = \alpha \varphi(x).$$

It follows that we have constructed a φ in $\mathfrak{B}^*$ such that $\varphi_\lambda \to \varphi$ (in w^*-topology). Thus every universal net is convergent, whence $\mathfrak{B}^*$ is compact by 1.6.2. $\qquad \square$

2.5.3. A *face* of a convex subset $\mathfrak{C}$ in a vector space $\mathfrak{X}$ is a nonempty, convex subset $\mathfrak{F}$ of $\mathfrak{C}$ with the property that $\lambda x + (1 - \lambda)y \in \mathfrak{F}$ implies $x \in \mathfrak{F}$ and $y \in \mathfrak{F}$, for all x and y in $\mathfrak{C}$ and $0 < \lambda < 1$.

An *extreme point* in $\mathfrak{C}$ is a one-point face, i.e., a point in $\mathfrak{C}$ that cannot be expressed as a nontrivial convex combination of elements from $\mathfrak{C}$. The *extremal boundary* of $\mathfrak{C}$ is the set of extreme points in $\mathfrak{C}$, denoted by $\partial \mathfrak{C}$ (and not to be confused with the equisymbolized topological boundary).

We are primarily interested in proving the *Krein–Milman theorem* (2.5.4) for convex, w^*-compact subsets in the dual space $\mathfrak{X}^*$ of a normed space $\mathfrak{X}$. However, it is convenient to have the more general result at our disposal.

2.5.4. Theorem. *Consider a vector space $\mathfrak{X}$ equipped with the weak topology induced by a separating space $\mathfrak{X}^*$ of functionals on $\mathfrak{X}$. Then for each convex, compact subset $\mathfrak{C}$ of $\mathfrak{X}$, the convex hull of the extremal boundary $\partial\mathfrak{C}$ of $\mathfrak{C}$ is dense in $\mathfrak{C}$.*

PROOF. Take a closed face $\mathfrak{C}_0$ of $\mathfrak{C}$, and consider the set Λ of closed faces of $\mathfrak{C}_0$. From the definition of a face in 2.5.3, we see that faces of $\mathfrak{C}_0$ are also faces of $\mathfrak{C}$. We order Λ under reverse inclusion, and claim that the order is inductive. Indeed, if $\{\mathfrak{F}_j | j \in J\}$ is a totally ordered subset of faces of $\mathfrak{C}_0$, then $\bigcap \mathfrak{F}_j \neq \emptyset$ because $\mathfrak{C}$ is compact, and clearly $\bigcap \mathfrak{F}_j$ is a face in Λ majorizing every $\mathfrak{F}_j$. By Zorn's lemma (1.1.3) we conclude that $\mathfrak{C}_0$ contains a minimal face $\mathfrak{F}$.

Take φ in $\mathfrak{X}^*$ and put

$$s = \inf\{\operatorname{Re}\langle x, \varphi\rangle \,|\, x \in \mathfrak{F}\}.$$

Since the function $x \to \operatorname{Re}\langle x, \varphi\rangle$ is weakly continuous, it attains its minimal value s on $\mathfrak{F}$ [use e.g. 1.6.2(v)], so that

$$\mathfrak{F}_\varphi = \{x \in \mathfrak{F} \,|\, \operatorname{Re}\langle x, \varphi\rangle = s\} \neq \emptyset. \tag{*}$$

Evidently $\mathfrak{F}_\varphi$ is a face of $\mathfrak{F}$, hence of $\mathfrak{C}_0$, and since $\mathfrak{F}$ is minimal, we conclude that $\mathfrak{F}_\varphi = \mathfrak{F}$. Since $\mathfrak{X}^*$ separates points in $\mathfrak{X}$, this implies that $\mathfrak{F}$ must be a one-point set, i.e., $\mathfrak{F}$ is an extreme point. We have thus shown that $\partial\mathfrak{C} \cap \mathfrak{C}_0 \neq \emptyset$ for every closed face $\mathfrak{C}_0$ of $\mathfrak{C}$.

Now let $\operatorname{conv}\{\partial\mathfrak{C}\}$ denote the convex hull of $\partial\mathfrak{C}$, i.e. the set of convex combinations of points from $\partial\mathfrak{C}$. Then $\operatorname{conv}\{\partial\mathfrak{C}\}$ is a convex subset of $\mathfrak{C}$, and its closure $\mathfrak{B}$ is therefore also convex, since the vector operations are continuous. If $x \in \mathfrak{C}\backslash\mathfrak{B}$, there is an open, convex neighborhood $\mathfrak{A}$ of x disjoint from $\mathfrak{B}$, cf. 2.4.2. Applying 2.4.7 we find φ in $\mathfrak{X}^*$ and t in $\mathbb{R}$, such that

$$\operatorname{Re}\varphi(x) \in \operatorname{Re}\varphi(\mathfrak{A}) < t \leq \operatorname{Re}\varphi(\mathfrak{B}).$$

Thus $s < t$, where s denotes the minimum of $\operatorname{Re}\varphi$ on $\mathfrak{C}$, and the face $\mathfrak{F}_\varphi$ defined as in (*) above (with $\mathfrak{F}$ replaced by $\mathfrak{C}$) satisfies $\mathfrak{F}_\varphi \cap \mathfrak{B} = \emptyset$. In particular, $\mathfrak{F}_\varphi \cap \partial\mathfrak{C} = \emptyset$, which contradicts the first result in the proof. Consequently, $\mathfrak{B} = \mathfrak{C}$, as desired. $\qquad\square$

2.5.5. The following strategy for the attack on a problem concerning a convex, compact set is often successful: First, find the extreme points of the set. These points are often simpler to handle, so that the problem can be solved for them. Now if the solution is stable under the formation of convex combinations, and stable under limits (i.e. continuous), then the Krein–Milman theorem asserts that the solution is valid on the whole set.

In order to use the strategy outlined above, it is necessary to have a catalogue of extremal boundaries for the most common convex sets, which are often unit balls in dual spaces. The catalogue follows. Except for the last, most important, item, the proofs are left to the reader.

2.5.6. Catalogue. (a) Consider $C(X)$, where X is an infinite compact Hausdorff space. The unit ball in $C(X)$ under ∞-norm is not compact. Even so, the ball is often well supplied with extreme points. These are the functions f in $C(X)$, such that $|f(x)| = 1$ for every x in X. If $\mathbb{F} = \mathbb{R}$ and X is connected, there are only two extreme points. However, if $\mathbb{F} = \mathbb{C}$, the convex hull of the extreme points (the unitary functions) is uniformly dense in the ball.

(b) Consider $L^1(X)$, where $X \subset \mathbb{R}^n$ (cf. 2.1.14). The unit ball is not compact, and there are no extreme points.

(c) Consider $L^p(X)$ for $1 < p < \infty$. Since $L^p(X) = (L^q(X))^*$ if $p^{-1} + q^{-1} = 1$ by 6.5.11, we know from 2.5.2 that the unit ball is w^*-compact. In this case, however, the extreme boundary coincides with the topological boundary, so that every unit vector is an extreme point. This corresponds to the geometrical fact that p-norms give "uniformly round" balls with no edges.

(d) Consider $L^\infty(X) = (L^1(X))^*$ (cf. 6.5.11). The extreme points in the unit ball are the functions f such that $|f(x)| = 1$ for (almost) all x in X.

(e) Consider the convex set of monotone increasing functions $f: [0,1] \to [0,1]$, which is compact in the topology of pointwise convergence. The extreme points are those functions that only take the values 0 and 1.

(f) Consider the convex set of holomorphic functions f on an open subset Ω of $\mathbb{C}$, such that $\|f\|_\infty \leq 1$. The extreme points are the functions $f(z) = \alpha(z - z_0)^{-1}$, where $z_0 \notin \Omega$ and $|\alpha| = d(z_0, \Omega)$. The strategy in 2.5.5 is capitalized in the Cauchy integral formula.

(g) Consider $\mathbf{M}_n$—the $n \times n$-matrices over $\mathbb{F}$. Note that $\mathbf{M}_n^* = \mathbf{M}_n$, because $\dim(\mathbf{M}_n) = n^2$. Identifying $\mathbf{M}_n$ with $\mathbf{B}(\mathbb{F}^n)$, we obtain an operator norm on $\mathbf{M}_n$ corresponding to the 2-norm on $\mathbb{F}^n$. The extreme points in the unit ball of $\mathbf{M}_n$ are the isometries on $\mathbb{F}^n$. For $\mathbb{F} = \mathbb{R}$ these are the orthogonal matrices, for $\mathbb{F} = \mathbb{C}$ the unitary matrices.

(h) Consider $\mathbf{B}(\mathfrak{H})$, where $\mathfrak{H}$ is an infinite-dimensional Hilbert space (see Chapter 3). By 3.4.13 we have $\mathbf{B}(\mathfrak{H}) = (\mathbf{B}^1(\mathfrak{H}))^*$, so that the unit ball in $\mathbf{B}(\mathfrak{H})$ is compact in the weak* topology (known in this case as the σ-weak topology). The extreme points are operators T such that either $T^*T = I$ (isometries) or $TT^* = I$ (co-isometries), possibly both (unitaries). Quite unexpectedly, the convex hull of the extreme points is the whole unit ball. In fact, already the convex hull of the unitary operators contains the open unit ball, see 3.2.23.

(i) Consider $\mathbf{B}(\mathfrak{H})_{sa}$, the self-adjoint operators in $\mathbf{B}(\mathfrak{H})$. This is a real Banach space (and a dual space in the σ-weak topology). The extreme points in the unit ball are the symmetries, i.e. the operators $S = S^*$ such that $S^2 = I$.

(j) Consider $\mathbf{B}(\mathfrak{H})_+$, the positive (definite) operators in $\mathbf{B}(\mathfrak{H})$. This is a closed, generating cone for $\mathbf{B}(\mathfrak{H})$. The extreme points in the unit ball of $\mathbf{B}(\mathfrak{H})_+$ are the (orthogonal) projections, i.e. the operators $P = P^*$ such that $P^2 = P$. The strategy in 2.5.5 is capitalized in the spectral theorem for compact operators (3.3.7) and for arbitrary (normal) operators (4.5.8).

2.5.7. Proposition. *Consider the Banach algebra $C(X)$, with X a compact Hausdorff space, equipped with the pointwise algebraic operations and ∞-norm. Let $M(X)$ denote the dual space of $C(X)$ and $P(X)$ denote the subset consisting* of those μ in $M(X)$ such that $\|\mu\| \leq 1$ and $\mu(1) = 1$. Then $P(X)$ is a convex, w*-compact set, whose extremal points are the Dirac measures δ_x, $x \in X$, given by $\delta_x(f) = f(x)$, for every f in $C(X)$.

PROOF. The unit ball in $M(X)$ is weak* compact (2.5.2), and $P(X)$ is a w*-closed face of it, because the map $\mu \to \mu(1)$ is a w*-continuous function. Thus $P(X)$ is a convex, w*-compact set.

If $\mu \in P(X)$ and $f = \bar{f}$ in $C(X)$, then $\mu(f) \in \mathbb{R}$. Indeed, if $\mu(f) = s + it$, then for each n,

$$s^2 + t^2(1 + n)^2 = |\mu(f + itn)|^2 \leq \|f + itn\|^2 = \|f\|^2 + t^2 n^2,$$

which only holds if $t = 0$. Moreover, if $f \geq 0$, then $\mu(f) \geq 0$. Indeed, as $0 \leq \|f\| - f \leq \|f\|$, we have

$$\|f\| - \mu(f) = \mu(\|f\| - f) \leq \|f\|,$$

whence $\mu(f) \geq 0$. We have therefore shown that $P(X)$ consists of positive, self-adjoint functionals. As a consequence we have, for each f in $C(X)$,

$$\mathrm{Re}\,\mu(f) = \mu(\mathrm{Re}\,f) \leq \mu(|f|),$$

which implies that $|\mu(f)| \leq \mu(|f|)$.

Suppose now that μ is an extreme point in $P(X)$, and take f in $C(X)$ with $0 \leq f \leq 1$. Then with $\alpha = \mu(f)$ we have $0 \leq \alpha \leq 1$. Assuming that $0 < \alpha < 1$ we define the functionals φ and ψ on $C(X)$ by

$$\varphi(g) = \alpha^{-1}\mu(fg), \qquad \psi(g) = (1 - \alpha)^{-1}\mu((1 - f)g)$$

for g in $C(X)$. Note that $\varphi(1) = \psi(1) = 1$ and that

$$|\mu(fg)| \leq \mu(|fg|) = \mu(f|g|) \leq \|g\|\mu(f) = \alpha\|g\|, \tag{$*$}$$

whence $\|\varphi\| \leq 1$; so that $\varphi \in P(X)$. Similarly $\psi \in P(X)$. Now by definition $\mu = \alpha\varphi + (1 - \alpha)\psi$, and since μ is extreme, this implies that $\varphi = \psi = \mu$. We conclude that $\mu(fg) = \mu(f)\mu(g)$ if $0 < \mu(f) < 1$; but of course that equation holds also when $\mu(f) = 0$ by $(*)$ above, and replacing f with $1 - f$ we see that it holds when $\mu(f) = 1$ as well. It is elementary to verify that the linear span of the positive functions in the unit ball of $C(X)$ is all of $C(X)$, and since μ is linear, we see that $\mu(fg) = \mu(f)\mu(g)$ for all f and g in $C(X)$; i.e. μ is a multiplicative functional, or a homomorphism of $C(X)$ onto $\mathbb{C}$.

Assume for a moment that to each x in X there is an f in $\ker\mu$ such that $f(x) \neq 0$. Then f is also nonzero in a neighborhood of x, so by a standard compactness argument we find $f_1, \ldots, f_n$ in $\ker\mu$ such that the open sets $\{x \in X | f_k(x) \neq 0\}$, $1 \leq k \leq n$, form a covering of X. The function $f = \sum \bar{f}_k f_k$ belongs to $\ker\mu$, because μ is multiplicative; but $f(x) > 0$ for every x in X, so that $1/f \in C(X)$. This leads to the contradiction

$$1 = \mu(1) = \mu(f \cdot 1/f) = \mu(f)\mu(1/f) = 0.$$

Consequently, there is an x in X such that (in the notation of 2.5.7) $\ker\delta_x \supset$

ker μ. Since $f - \mu(f)1 \in \ker \mu$ for every f in $C(x)$, we see that $f(x) = \mu(f)$, i.e. $\delta_x = \mu$.

Conversely, if we have a convex combination $\alpha\varphi + (1 - \alpha)\psi = \delta_x$ for some x in X, $0 < \alpha < 1$, and φ, ψ in $P(X)$, then from the inequalities

$$\alpha|\varphi(f)| \leq \alpha\varphi(|f|) \leq \delta_x(|f|) = |f|(x) = |f(x)|,$$

we see that $\ker \varphi \supset \ker \delta_x$. As above this implies that $\varphi = \delta_x$, whence also $\psi = \delta_x$; so that δ_x is an extreme point in $P(X)$.   □

2.5.8. Remark. To appreciate the result in 2.5.7 one should know that $M(X)$ is the set of Radon charges on X (signed measures, see 6.5.5); whereas $P(X)$ are the probability measures on X. Thus we learn from the Krein–Milman theorem that every probability measure on X can be approximated pointwise on $C(X)$ by measures with finite support on X.

2.5.9. Theorem. *Let $\mathfrak{X}$ be a Banach space, and denote by $\mathfrak{B}^*$ the unit ball in $\mathfrak{X}^*$. A convex set $\mathfrak{C}$ in $\mathfrak{X}^*$ is then w^*-closed if each of the sets $r\mathfrak{B}^* \cap \mathfrak{C}$, $r > 0$, is w^*-compact.*

PROOF. Under the assumptions on $\mathfrak{C}$, it is easy to see that it must at least be norm closed. Indeed, every norm convergent sequence contained in $\mathfrak{C}$ is bounded, and thus contained in a ball $r\mathfrak{B}^*$ for some $r > 0$; and $r\mathfrak{B}^* \cap \mathfrak{C}$ is w^*-closed and, a fortiori, norm closed.

Therefore, if $\varphi \notin \mathfrak{C}$, there is an $r > 0$ such that $r\mathfrak{B}^* \cap (\mathfrak{C} - \varphi) = \emptyset$. We may as well assume that $r = 1$, and replacing $\mathfrak{C}$ with $\mathfrak{C} - \varphi$, we may assume that $\mathfrak{B}^* \cap \mathfrak{C} = \emptyset$. We have then to prove that 0 is not in the w^*-closure of $\mathfrak{C}$.

For any subset $\mathfrak{F}$ of $\mathfrak{X}$ we define the *polar* of $\mathfrak{F}$ as the w^*-closed, convex subset

$$P(\mathfrak{F}) = \bigcap_{x \in \mathfrak{F}} \{\varphi \in \mathfrak{X}^* | \mathrm{Re}\langle x, \varphi \rangle \geq -1\}.$$

We claim that $P(r\mathfrak{B}) = r^{-1}\mathfrak{B}^*$, where $\mathfrak{B}$ as usual denotes the closed unit ball in $\mathfrak{X}$. Clearly, $r^{-1}\mathfrak{B}^* \subset P(r\mathfrak{B})$, and if $\varphi \notin r^{-1}\mathfrak{B}^*$, there is by 2.4.7 an x in $\mathfrak{X}$ and a t in $\mathbb{R}$ such that

$$\mathrm{Re}\langle x, \varphi \rangle < t \leq \mathrm{Re}\langle x, r^{-1}\mathfrak{B}^* \rangle.$$

As $\mathrm{Re}\langle x, r^{-1}\mathfrak{B}^* \rangle = [-r^{-1}\|x\|, r^{-1}\|x\|]$ by 2.3.4, we obtain by the normalization $\|x\| = r$ an element in $r\mathfrak{B}$ such that $\mathrm{Re}\langle x, \varphi \rangle < -1$, whence $\varphi \notin P(r\mathfrak{B})$. Thus $P(r\mathfrak{B}) = r^{-1}\mathfrak{B}^*$, as claimed.

We now by induction define a sequence $\mathfrak{F}_1, \mathfrak{F}_2, \ldots$ of finite subsets of $\mathfrak{X}$ such that $\mathfrak{F}_1 = \{0\}$ and

$$\mathfrak{F}_n \subset (n - 1)^{-1}\mathfrak{B}, \tag{$*$}$$

$$n\mathfrak{B}^* \cap \mathfrak{C} \cap P(\mathfrak{F}_1) \cap \cdots \cap P(\mathfrak{F}_n) = \emptyset, \tag{$**$}$$

for every n. The case $n = 1$ is given in advance, and we assume that $\mathfrak{F}_1, \mathfrak{F}_2, \ldots, \mathfrak{F}_n$ have been chosen. Put

$$\mathfrak{D} = (n+1)\mathfrak{B}^* \cap \mathfrak{C} \cap P(\mathfrak{F}_1) \cap \cdots \cap P(\mathfrak{F}_n),$$

and note that $\mathfrak{D}$ is a convex, w^*-compact subset of $\mathfrak{X}^*$ such that $\mathfrak{D} \cap n\mathfrak{B}^* = \emptyset$. This means that

$$\emptyset = \mathfrak{D} \cap P(n^{-1}\mathfrak{B}) = \bigcap_{x \in n^{-1}\mathfrak{B}} \mathfrak{D} \cap P(\{x\}),$$

and since $\mathfrak{D}$ is compact, there is a finite subset $\mathfrak{F}_{n+1}$ in $n^{-1}\mathfrak{B}$ such that $\mathfrak{D} \cap P(\mathfrak{F}_{n+1}) = \emptyset$. Thus $\mathfrak{F}_{n+1}$ satisfies (*) and (**), and the induction proceeds.

The elements in $\bigcup \mathfrak{F}_n$ can be ordered in a sequence (x_n), which converges to 0 by (*). We can therefore define the bounded operator $T: \mathfrak{X}^* \to c_0$ (cf. 2.1.18) by $T(\varphi) = (\langle x_n, \varphi \rangle)$. If $\varphi \in \mathfrak{C}$, then

$$m\mathfrak{B}^* \cap \{\varphi\} \cap P(\{x_n \mid n \in \mathbb{N}\}) = \emptyset$$

by (**) for every m (in particular for $m > \|\varphi\|$), which means that $\inf \operatorname{Re}\langle x_n, \varphi \rangle \le -1$, whence $\|T\varphi\|_\infty \ge 1$. This shows that the convex set $T(\mathfrak{C})$ does not intersect the open unit ball $\mathscr{B}_0$ in c_0. As $c_0^* = \ell^1$, we can apply 2.4.7 to obtain an element $\lambda = (\lambda_n)$ in ℓ^1 such that

$$\operatorname{Re}\langle \mathscr{B}_0, \lambda \rangle < t \le \operatorname{Re}\langle T(\mathfrak{C}), \lambda \rangle$$

for some t in $\mathbb{R}$. Normalizing $\|\lambda\|_1 = 1$ we have $1 \le t$. Put $x = \sum \lambda_n x_n$ in (the completion of) $\mathfrak{X}$ and note that

$$\operatorname{Re}\langle x, \varphi \rangle = \sum \operatorname{Re}\langle \lambda_n x_n, \varphi \rangle = \operatorname{Re}\langle T\varphi, \lambda \rangle \ge 1$$

for every φ in $\mathfrak{C}$; which shows that 0 does not belong to the w^*-closure of $\mathfrak{C}$, as desired. This is the *Krein–Smulian theorem*. $\qquad\square$

2.5.10. Corollary. *A subspace $\mathfrak{Z}$ in $\mathfrak{X}^*$ is w^*-closed iff $\mathfrak{Z} \cap \mathfrak{B}^*$ is w^*-closed ($= w^*$-compact).*

2.5.11. Corollary. *A functional x on $\mathfrak{X}^*$ is w^*-continuous (and therefore belongs to $\mathfrak{X}$) iff the restriction $x|\mathfrak{B}^*$ is w^*-continuous.*

2.5.12. Consider a function $f: X \to \mathfrak{X}$, where X is a locally compact Hausdorff space, and $\mathfrak{X}$ is a Banach space. If $\int$ is a Radon integral on X, cf. 6.1, we may ask for suitable conditions on f, which will ensure the existence of an element in $\mathfrak{X}$, worthy of the symbol $\int f$. Such a *vector-valued* integral should at least be consistent with the coherent family of scalar-valued integrals obtained by composing f with a continuous functional. Thus, we expect that

$$\left\langle \int f, \varphi \right\rangle = \int \langle f(\cdot), \varphi \rangle \tag{*}$$

for every φ in $\mathfrak{X}^*$. We shall see that condition (*) can actually be used to define the integral.

As usual in integration theory there is no loss of complexity (but maybe some gain in familiarity) by assuming that X is an open or a closed subset of $\mathbb{R}^n$, and that $\int$ is the Lebesgue integral.

To simplify matters, we shall assume throughout that the scalar function $\|f(\cdot)\|$ is integrable on X.

2.5.13. Lemma. *Given $f: X \to \mathfrak{X}$ as in 2.5.12, let $\mathfrak{Y}_f$ denote the set of elements φ in $\mathfrak{X}^*$, for which the function $\langle f(\cdot), \varphi \rangle$ is measurable. Then $\mathfrak{Y}_f$ is a norm closed subspace of $\mathfrak{X}^*$; and if $\mathfrak{Y}_f$ is separating for $\mathfrak{X}$, there is at most one element $\int f$ in $\mathfrak{X}$ for which*

$$\left\langle \int f, \varphi \right\rangle = \int \langle f(\cdot), \varphi \rangle, \quad \varphi \in \mathfrak{Y}_f.$$

PROOF. If (φ_n) is a sequence in $\mathfrak{Y}_f$ that is w^*-convergent to some φ in $\mathfrak{X}^*$, then $(\langle f(\cdot), \varphi_n \rangle)$ converges pointwise to $\langle f(\cdot), \varphi \rangle$ as functions on X. Since a sequential limit of measurable functions is measurable by 6.2.14, it follows that $\varphi \in \mathfrak{Y}_f$. In particular, $\mathfrak{Y}_f$, which is clearly a linear subspace, is norm closed in $\mathfrak{X}^*$. Note, moreover, that since $|\langle f(\cdot), \varphi \rangle| \leq \|f(\cdot)\| \|\varphi\|$, which is an integrable majorant by assumption, we may describe $\mathfrak{Y}_f$ as the functionals φ in $\mathfrak{X}^*$ for which $\langle f(\cdot), \varphi \rangle \in \mathscr{L}^1(X)$; cf. 6.2.16.

If $\mathfrak{Y}_f$ is w^*-dense in $\mathfrak{X}^*$, the unicity of $\int f$ is evident.   □

2.5.14. Proposition. *If $\mathfrak{Y}$ is a Banach space, and $f: X \to \mathfrak{Y}^*$ is a w^*-measurable function with integrable norm, there is a unique element $\int f$ in $\mathfrak{Y}^*$ such that*

$$\left\langle y, \int f \right\rangle = \int \langle y, f(\cdot) \rangle, \quad y \in \mathfrak{Y}.$$

PROOF. Taking $\mathfrak{X} = \mathfrak{Y}^*$ we have $\mathfrak{Y}_f \supset \mathfrak{Y}$ by assumption. For each y in $\mathfrak{Y}$ we can form the integral $\int \langle y, f(\cdot) \rangle$, and this clearly defines a linear functional $\int f$ on $\mathfrak{Y}$. Since

$$\left| \left\langle y, \int f \right\rangle \right| = \left| \int \langle y, f(\cdot) \rangle \right| \leq \|y\| \int \|f(\cdot)\|,$$

the functional $\int f$ is bounded (by $\|f(\cdot)\|_1$), whence $\int f \in \mathfrak{Y}^*$. The uniqueness follows from 2.5.13.   □

2.5.15. Proposition. *Let $f: X \to \mathfrak{X}$ be a weakly measurable function from a locally compact Hausdorff space into a separable Banach space, and assume that $\|f(\cdot)\| \in \mathscr{L}^1(X)$. There is then a unique element $\int f$ in $\mathfrak{X}$ such that*

$$\left\langle \int f, \varphi \right\rangle = \int \langle f(\cdot), \varphi \rangle, \quad \varphi \in \mathfrak{X}^*. \qquad (*)$$

PROOF. By assumption $\mathfrak{Y}_f = \mathfrak{X}^*$; cf. 2.5.13. Regarding $\mathfrak{X}$ as a subspace of $\mathfrak{X}^{**}$ there is therefore by 2.5.14 a unique element $\int f$ in $\mathfrak{X}^{**}$ satisfying $(*)$. Since $\mathfrak{X}$

is separable, the *w**-topology on the unit ball $\mathfrak{B}^*$ of $\mathfrak{X}^*$ is second countable. To show that $\int f$ is *w**-continuous on $\mathfrak{B}^*$ it therefore suffices to check its behavior on a *sequence* (φ_n) in $\mathfrak{B}^*$, that is *w**-convergent to an element φ. This means that we have a sequence of functions $(\langle f(\cdot), \varphi_n \rangle)$ in $\mathscr{L}^1(X)$ that is pointwise convergent to $\langle f(\cdot), \varphi \rangle$. The function $\|f(\cdot)\|$ being a common integrable majorant, we see from Lebesgue's dominated convergence theorem (6.1.15) that

$$\int \langle f(\cdot), \varphi_n \rangle \to \int \langle f(\cdot), \varphi \rangle.$$

But this means that $\langle \int f, \varphi_n \rangle \to \langle \int f, \varphi \rangle$, so that $\int f$ is *w**-continuous on $\mathfrak{B}^*$. It follows from 2.5.11 that $\int f \in \mathfrak{X}$, as desired. $\qquad\square$

EXERCISES

E 2.5.1. Let C denote the convex hull in $\mathbb{R}^3$ of the points $(0, 0, 1)$, $(0, 0, -1)$, and the circle

$$\{(1 + \cos \theta, \sin \theta, 0) | 0 \le \theta \le 2\pi\}.$$

Show that the set of extreme points in C is not closed in C.

E 2.5.2. A Banach space $\mathfrak{X}$ is *strictly convex* if the equality $\|x + y\| = \|x\| + \|y\|$ for x and y in $\mathfrak{X}$ always implies that x and y are proportional. Show that $\mathfrak{X}$ is strictly convex iff every x on the unit sphere of $\mathfrak{X}$ is an extreme point of the unit ball. Compare strict convexity with uniform convexity defined in E 2.4.9.

E 2.5.3. Let $\mathfrak{B}^*$ denote the closed unit ball in $\mathfrak{X}^*$, for a separable Banach space $\mathfrak{X}$. Show that the *w**-topology on $\mathfrak{B}^*$ is metrizable, and deduce that $\mathfrak{X}^*$ is separable in the *w**-topology.

Hint: Show that a basis for the *w**-topology on $\mathfrak{B}^*$ can be obtained from a countable number of seminorms. Note also that a compact, metric space is separable.

E 2.5.4. Let $\mathfrak{C}$ be a convex, *w**-compact subset of $\mathfrak{X}^*$, where $\mathfrak{X}$ is a separable, normed space. Show that the extremal boundary $\partial\mathfrak{C}$ of $\mathfrak{C}$ can be written as a countable intersection of open subsets of $\mathfrak{C}$ (known as a G_δ-set).

Hint: Note first that $\mathfrak{C}$ is bounded and the *w**-topology is therefore metrizable by E 2.5.3. Then use the fact that $\varphi \in \mathfrak{C} \backslash \partial\mathfrak{C}$ iff φ belongs to one of the closed sets $f(\mathfrak{F}_n)$, $n \in \mathbb{N}$, where

$$\mathfrak{F}_n = \{(\varphi_1, \varphi_2) \in \mathfrak{C} \times \mathfrak{C} | d(\varphi_1, \varphi_2) \ge n^{-1}\},$$

and $f: \mathfrak{C} \times \mathfrak{C} \to \mathfrak{C}$ is the continuous function given by $f(\varphi_1, \varphi_2) = \frac{1}{2}(\varphi_1 + \varphi_2)$.

E 2.5.5. Prove some of the assertions in the catalogue 2.5.6.

E 2.5.6. Let (f_n) be a bounded sequence in $C(X)$, for a compact Hausdorff
space X, such that $f_n(x) \to f(x)$ for every x in X, where $f \in C(X)$. Show
that for each $\varepsilon > 0$ there is a convex combination $g = \sum \lambda_n f_n$ such
that $\| f - g \|_\infty < \varepsilon$.

Hint: Use Lebesgue's dominated convergence theorem (6.1.15) and
6.5.9 to conclude that $f_n \to f$ weakly, cf. 2.4.8. Then apply E 2.4.6.

E 2.5.7. (*Markov's fixed point theorem.*) Let $\mathfrak{C}$ be a convex, compact subset of
a vector space $\mathfrak{X}$, equipped with the weak topology induced by a
separating subspace $\mathfrak{X}^*$ of functionals on $\mathfrak{X}$. Assume that $\mathfrak{T}$ is a family
of mutually commuting, weakly continuous operators $T: \mathfrak{X} \to \mathfrak{X}$, such
that $T(\mathfrak{C}) \subset \mathfrak{C}$ for every T in $\mathfrak{T}$. Show that there exists a point x in $\mathfrak{C}$
such that $Tx = x$ for all T.

Hint: Put $T_n = n^{-1}(I + T + \cdots + T^{n-1})$ for each T in $\mathfrak{T}$ and n in
$\mathbb{N}$. Show that $T_n(\mathfrak{R})$ is convex and compact for every convex, compact
subset $\mathfrak{R}$. Show further that

$$T_n S_m(\mathfrak{C}) \subset T_n(\mathfrak{C}) \cap S_m(\mathfrak{C})$$

for all S and T in $\mathfrak{T}$ and all n and m, and deduce that $\mathfrak{F} = \bigcap T_n(\mathfrak{C})$
(intersection over $\mathfrak{T} \times \mathbb{N}$) is nonempty. If $x \in \mathfrak{F}$ and $Tx \neq x$ for some
T in $\mathfrak{T}$, choose φ in $\mathfrak{X}^*$ with $\varphi(Tx - x) = 1$. Given n there is a y in $\mathfrak{C}$
with $T_n y = x$. This implies that

$$n = \varphi(n(TT_n y - T_n y)) = \varphi(T^n y - y).$$

But $\mathfrak{C} - \mathfrak{C}$ is compact (as the continuous image of the compact
subset $\mathfrak{C} \times \mathfrak{C}$ in $\mathfrak{X} \times \mathfrak{X}$), so φ is bounded on $\mathfrak{C} - \mathfrak{C}$, and we have a
contradiction.

E 2.5.8. Let $f: X \to \mathfrak{X}$ be a continuous function from a compact Hausdorff
space into a Banach space. Let μ be a Radon measure on X and
consider elements of the form

$$I_\lambda(f) = \sum_{k=1}^n f(s_k)\mu(E_k),$$

where the E_k's form a partition of X into disjoint Borel subsets, and

$$s_k \in E_k \subset \{s \in X \,|\, \| f(s) - f(s_k)\| \leq \varepsilon\}.$$

With $\lambda = \{E_1, \ldots, E_n, \varepsilon\}$, show that $(I_\lambda(f))_{\lambda \in \Lambda}$ is a uniformly convergent net in $\mathfrak{X}$. If $\int f(s)\,d\mu(s)$ denotes the limit of the net, show that

$$\varphi\left(\int f(s)\,d\mu(s)\right) = \int \varphi(f(s))\,d\mu(s)$$

for every φ in $\mathfrak{X}^*$.

Hint 1: Proceed as if you were constructing the Riemann integral.
Hint 2: Apply 2.5.15.

CHAPTER 3

Hilbert Spaces

The geometry of infinite-dimensional Banach spaces offers quite a few surprises from the viewpoint of finite-dimensional euclidean spaces. Thus, the unit ball may have corners, and closed convex sets may fail to have elements of minimal norm. Even more alienating, there may be no notion of perpendicular vectors and no good notion of a basis. By contrast, the Hilbert spaces are perfect generalizations of euclidean spaces, to the point of being almost trivial as geometrical objects. The deeper theory (and the fruitful applications) is, however, concerned with the *operators* on Hilbert space. Accordingly, we devote only a single section to Hilbert spaces as such, centered around the notions of sesquilinear forms, orthogonality, and self-duality. We then develop the elementary theory of bounded linear operator on a Hilbert space $\mathfrak{H}$, i.e. we initiate the study of the Banach *-algebra $\mathbf{B}(\mathfrak{H})$—to be continued in later chapters.

In order to present some advanced material, without yet having access to the spectral theorem, we study the compact operators in some detail. For these special operators we establish the spectral theorem and then proceed to the notion of index for Fredholm operators and its invariance properties. Finally, we introduce the trace on $\mathbf{B}(\mathfrak{H})$, define the trace class operators and the Hilbert–Schmidt operators, and prove the duality theorems that arise from the trace. We conclude with the basic theory of integral operators.

3.1. Inner Products

Synopsis. Sesquilinear forms and inner products. Polarization identities and the Cauchy–Schwarz inequality. Parallelogram law. Orthogonal sum. Orthogonal complement. Conjugate self-duality of Hilbert spaces. Weak topology.

Orthonormal basis. Orthonormalization. Isomorphism of Hilbert spaces. Exercises.

3.1.1. A *sesquilinear form* on a vector space $\mathfrak{X}$ is a map

$$(\cdot|\cdot): \mathfrak{X} \times \mathfrak{X} \to \mathbb{F}$$

that is linear in the first variable and conjugate linear in the second. The word means one and a half (*semis qui* $\sim$ a half more) which is nonsense; but only when $\mathbb{F} = \mathbb{R}$ is the form honestly bilinear, and we are primarily interested in the case $\mathbb{F} = \mathbb{C}$. (A mathematical physicist is a mathematician believing that a sesquilinear form is conjugate linear in the first variable and linear in the second.)

To each sesquilinear form $(\cdot|\cdot)$ we define the *adjoint* form $(\cdot|\cdot)^*$ by

$$(x|y)^* = \overline{(y|x)}, \quad x, y \in \mathfrak{X}.$$

We say that the form is *self-adjoint* if $(\cdot|\cdot)^* = (\cdot|\cdot)$ (but if $\mathbb{F} = \mathbb{R}$, the term *symmetric* is often used). For $\mathbb{F} = \mathbb{C}$ a straightforward calculation shows that

$$4(x|y) = \sum_{k=0}^{3} i^k(x + i^k y | x + i^k y). \qquad (*)$$

It is immediate from $(*)$ that the form $(\cdot|\cdot)$ is self-adjoint iff $(x|x) \in \mathbb{R}$ for every x in $\mathfrak{X}$.

We say that a sesquilinear form $(\cdot|\cdot)$ is *positive* if $(x|x) \geq 0$ for every x in $\mathfrak{X}$. Thus, for $\mathbb{F} = \mathbb{C}$ a positive form is automatically self-adjoint. On a real space this is no longer true.

An *inner product* on $\mathfrak{X}$ is a positive, self-adjoint sesquilinear form, such that $(x|x) = 0$ implies $x = 0$ for every x in $\mathfrak{X}$.

3.1.2. For a positive, self-adjoint sesquilinear form $(\cdot|\cdot)$ we define the homogeneous function $\|\cdot\|: \mathfrak{X} \to \mathbb{R}_+$ by

$$\|x\| = (x|x)^{1/2}, \quad x \in \mathfrak{X}. \qquad (*)$$

From $(*)$ in 3.1.1 and a similar computation in the real case we obtain the following two *polarization identities*, the first for $\mathbb{F} = \mathbb{C}$, the second for $\mathbb{F} = \mathbb{R}$:

$$4(x|y) = \sum_{k=0}^{3} i^k \|x + i^k y\|^2; \qquad (**)$$

$$4(x|y) = \|x + y\|^2 - \|x - y\|^2. \qquad (***)$$

3.1.3. With $(\cdot|\cdot)$ as in 3.1.2 and α in $\mathbb{F}$ the formula

$$|\alpha|^2 \|x\|^2 + 2 \operatorname{Re} \alpha(x|y) + \|y\|^2 = \|\alpha x + y\|^2 \geq 0, \qquad (*)$$

for x and y in $\mathfrak{X}$, immediately leads to the *Cauchy–Schwarz inequality*:

$$|(x|y)| \leq \|x\| \|y\|. \qquad (**)$$

Inserting this in (∗) it follows that $\|\cdot\|$ is a subadditive function, and therefore a seminorm on $\mathfrak{X}$. In the case where $(\cdot|\cdot)$ is an inner product, the definition (∗) in 3.1.2 therefore gives a norm on $\mathfrak{X}$. In this case we see from (∗) that equality holds in the Cauchy–Schwarz inequality only when x and y are proportional.

Elementary computations show that the norm arising from an inner product satisfies the *parallellogram law*:

$$\|x + y\|^2 + \|x - y\|^2 = 2(\|x\|^2 + \|y\|^2). \qquad (\ast\ast\ast)$$

Conversely one may verify that if a norm satisfies (∗∗∗), then the equation (∗∗) in 3.1.2 [respectively (∗∗∗) in 3.1.2 for $\mathbb{F} = \mathbb{R}$] will define an inner product on $\mathfrak{X}$.

Two vectors x and y are *orthogonal*, in symbols $x \perp y$, if $(x|y) = 0$. It follows from (∗) that orthogonal elements satisfy the *Pythagoras identity*

$$\|x + y\|^2 = \|x\|^2 + \|y\|^2,$$

which conversely implies orthogonality in real spaces. Two subsets $\mathfrak{Y}$ and $\mathfrak{Z}$ are *orthogonal*, in symbols $\mathfrak{Y} \perp \mathfrak{Z}$, if $y \perp z$ for every y in $\mathfrak{Y}$ and z in $\mathfrak{Z}$.

3.1.4. A *Hilbert space* is a vector space $\mathfrak{H}$ with an inner product, such that $\mathfrak{H}$ is a Banach space with the associated norm. For this reason a space with an inner product is also called a *pre-Hilbert space*.

The euclidean spaces $\mathbb{F}^n$, $n \in \mathbb{N}$, are Hilbert spaces with the usual inner product $(x|y) = \sum x_k \bar{y}_k$. The associated norm is the 2-norm (2.1.13).

The spaces $C_c(\mathbb{R}^n)$ (2.1.14) have the inner product

$$(f|g) = \int f(x)\overline{g(x)}\, dx.$$

The associated norm is the 2-norm. By completion we obtain the Hilbert space $L^2(\mathbb{R}^n)$. More generally, the Banach space $L^2(X)$, corresponding to a Radon integral $\int$ on a locally compact Hausdorff space X, is a Hilbert space with the inner product $(f|g) = \int f\bar{g}$; cf. 2.1.15.

3.1.5. If $\{\mathfrak{H}_j | j \in J\}$ is a family of Hilbert spaces, we form the algebraic direct sum $\sum \mathfrak{H}_j$ as in 2.1.16, and define the inner product by

$$(x|y) = \sum (P_j x | P_j y), \quad x, y \in \sum \mathfrak{H}_j.$$

The associated norm is the 2-norm, and the completion of $\sum \mathfrak{H}_j$ is called the *orthogonal sum* of the $\mathfrak{H}_j$'s, denoted by $\bigoplus \mathfrak{H}_j$. We may identify each $\mathfrak{H}_j$ with a closed subspace of $\bigoplus \mathfrak{H}_j$, such that $\mathfrak{H}_i \perp \mathfrak{H}_j$ for $i \neq j$. As shown in 2.1.17 the elements in $\bigoplus \mathfrak{H}_j$ are exactly those x in $\prod \mathfrak{H}_j$ for which $\sum \|P_j x\|^2 < \infty$. In particular, $P_j x = 0$ except for countably many j's.

3.1.6. Lemma. *If $\mathfrak{C}$ is a closed, nonempty, convex subset of a Hilbert space $\mathfrak{H}$, there is for each y in $\mathfrak{H}$ a unique x in $\mathfrak{C}$ that minimizes the distance from y to $\mathfrak{C}$.*

PROOF. Replacing $\mathfrak{C}$ by $\mathfrak{C} - y$ we may assume that $y = 0$. Put $\alpha = \inf\{\|x\| \mid x \in \mathfrak{C}\}$ and choose a sequence (x_n) in $\mathfrak{C}$ such that $\|x_n\| \to \alpha$. For any y and z in $\mathfrak{C}$ the parallellogram law gives

$$2(\|y\|^2 + \|z\|^2) = \|y + z\|^2 + \|y - z\|^2 \geq 4\alpha^2 + \|y - z\|^2, \qquad (*)$$

because $\frac{1}{2}(y + z) \in \mathfrak{C}$. Replacing y and z by x_n and x_m in $(*)$ shows that (x_n) is a Cauchy sequence and, consequently, is convergent to an element x in $\mathfrak{C}$ with $\|x\| = \alpha$. If z is any other element in $\mathfrak{C}$ with $\|z\| = \alpha$, then $(*)$ immediately shows that $\|x - z\|^2 = 0$, i.e. $x = z$. $\square$

3.1.7. Theorem. *For a closed subspace $\mathfrak{X}$ of a Hilbert space $\mathfrak{H}$ set $\mathfrak{X}^\perp = \{x^\perp \in \mathfrak{H} \mid x^\perp \perp \mathfrak{X}\}$. Then each vector y in $\mathfrak{H}$ has a unique decomposition $y = x + x^\perp$, with x in $\mathfrak{X}$ and $x^\perp$ in $\mathfrak{X}^\perp$. The element x (respectively $x^\perp$) is the nearest point in $\mathfrak{X}$ (respectively $\mathfrak{X}^\perp$) to y. Moreover, $\mathfrak{H} = \mathfrak{X} \oplus \mathfrak{X}^\perp$ and $(\mathfrak{X}^\perp)^\perp = \mathfrak{X}$.*

PROOF. Given y in $\mathfrak{H}$ we take x to be the nearest point in $\mathfrak{X}$ to y, cf. 3.1.6, and put $x^\perp = y - x$. For every z in $\mathfrak{X}$ and $\varepsilon > 0$ we then have

$$\|x^\perp\|^2 = \|y - x\|^2 \leq \|y - (x + \varepsilon z)\|^2$$
$$= \|x^\perp - \varepsilon z\|^2 = \|x^\perp\|^2 - 2\varepsilon \operatorname{Re}(x^\perp \mid z) + \varepsilon^2 \|z\|^2.$$

It follows that $2\operatorname{Re}(x^\perp \mid z) \leq \varepsilon \|z\|^2$ for every $\varepsilon > 0$, whence $\operatorname{Re}(x^\perp \mid z) \leq 0$ for every z. Since $\mathfrak{X}$ is a linear subspace, this implies that $(x^\perp \mid z) = 0$, i.e. $x^\perp \in \mathfrak{X}^\perp$. Clearly $\mathfrak{X}^\perp$ is a closed subspace of $\mathfrak{H}$, and it follows from the Pythagoras identity that $\mathfrak{H}$ is isometrically isomorphic to the orthogonal sum $\mathfrak{X} \oplus \mathfrak{X}^\perp$ defined in 3.1.5.

If $y = z + z^\perp$ was another decomposition of y in $\mathfrak{X} \oplus \mathfrak{X}^\perp$, then $0 = (x - z) + (x^\perp - z^\perp)$, whence $0 = \|x - z\|^2 + \|x^\perp - z^\perp\|^2$, so that $x = z$ and $x^\perp = z^\perp$.

Take y in $(\mathfrak{X}^\perp)^\perp$ and decompose it as $y = x + x^\perp$ in $\mathfrak{X} \oplus \mathfrak{X}^\perp$. Since $\mathfrak{X} \subset (\mathfrak{X}^\perp)^\perp$, it follows that $x^\perp \in \mathfrak{X}^\perp \cap (\mathfrak{X}^\perp)^\perp$, whence $x^\perp = 0$ and $\mathfrak{X} = (\mathfrak{X}^\perp)^\perp$. Therefore, replacing $\mathfrak{X}$ by $\mathfrak{X}^\perp$ in the previous arguments we see that if $y = x^\perp + x$ in $\mathfrak{X}^\perp \oplus \mathfrak{X}$, then $x^\perp$ is the nearest point in $\mathfrak{X}^\perp$ to y. $\square$

3.1.8. Corollary. *For every subset $\mathfrak{X} \subset \mathfrak{H}$, the smallest closed subspace of $\mathfrak{H}$ containing $\mathfrak{X}$ is $(\mathfrak{X}^\perp)^\perp$. In particular, if $\mathfrak{X}$ is a subspace of $\mathfrak{H}$, then $\mathfrak{X}^= = (\mathfrak{X}^\perp)^\perp$.*

3.1.9. Proposition. *The map Φ given by $\Phi(x) = (\cdot \mid x)$ is a conjugate linear isometry of $\mathfrak{H}$ onto $\mathfrak{H}^*$.*

PROOF. Evidently Φ is a conjugate linear map of $\mathfrak{H}$ into $\mathfrak{H}^*$, and is norm decreasing by the Cauchy–Schwarz inequality. Since $\langle x, \Phi(x) \rangle = (x \mid x) = \|x\|^2$, we see that Φ is an isometry.

Now take φ in $\mathfrak{H}^* \backslash \{0\}$, and put $\mathfrak{X} = \ker \varphi$. Then $\mathfrak{X}$ is a proper, closed subspace of $\mathfrak{H}$, so by 3.1.7 there is a vector x in $\mathfrak{X}^\perp$ with $\varphi(x) = 1$. For every y in $\mathfrak{H}$ we see that $y - \varphi(y)x \in \mathfrak{X}$, whence

$$(y|x) = (y - \varphi(y)x + \varphi(y)x|x) = \varphi(y)\|x\|^2.$$

It follows that $\varphi = \Phi(\|x\|^{-2}x)$.   □

3.1.10. We define the *weak topology* on a Hilbert space $\mathfrak{H}$ as the initial topology corresponding to the family of functionals $x \to (x|y)$, where y ranges over $\mathfrak{H}$; see 2.4.1. It follows from 3.1.9 that the weak topology on $\mathfrak{H}$ is the w^*-topology on $\mathfrak{H}^*$ (2.4.8) pulled back to $\mathfrak{H}$ via the map Φ. In particular, the unit ball in $\mathfrak{H}$ is weakly compact by 2.5.2.

It will follow immediately from 3.2.3 that every operator T in $\mathbf{B}(\mathfrak{H})$ is continuous as a function $T: \mathfrak{H} \to \mathfrak{H}$, when both copies of $\mathfrak{H}$ are endowed with the weak topology. We say that T is *weak–weak continuous*. Conversely, if T is a weak–weak continuous operator on $\mathfrak{H}$, we conclude from the closed graph theorem 2.2.7 that $T \in \mathbf{B}(\mathfrak{H})$. Indeed, if $x_n \to x$ and $Tx_n \to y$ (norm convergence), then $x_n \to x$ and $Tx_n \to y$ weakly, because the norm topology is stronger than the weak topology. By assumption $Tx_n \to Tx$ weakly, whence $Tx = y$ as desired.

A similar argument shows that every norm–weak continuous operator on $\mathfrak{H}$ is bounded. An operator that is weak–norm continuous must have finite rank (3.3.1), which is a very special demand. A small variation along these lines, however, gives a characterization of the compact operators, see 3.3.2.

3.1.11. A subset $\{e_j | j \in J\}$ of a Hilbert space $\mathfrak{H}$ is said to be *orthonormal* if $\|e_j\| = 1$ for every j, and $(e_j|e_i) = 0$ for all $i \neq j$. Furthermore, if the subspace spanned by the family $\{e_j | j \in J\}$ is dense in $\mathfrak{H}$, we call it an *orthonormal basis*. By 3.1.5 this implies that $\mathfrak{H}$ is the orthogonal sum of the one-dimensional subspaces $\mathbb{F}e_j, j \in J$. Consequently (2.1.17), each element x in $\mathfrak{H}$ has the form

$$x = \sum \alpha_j e_j,$$

where the sum converges in (2)-norm. Taking inner products with the e_j's we see that the coordinates for x are determined by $\alpha_j = (x|e_j)$. Finally, we note the *Parseval identity* (a generalization of Pythagoras')

$$\|x\|^2 = \sum |\alpha_j|^2, \tag{$*$}$$

obtained by computing $(x|x)$.

3.1.12. Proposition. *Every orthonormal set in a Hilbert space $\mathfrak{H}$ can be enlarged to an orthonormal basis for $\mathfrak{H}$.*

PROOF. Let $\{e_j | j \in J_0\}$ be an orthonormal set in $\mathfrak{H}$. Zorn's lemma (1.1.3), applied to the inductively ordered system of orthonormal subsets of $\mathfrak{H}$ containing the given one, shows the existence of a maximal orthonormal set $\{e_j | j \in J\}$, where $J_0 \subset J$.

Let $\mathfrak{X}$ be the closed subspace of $\mathfrak{H}$ generated by $\{e_j | j \in J\}$. If $\mathfrak{X} \neq \mathfrak{H}$, there is by 3.1.7 a unit vector e in $\mathfrak{X}^\perp$, contradicting the maximality of the system $\{e_j | j \in J\}$. Thus, $\mathfrak{X} = \mathfrak{H}$ and we have a basis.   □

3.1.13. It is often possible (and desirable) to prove 3.1.12 in a more constructive manner. Assume that we are given a sequence (x_n) in $\mathfrak{H}$ with dense linear span (known as a *total* family). In particular, $\mathfrak{H}$ is separable. Without loss of generality we may assume that the vectors (x_n) are linearly independent. There is then an orthonormal basis (e_n) for $\mathfrak{H}$ such that

$$\text{span}\{e_1,\dots,e_n\} = \text{span}\{x_1,\dots,x_n\}$$

for every n. To construct this basis, put $e_1 = \|x_1\|^{-1}x_1$ and then inductively $e_{n+1} = \|y_{n+1}\|^{-1}y_{n+1}$, where

$$y_{n+1} = x_{n+1} - \sum_{k=1}^{n}(x_{n+1}|e_k)e_k.$$

The method above is called the *Gram–Schmidt orthonormalization process*. Its main advantage is that, when the Hilbert space has a dense subspace of "interesting" vectors, the basis may be chosen among these vectors. For a concrete Hilbert space of (equivalence classes of) square integrable functions (cf. 3.1.4), the "interesting" vectors could be continuous functions, C^∞-functions, polynomials, et cetera.

3.1.14. Proposition. *If $\mathfrak{H}$ and $\mathfrak{K}$ are Hilbert spaces with orthonormal bases $\{e_i|i \in I\}$ and $\{f_j|j \in J\}$, and if I and J have the same cardinality (i.e. there is a bijective map $\gamma: I \to J$), then there is an isometric operator U of $\mathfrak{H}$ onto $\mathfrak{K}$ such that $(Ux|Uy) = (x|y)$ for all x and y in $\mathfrak{H}$.*

PROOF. For every $x = \sum \alpha_i e_i$ in the algebraic direct sum $\sum \mathbb{F}e_i$ we define

$$U_0 x = \sum \alpha_i f_{\gamma(i)}.$$

Then U_0 is linear and isometric by the Parseval identity (3.1.11); and it maps $\sum \mathbb{F}e_i$ onto $\sum \mathbb{F}f_j$ because γ is surjective. Extending U_0 by continuity (2.1.11) we obtain an isometry U of $\mathfrak{H}$ onto $\mathfrak{K}$. For each such we have by the polarization identities (**) and (***) in 3.1.2 that

$$4(Ux|Uy) = \sum i^k \|U(x + i^k y)\|^2$$
$$= \sum i^k \|x + i^k y\|^2 = 4(x|y). \qquad \square$$

3.1.15. We see from 3.1.14 that all infinite-dimensional, separable Hilbert spaces are isomorphic, because they have a countable orthonormal basis. They can therefore all be identified with the sequence space ℓ^2 defined in 2.1.18. The real interest of Hilbert spaces, however, arises not so much from the space (a rather trivial generalization of euclidean spaces), but from the operators associated with it (translation operators, multiplication operators, integral operators, and differential operators). A given operator is often intimately linked with a certain realization of the Hilbert space, and an insensitive choice of basis can make even a simple problem quite incomprehensible. The further theory of the Hilbert space concerns the operators on it.

EXERCISES

E 3.1.1. Given n vectors $x_1, x_2, \ldots, x_n$ in the Hilbert space $\mathfrak{H}$, consider the $n \times n$ matrix $A = (a_{ij})$ given by $a_{ij} = (x_i | x_j)$. Show that the vectors are linearly independent iff $\det(A) \neq 0$.

E 3.1.2. Let Ω be an open subset of $\mathbb{C}$ and denote by $A^2(\Omega)$ the linear space of complex functions that are holomorphic in Ω and square integrable (with respect to Lebesgue measure). Show that if $B(z, r)$ is a closed disk contained in Ω (with center z and radius r), then

$$f(z) = \pi^{-1} r^{-2} \int_{B(z,r)} f(x) \, dx \qquad (*)$$

for every f in $A^2(\Omega)$.

Hint: Consider the power series expansion of f in $B(z, r)$, and integrate term by term.

E 3.1.3. On the space $A^2(\Omega)$ defined in E 3.1.2 consider the inner product

$$(f | g) = \int f(x) \overline{g(x)} \, dx,$$

and the corresponding 2-norm. Show that $A^2(\Omega)$ is a Hilbert space.

Hint: Use $(*)$ in E 3.1.2 and the Cauchy–Schwarz inequality in $L^2(B(z, r))$ to show that

$$|f(z)| \leq \pi^{-1/2} r^{-1} \left(\int_{B(z,r)} |f(x)|^2 \, dx \right)^{1/2}.$$

Conclude that convergence in 2-norm implies uniform convergence on compact subsets of Ω.

E 3.1.4. Let $\Omega = \{ z \in \mathbb{C} \, | \, |z| < 1 \}$ and consider the Hilbert space $A^2 = A^2(\Omega)$; cf. E 3.1.2 and E 3.1.3. Show that the sequence

$$\tilde{e}_n(z) = (n + 1)^{1/2} \pi^{-1/2} z^n, \quad n = 0, 1, \ldots$$

is an orthonormal basis for A^2.

Hint: Show that the power series for each f in A^2 converges in 2-norm toward f.

E 3.1.5. Let $\mathbb{T}$ denote the unit circle in $\mathbb{C}$ and consider the Hilbert space $L^2(\mathbb{T})$ (with respect to Lebesgue measure on the circle). Show that the functions

$$e_n(z) = (2\pi)^{-1/2} z^n, \quad n \in \mathbb{Z},$$

form an orthonormal basis for $L^2(\mathbb{T})$.

E 3.1.6. (*The Hardy space.*) Let H^2 denote the closed subspace of $L^2(\mathbb{T})$ (cf. E 3.1.5) spanned by the vectors e_n, $n \geq 0$. For each vector $f =$

$\sum \alpha_n e_n$ in H^2 define $\tilde{f}(z) = (2\pi)^{-1/2} \sum \alpha_n z^n$ for $|z| < 1$. Show that $\tilde{f} \in A^2$ (cf. E 3.1.4). Conversely, show that if g is holomorphic in the open unit disk with Taylor series $g(z) = \sum \beta_n z^n$, such that $\sum |\beta_n|^2 < \infty$, then $g = \tilde{f}$ for some (unique) f in H^2. Show that if $g \in A^2$ and $g_t(z) = g(tz)$, $0 < t < 1$, then $g_t = \tilde{f_t}$ for some f_t in H^2. Finally show that $g = \tilde{f}$ for some f in H^2 iff $\sup \|f_t\| < \infty$; in which case $f_t \to f$ in H^2.

   Hint: Define $T: H^2 \to A^2$ by $Te_n = (2n + 2)^{-1/2} \tilde{e}_n$ and show that $\tilde{f} = Tf$.

E 3.1.7. Let $\mathfrak{X}$ and $\mathfrak{Y}$ be closed subspaces of a Hilbert space $\mathfrak{H}$. Assume that $\dim \mathfrak{X} < \infty$ and that $\dim \mathfrak{X} < \dim \mathfrak{Y}$. Show that $\mathfrak{X}^\perp \cap \mathfrak{Y} \neq \{0\}$.

E 3.1.8. Let $\{e_n | n \in \mathbb{N}\}$ be an orthonormal basis for the Hilbert space $\mathfrak{H}$, and put

$$\mathfrak{C} = \left\{ x \in \mathfrak{H} \mid \sum \left(1 + \frac{1}{n}\right)^2 |(x|e_n)|^2 \leq 1 \right\}.$$

Show that $\mathfrak{C}$ is a closed, bounded, and convex set, but that it contains no vector with maximal norm.

   Hint: Define $Tx = \sum (1 + n^{-1})(x|e_n)e_n$. Then $T \in \mathbf{B}(\mathfrak{H})$ and $\mathfrak{C} = \{x \in \mathfrak{H} | \|Tx\| \leq 1\}$, which proves that $\mathfrak{C}$ is closed and convex. Furthermore,

$$\|x\|^2 = \sum |(x|e_n)|^2 < \sum \left(1 + \frac{1}{n}\right)^2 |(x|e_n)|^2 \leq 1$$

for every x in $\mathfrak{C}$.

E 3.1.9. Let $\{e_n | n \in \mathbb{N}\}$ be an orthonormal basis for the Hilbert space $\mathfrak{H}$, and put $\mathfrak{X} = \{n^{1/2} e_n | n \in \mathbb{N}\}$. Show that 0 belongs to the weak closure of $\mathfrak{X}$, but that no sequence from $\mathfrak{X}$ is weakly convergent to 0. Conclude from this that the weak topology on $\mathfrak{H}$ does not satisfy the first axiom of countability and, hence, is not metrizable.

   Hint: If $x = \sum \alpha_n e_n \in \mathfrak{H}$, there is no $\varepsilon > 0$ such that $|(y|x)| > \varepsilon$ for every y in $\mathfrak{X}$. Thus $0 \in \mathfrak{X}^{-w}$. On the other hand, every weakly convergent sequence in $\mathfrak{H}$ is bounded by the principle of uniform boundedness (2.2.10 and 2.2.11).

E 3.1.10. (*The Hilbert cube.*) Let $\mathfrak{C}$ denote the set of vectors $x = (x_n)$ in the real Hilbert space ℓ^2, such that $|x_n| \leq n^{-1}$ for every n. Show directly that $\mathfrak{C}$ is convex and (norm) compact. [A more sophisticated proof would use 3.3.2 and 3.3.8, because $\mathfrak{C} = T(\mathfrak{B})$, where $\mathfrak{B}$ is the unit ball in ℓ^2 and T is the compact operator given by $(Tx)_n = n^{-1} x_n$, $x = (x_n) \in \ell^2$.]

E 3.1.11. Show that every infinite, orthonormal sequence in a Hilbert space converges weakly to 0.

E 3.1.12. (*The Riemann–Lebesgue lemma.*) Consider a function f in $L^1([0, 2\pi])$ with Fourier coefficients

$$\hat{f}(n) = (2\pi)^{-1} \int_0^{2\pi} f(x) \exp(-inx)\, dx.$$

Show that $\hat{f}(n) \to 0$ as $n \to \pm\infty$.

    Hint: Given $\varepsilon > 0$ find g in $L^2([0, 2\pi])$ such that $\|f - g\|_1 < \varepsilon$. Then apply E 3.1.11.

E 3.1.13. Let $\mathfrak{X}$ be a closed subspace of the Hilbert space $L^2([0, 1])$, and assume that every element in $\mathfrak{X}$ is essentially bounded (i.e. belongs to $L^\infty([0, 1])$). Prove that $\dim \mathfrak{X} < \infty$.

    Hint: As the identity map of $\mathfrak{X}$ (with 2-norm) into $L^\infty([0, 1])$ (with ∞-norm) has closed graph, there is an $\alpha > 0$ such that $\|f\|_\infty \leq \alpha \|f\|_2$ for every f in $\mathfrak{X}$. Let $\{f_1, f_2, \ldots, f_n\}$ be an orthonormal set in $\mathfrak{X}$ and for each $x = (x_1, \ldots, x_n)$ in $\mathbb{C}^n$ set $f_x = \sum x_k f_k$. Then $|f_x(t)| \leq \alpha \|f_x\|_2 \leq \alpha \|x\|_2$ for almost all t in $[0, 1]$, so if Λ is a countable dense subset of $\mathbb{C}^n$, there is a null set N in $[0, 1]$ such that $|f_x(t)| \leq \alpha \|x\|_2$ for every x in Λ and t in $[0, 1] \setminus N$. Each map $x \to f_x(t)$ from $\mathbb{C}^n$ to $\mathbb{C}$ is linear, hence continuous, whence $|f_x(t)| \leq \alpha \|x\|_2$ for all x in $\mathbb{C}^n$ and $t \notin N$. In particular,

$$\sum |f_k(t)|^2 = \sum \overline{f_k(t)} f_k(t) \leq \alpha \left(\sum |f_k(t)|^2 \right)^{1/2},$$

whence $\sum |f_k(t)|^2 \leq \alpha^2$ for $t \notin N$. Integration gives

$$n = \left\| \sum f_k \right\|_2^2 = \sum \int |f_k(t)|^2\, dt \leq \alpha^2.$$

E 3.1.14. (*Fourier transform.*) In this and the next two exercises we use the normalized Lebesgue integral on $\mathbb{R}$ [i.e. the usual one divided by $(2\pi)^{1/2}$]. One consequence of this is that $\int g_0(x)\, dx = 1$, where $g_0(x) = \exp(-\frac{1}{2}x^2)$. We shall use the abbreviated notation L^p for $L^p(\mathbb{R})$.

    For each f in L^1 define $\hat{f}(y) = \int f(x) \exp(-ixy)\, dx$.

  (a) Show that if $f \in L^1$ and $id \cdot f \in L^1$ [where $id(x) = x$], then $\hat{f}$ is a differentiable function and $\hat{f}'(y) = -i(id \cdot f)^\wedge(y)$.

  (b) Show that if f is differentiable and both f and f' belong to L^1, then $(f')^\wedge(y) = iy\hat{f}(y)$.

  (c) Let $\mathscr{S}$ (the *Schwartz space*) denote the space of C^∞-functions on $\mathbb{R}$ such that $(id)^n f^{(m)} \in L^1$ for all n and m. Use (a) and (b) to show that $\hat{f} \in \mathscr{S}$ for every f in $\mathscr{S}$.

  (d) Show that the map $F: f \to \hat{f}$ belongs to $\mathbf{B}(L^1, C_0(\mathbb{R}))$.

    Hints: (a) Use Lebesgue's theorem on majorized convergence (6.1.15) on the difference quotient of $\hat{f}$. (b) Use integration by parts and the fact that $f \in C_0(\mathbb{R})$. (d) For any polynomial p we have $pg_0 \in \mathscr{S}$. Use this to show that $\mathscr{S}$ is dense in L^1.

E 3.1.15. (The *inversion theorem.*) (a) Show that $\hat{g}_0 = g_0$, where $g_0(x) = \exp(-\frac{1}{2}x^2)$.

(b) Take g in L^1 and define g_n by $g_n(x) = g(n^{-1}x)$. Show that $\hat{g}_n(y) = n\hat{g}(ny)$.

(c) For f and g in L^1 and n in $\mathbb{N}$, show that

$$\int \hat{f}(y)g(n^{-1}y)\,dy = \int f(n^{-1}x)\hat{g}(x)\,dx.$$

(d) Show that if $f \in L^1$ and $\hat{f} \in L^1$, then $\hat{\hat{f}}(x) = f(-x)$ for almost all x in $\mathbb{R}$.

Hints: (a) Use 3.1.14 to show that g_0 and $\hat{g}_0$ both are solutions to the differential equation $g'(x) + xg(x) = 0$; and therefore proportional. Observe next that $\hat{g}_0(0) = g_0(0)$. (c) For $n = 1$ the equation follows by applying Fubini's theorem (6.6.6) to the function $h(x, y) = f(x)g(y)\exp(-ixy)$. Replacing g by g_n and using (b) we get the general formula. (d) If $f \in \mathscr{S}$ [cf. E 3.1.14(c)], we replace g by g_0 in (c) and let $n \to \infty$ to obtain $\int \hat{f}(y)\,dy = f(0)$. Now replace f with the function $y \to f(x + y)$. In the general case, put $f_0(x) = \hat{\hat{f}}(-x)$. The inversion formula for a g in $\mathscr{S}$ in conjunction with (c) (for $n = 1$) then gives

$$\int f_0(x)g(x)\,dx = \int \int \hat{f}(y)\exp(i\,xy)g(x)\,dx\,dy$$

$$= \int \hat{f}(y)\hat{g}(-y)\,dy = \int f(x)\hat{\hat{g}}(-x)\,dx = \int f(x)g(x)\,dx.$$

As $\mathscr{S}$ is dense in L^1 it follows that $f_0(x) = f(x)$ for almost all x in $\mathbb{R}$.

E 3.1.16. (*Plancherel's theorem.*) Show that there is a unitary operator F on L^2 with $F^4 = I$, such that

$$Ff(x) = \int f(y)\exp(-ixy)\,dy$$

for every f in $L^1 \cap L^2$.

Hint: Take f and g in $\mathscr{S}$ [cf. E 3.1.14(c)] and use E 3.1.15 to prove that $(f|g) = (\hat{f}|\hat{g})$. Set $F_0 f = \hat{f}$ for f in $L^1 \cap L^2$ and extend it by continuity.

3.2. Operators on Hilbert Space

Synopsis. The correspondence between sesquilinear forms and operators. Adjoint operator and involution in $\mathbf{B}(\mathfrak{H})$. Invertibility, normality, and positivity in $\mathbf{B}(\mathfrak{H})$. The square root. Projections and diagonalizable operators.

Unitary operators and partial isometries. Polar decomposition. The Russo–Dye–Gardner theorem. Numerical radius. Exercises.

3.2.1. The first lemma is of historical interest (apart from being quite a useful result). It should be recalled that the spectral theory was developed by Hilbert (in the years 1904–1910) as a theory for quadratic and bilinear forms. Composition of bilinear forms (by convolution product) entailed by necessity the selection of a basis for the space, and the expression of the forms (and their product) as infinite matrices. Even the simplest computation had a tendency to drown in indices under these circumstances. One of the reasons (but not the only one) to von Neumann's success was his consistent use of the operator concept to tackle problems on Hilbert space.

Throughout this section $\mathfrak{H}$ will denote a Hilbert space, and I will denote the identical map on $\mathfrak{H}$, so that I is the unit in $\mathbf{B}(\mathfrak{H})$.

3.2.2. Lemma. *There is a bijective, isometric correspondence between operators in $\mathbf{B}(\mathfrak{H})$ and bounded, sesquilinear forms on $\mathfrak{H}$, given by $T \to B_T$, where*

$$B_T(x, y) = (x \,|\, Ty).$$

PROOF. If $T \in \mathbf{B}(\mathfrak{H})$, then clearly B_T is a sesquilinear form on $\mathfrak{H}$, bounded by $\|T\|$, since

$$\|B_T\| = \sup\{|B_T(x, y)| \,|\, \|x\| \le 1, \|y\| \le 1\}$$
$$= \sup\{|(x\,|\,Ty)| \,|\, \|x\| \le 1, \|y\| \le 1\} \le \|T\|.$$

On the other hand,

$$\|Tx\|^2 = (Tx \,|\, Tx) = B_T(Tx, x)$$
$$\le \|B_T\| \|Tx\| \|x\| \le \|B_T\| \|T\| \|x\|^2,$$

which shows that $\|T\| \le \|B_T\|$, whence $\|T\| = \|B_T\|$.

Assume now that B is a bounded, sesquilinear form on $\mathfrak{H}$. Then $B(\cdot, y) \in \mathfrak{H}^*$ for each y in $\mathfrak{H}$. By 3.1.9 there is therefore a unique vector in $\mathfrak{H}$, denoted by Ty, such that

$$(x \,|\, Ty) = B(x, y), \quad x \in \mathfrak{H}.$$

The map $y \to Ty$ is linear (we conjugate twice) and bounded by $\|B\|$, whence $T \in \mathbf{B}(\mathfrak{H})$ and $B = B_T$. □

3.2.3. Theorem. *To each T in $\mathbf{B}(\mathfrak{H})$ there is a unique T^* in $\mathbf{B}(\mathfrak{H})$ such that*

$$(Tx \,|\, y) = (x \,|\, T^*y), \quad x, y \in \mathfrak{H}. \tag{$*$}$$

The map $T \to T^*$ is a conjugate linear, antimultiplicative isometry of $\mathbf{B}(\mathfrak{H})$ onto itself of period two, and satisfies the identity

$$\|T^*T\| = \|T\|^2. \tag{$**$}$$

PROOF. Corresponding to the bounded sesquilinear form $x, y \to (Tx|y)$ on $\mathfrak{H}$ there is by 3.2.2 a uniquely determined operator T^* in $\mathbf{B}(\mathfrak{H})$ satisfying (∗). As the inner product is self-adjoint (cf. 3.1.1), we have

$$(x|T^{**}y) = (T^*x|y) = \overline{(y|T^*x)}$$
$$= \overline{(Ty|x)} = (x|Ty).$$

Thus T^{**} and T correspond to the same sesquilinear form, whence $T^{**} = T$ by 3.2.2. In the same manner we show that the involution $T \to T^*$ is conjugate linear. Furthermore,

$$(x|(ST)^*y) = (STx|y) = (Tx|S^*y) = (x|T^*S^*y),$$

whence $(ST)^* = T^*S^*$, so that the involution is antimultiplicative.

Since the operator norm is submultiplicative we have $\|T^*T\| \le \|T\|\|T^*\|$. On the other hand, applying (∗) to T^* we get

$$\|Tx\|^2 = (Tx|Tx) = (T^*Tx|x) \le \|T^*T\|\|x\|^2,$$

which shows that $\|T\|^2 \le \|T^*T\|$. Combining these inequalities we see that $\|T\| \le \|T^*\|$, whence $\|T\| = \|T^*\|$ (since $T^{**} = T$). All taken together show that the involution is isometric and satisfies (∗∗).   □

3.2.4. The *involution* defined in 3.2.3 is clearly a generalization of the well-known process of taking adjoints of matrices, where the matrix $A = (\alpha_{ij})$ is transformed into the matrix $A^* = (\alpha_{ij}^*)$, with $\alpha_{ij}^* = \bar{\alpha}_{ji}$. We will use the same terminology as developed in linear algebra. Thus, we say that T^* is the *adjoint* of the operator T and that T is *self-adjoint* if $T = T^*$. [The word *hermitian* (after C. Hermite) is also used.] Since T^* corresponds to the adjoint of the form corresponding to T, it follows from (∗) in 3.1.1 that for $\mathbb{F} = \mathbb{C}$ we have $T = T^*$ iff $(Tx|x) \in \mathbb{R}$ for every x in $\mathfrak{H}$.

3.2.5. Proposition. *For every T in $\mathbf{B}(\mathfrak{H})$ we have*

$$\ker T^* = (T(\mathfrak{H}))^{\perp}.$$

PROOF. From the defining identity $(Tx|y) = (x|T^*y)$ we see that if $y \in \ker T^*$, then $y \in (T(\mathfrak{H}))^{\perp}$. Conversely, if $y \in (T(\mathfrak{H}))^{\perp}$, then $T^*y \in \mathfrak{H}^{\perp} = \{0\}$.   □

3.2.6. Proposition. *For an operator T in $\mathbf{B}(\mathfrak{H})$ the following conditions are equivalent:*

  (i) *T is invertible [i.e. $T^{-1} \in \mathbf{B}(\mathfrak{H})$].*
  (ii) *T^* is invertible.*
 (iii) *Both T and T^* are bounded away from zero.*
 (iv) *Both T and T^* are injective, and $T(\mathfrak{H})$ is closed.*
  (v) *T is injective and $T(\mathfrak{H}) = \mathfrak{H}$.*
 (vi) *Both T and T^* are surjective [i.e. $T(\mathfrak{H}) = T^*(\mathfrak{H}) = \mathfrak{H}$].*

PROOF. By definition of the inverse we have $T^{-1}T = TT^{-1} = I$, and therefore $(T^{-1})^*$ is the inverse of T^*. Thus (i) ⇔ (ii).

(i) ⇒ (iii). For every x in $\mathfrak{H}$ we have

$$\|x\| = \|T^{-1}Tx\| \leq \|T^{-1}\| \|Tx\|,$$

so that T is bounded away from zero by $\|T^{-1}\|^{-1}$. Since (i) ⇔ (ii) we see that also T^* is bounded away from zero.

(iii) ⇒ (iv). If $\|Tx\| \geq \varepsilon\|x\|$ and $\|T^*x\| \geq \varepsilon\|x\|$ for some $\varepsilon > 0$ and all x in $\mathfrak{H}$, then evidently T and T^* are injective. Moreover, $\|Tx - Ty\| \geq \varepsilon\|x - y\|$, which shows that $T(\mathfrak{H})$ is complete and therefore a closed subspace of $\mathfrak{H}$.

(iv) ⇒ (v). By 3.1.8 and 3.2.5 we have

$$(T(\mathfrak{H}))^= = (T(\mathfrak{H})^\perp)^\perp = (\ker T^*)^\perp = \{0\}^\perp = \mathfrak{H},$$

since T^* is injective.

(v) ⇒ (i). This is the open mapping theorem (2.2.5).

(vi) ⇒ (v). Since $\ker T = (T^*(\mathfrak{H}))^\perp = \{0\}$ this is obvious, and so is (i) ⇒ (vi). □

3.2.7. An operator T in $\mathbf{B}(\mathfrak{H})$ is *normal* if it commutes with its adjoint, i.e. if $T^*T = TT^*$. Evidently this entails that

$$\|Tx\| = (T^*Tx|x)^{1/2} = (TT^*x|x)^{1/2} = \|T^*x\|$$

for every x in $\mathfrak{H}$. We say that T and T^* are *metrically identical*. Conversely, if $T \in \mathbf{B}(\mathfrak{H})$ such that T and T^* are metrically identical, the polarization identity (3.1.2) shows that $(T^*Tx|y) = (TT^*x|y)$ for all x and y, whence T is normal by 3.2.2. Note from 3.2.6 that a normal operator is invertible iff it is bounded away from zero.

3.2.8. An operator T in $\mathbf{B}(\mathfrak{H})$ is *positive*, in symbols $T \geq 0$, if $T = T^*$ and $(Tx|x) \geq 0$ for every x in $\mathfrak{H}$. If $\mathbb{F} = \mathbb{C}$, the positivity condition alone implies self-adjointness by the polarization identity.

Clearly, the sum of two positive operators is positive [the positive operators form a cone in $\mathbf{B}(\mathfrak{H})$]; but the product need not be positive, need not even be self-adjoint. However, if the operators commute (i.e. if the product is self-adjoint), we see from 3.2.11 that their product *is* positive (since $ST = S^{1/2}TS^{1/2} \geq 0$ by 3.2.9).

The self-adjoint operators in $\mathbf{B}(\mathfrak{H})$ form a closed, real subspace, denoted by $\mathbf{B}(\mathfrak{H})_{sa}$. On this subspace the cone of positive operators defines an order: $S \leq T$ if $T - S \geq 0$. As we know already from the 2×2 matrices, this order is not total; it is not even a lattice order. For the deeper properties of (positive) operators we need the *square root lemma* (3.2.11). We include a direct proof, but must emphasize that the result is an immediate consequence of the spectral theorem (4.4.1). It will therefore be reproved in that context; see 4.4.8.

3.2.9. Proposition. *If $S \leq T$ in $\mathbf{B}(\mathfrak{H})_{sa}$, then $A^*SA \leq A^*TA$ for every A in $\mathbf{B}(\mathfrak{H})$. If, moreover, $0 \leq S$, then $\|S\| \leq \|T\|$. In particular, $S \leq I$ iff $\|S\| \leq 1$. Also, for $T = T^*$ we have $-I \leq T \leq I$ iff $\|T\| \leq 1$.*

PROOF. We know that $((T - S)y|y) \geq 0$ for every y in $\mathfrak{H}$. Replacing y with Ax, $x \in \mathfrak{H}$, it follows that $A^*(T - S)A \geq 0$, i.e. $A^*SA \leq A^*TA$.

If $0 \leq S \leq T$, we use the Cauchy–Schwarz inequality on the positive sesquilinear form $(S \cdot | \cdot)$ to obtain, for every pair of unit vectors x and y in $\mathfrak{H}$,

$$|(Sx|y)|^2 \leq (Sx|x)(Sy|y) \leq (Tx|x)(Ty|y) \leq \|T\|^2,$$

whence $\|S\|^2 \leq \|T\|^2$ by 3.2.2. Taking $T = I$ we see that $S \leq I$ iff $\|S\| \leq 1$, whenever $0 \leq S$.

For the last assertion, taking $T = T^*$ with $-I \leq T \leq I$. Then with x and y unit vectors in $\mathfrak{H}$ we have

$$(T(x + y)|x + y) \leq \|x + y\|^2$$

$$-(T(x - y)|x - y) \leq \|x - y\|^2.$$

Adding these inequalities and using the parallellogram law gives $4 \operatorname{Re}(Tx|y) \leq 4$, and since x and y are arbitrary this implies that $\|T\| \leq 1$ by 3.2.2. Conversely, if $\|T\| \leq 1$ then evidently $|(Tx|x)| \leq \|x\|^2$, whence $-I \leq T \leq I$. □

3.2.10. Lemma. *There is a sequence (p_n) of polynomials with positive coefficients, such that $\sum p_n$ converges uniformly on the interval $[0, 1]$ to the function $t \to 1 - (1 - t)^{1/2}$.*

PROOF. Define a sequence (q_n) of polynomials inductively by $q_0 = 0$, $q_n(t) = \frac{1}{2}(t + q_{n-1}(t)^2)$. It is easily verified (by induction) that each q_n has positive coefficients and that $q_n(t) \leq 1$ for every t in $[0, 1]$. Moreover, since

$$2(q_{n+1} - q_n) = q_n^2 - q_{n-1}^2 = (q_n - q_{n-1})(q_n + q_{n-1}),$$

it follows, again by induction, that each polynomial $p_n = q_n - q_{n-1}$ has positive coefficients.

Regarding the q_n's as elements in $C([0, 1])$, we see that they converge pointwise on $[0, 1]$ to a function q such that $2q(t) = t + q(t)^2$. Thus $q(t) = 1 - (1 - t)^{1/2}$. Since $q \in C([0, 1])$ and $[0, 1]$ is compact, the monotone convergence $q_n \nearrow q$ is in fact uniform by Dini's lemma (E 1.6.6). Thus $\sum p_n = \lim q_n = q$, as claimed. □

3.2.11. Proposition. *To each positive operator T in $\mathbf{B}(\mathfrak{H})$ there is a unique positive operator, denoted by $T^{1/2}$, satisfying $(T^{1/2})^2 = T$. Moreover, $T^{1/2}$ commutes with every operator commuting with T.*

PROOF. Since $(\alpha T)^{1/2} = \alpha^{1/2} T^{1/2}$ if $\alpha \geq 0$, we may evidently assume that $T \leq I$. Thus, with $S = I - T$ we have $0 \leq S \leq I$. Now take (p_n) as in 3.2.10 and define $S_n = p_n(S)$. Given $\varepsilon > 0$ we can find n_0 such that

$$0 \leq \sum_{n=n_0}^{m} p_n(t) = \sum \alpha_k t^k \leq \varepsilon$$

for all $m > n_0$ and every t in $[0, 1]$. Since all the coefficients α_k in the polynomial above are positive, this means that

$$\left\| \sum_{n=n_0}^{m} S_n \right\| \le \sum \alpha_k \|S^k\| \le \sum \alpha_k \le \varepsilon.$$

It follows that $\sum S_n$ converges uniformly to an operator R with $0 \le R \le I$. With q_n as in the proof of 3.2.10 we have

$$(I - R)^2 = (I - \sum S_n)^2 = \lim(I - q_n(S))^2$$
$$= \lim(I - 2q_n(S) + q_n(S)^2) = \lim(I - 2q_n(S) + 2q_{n+1}(S) - S)$$
$$= I - S + 2\lim p_{n+1}(S) = I - S = T.$$

We may therefore take $T^{1/2} = I - R$.

Since $T^{1/2}$ is a uniform limit of polynomials in the variables I and T, it commutes with every operator commuting with T. Finally, if $A \in \mathbf{B}(\mathfrak{H})$, such that $A \ge 0$ and $A^2 = T$, then

$$(T^{1/2} - A)(T^{1/2} + A)x = ((T^{1/2})^2 - A^2)x = 0$$

for every x in $\mathfrak{H}$, since $T^{1/2}$ commutes with A. Thus, $(T^{1/2} - A)y = 0$ for every y in $\mathfrak{Y} = ((T^{1/2} + A)\mathfrak{H})^=$. Since $\mathfrak{Y}^\perp = \ker(T^{1/2} + A)$ by 3.2.5, it follows that

$$(T^{1/2}z|z) \le ((T^{1/2} + A)z|z) = 0$$

for every z in $\mathfrak{Y}^\perp$, whence $\|(T^{1/2})^{1/2}z\|^2 = 0$, so that $T^{1/2}z = (T^{1/2})^{1/2}(T^{1/2})^{1/2}z = 0$. Similarly, $Az = 0$ for every z in $\mathfrak{Y}^\perp$, whence $T^{1/2} = A$, as desired. $\square$

3.2.12. Proposition. *A positive operator T is invertible in $\mathbf{B}(\mathfrak{H})$ iff $T \ge \varepsilon I$ for some $\varepsilon > 0$. In that case $T^{-1} \ge 0$ and $T^{1/2}$ is invertible, and $(T^{-1})^{1/2} = (T^{1/2})^{-1}$. If, moreover, $T \le S$, then $S^{-1} \le T^{-1}$.*

PROOF. If $T \ge \varepsilon I$, then $T - \varepsilon I$ and $T + \varepsilon I$ are two commuting, positive operators. Their product, $T^2 - \varepsilon^2 I$, is therefore also positive, so $T^2 \ge \varepsilon^2 I$. This means that

$$\|Tx\|^2 = (T^2x|x) \ge \varepsilon^2(x|x) = \varepsilon^2\|x\|^2,$$

so T is bounded away from zero, hence invertible by 3.2.6. Moreover, $(T^{-1}Tx|Tx) = (Tx|x) \ge 0$, whence $T^{-1} \ge 0$ since $T(\mathfrak{H}) = \mathfrak{H}$.

Conversely, if T is invertible, hence bounded away from zero by some $\varepsilon > 0$, then

$$(T^2x|x) = \|Tx\|^2 \ge \varepsilon^2\|x\|^2 = \varepsilon^2(x|x)$$

for every x in $\mathfrak{H}$, whence $T^2 \ge \varepsilon^2 I$. From the first part of the proof we know that $T + \varepsilon I$ is invertible with a positive inverse, so $T - \varepsilon I \ge 0$, being the product of the two positive, commuting operators $T^2 - \varepsilon^2 I$ and $(T + \varepsilon I)^{-1}$. Thus, $T \ge \varepsilon I$. Iterating this argument we see that $T^{1/2} \ge \varepsilon^{1/2} I$, so that $T^{1/2}$ is invertible with a positive inverse $(T^{1/2})^{-1}$. Since the square of this element is T^{-1}, we see from the uniqueness part of 3.2.11 that $(T^{1/2})^{-1} = (T^{-1})^{1/2}$. This element will from now on be denoted $T^{-1/2}$.

Suppose now that $T \leq S$. Then $S^{-1/2}$ exists by the previous arguments, so $S^{-1/2} T S^{-1/2} \leq I$ by 3.2.9. Combining 3.2.3 and 3.2.9 this gives

$$\|T^{1/2} S^{-1/2}\|^2 = \|S^{-1/2} T S^{-1/2}\| \leq 1.$$

But, again by 3.2.3, this means that

$$\|T^{1/2} S^{-1} T^{1/2}\| = \|S^{-1/2} T^{1/2}\|^2 = \|T^{1/2} S^{-1/2}\|^2 \leq 1,$$

so that $T^{1/2} S^{-1} T^{1/2} \leq I$. Multiplication by $T^{-1/2}$ as in 3.2.9 finally shows that $S^{-1} \leq T^{-1/2} I T^{-1/2} = T^{-1}$. $\square$

3.2.13. If $\mathfrak{X}$ is a closed subspace of the Hilbert space $\mathfrak{H}$, we have, for each y in $\mathfrak{H}$, the orthogonal decomposition $y = x + x^\perp$ in $\mathfrak{X} + \mathfrak{X}^\perp$; cf. 3.1.7. Since

$$\alpha y_1 + \beta y_2 = (\alpha x_1 + \beta x_2) + (\alpha x_1^\perp + \beta x_2^\perp)$$

is the decomposition of a linear combination, it follows that the map $P \colon \mathfrak{H} \to \mathfrak{H}$ defined by $Py = x$ is an operator in $\mathbf{B}(\mathfrak{H})$ with $\|P\| \leq 1$. Note that P is idempotent ($P^2 = P$), self-adjoint, and positive, since

$$(Py_1|y_2) = (x_1|x_2 + x_2^\perp) = (x_1|x_2) = (x_1 + x_1^\perp|x_2) = (y_1|Py_2)$$

and

$$(Py|y) = (x|x + x^\perp) = \|x\|^2 \geq 0.$$

We say that P is an (orthogonal) *projection*. If, conversely, P is a self-adjoint idempotent in $\mathbf{B}(\mathfrak{H})$, then

$$\mathfrak{X} = \{x \in \mathfrak{H} | Px = x\} = P(\mathfrak{H})$$

is a closed subspace of $\mathfrak{H}$; and if $x^\perp \in \mathfrak{X}^\perp$, we have (since $P = P^*$) that

$$\|Px^\perp\|^2 = (x^\perp|P^2 x^\perp) = 0.$$

Thus, P is the orthogonal projection of $\mathfrak{H}$ onto $\mathfrak{X}$. Note that $I - P$ is the projection on $\mathfrak{X}^\perp$.

3.2.14. The projections are just about the most elementary operators one can imagine, and correspond to the characteristic functions in the theory of functions of a real variable. The analogue of a simple function is obtained by decomposing $\mathfrak{H}$ as a finite orthogonal sum $\mathfrak{H} = \mathfrak{X}_1 \oplus \mathfrak{X}_2 \oplus \cdots \oplus \mathfrak{X}_n$, corresponding to the projections $P_1, P_2, \ldots, P_n$ (which are pairwise orthogonal, i.e. $P_i P_j = 0$ for $i \neq j$, and satisfy $\sum P_i = I$). Then one considers $T = \sum \lambda_n P_n$ for some scalars $\lambda_1, \lambda_2, \ldots, \lambda_n$ in $\mathbb{F}$. More generally, we say that an operator T is *diagonalizable* if there is an orthonormal basis $\{e_j | j \in J\}$ for $\mathfrak{H}$ and a bounded set $\{\lambda_j | j \in J\}$ in $\mathbb{F}$ such that

$$Tx = \sum \lambda_j(x|e_j)e_j \qquad (*)$$

for every x in $\mathfrak{H}$. Note that the numbers $(x|e_j)$ are the coordinates for x in the basis $\{e_j\}$ and that each λ_j is an eigenvalue for T corresponding to the eigenvector e_j. Denoting by P_j the projection of $\mathfrak{H}$ on the subspace $\mathbb{F}e_j$, we have $P_j x = (x|e_j)e_j$, so that we may formally write

$$T = \sum \lambda_j P_j$$

in analogy with the finite case above. However, the sum is no longer necessarily convergent in $\mathbf{B}(\mathfrak{H})$, but only pointwise convergent (i.e. convergent in the strong operator topology defined in 4.6.1).

It follows from $(*)$ that $T^*x = \sum \bar{\lambda}_j(x|e_j)e_j$, so that T^* is diagonalizable along the basis $\{e_j\}$ with eigenvalues $\{\bar{\lambda}_j\}$. Moreover, $T^*T = TT^*$ (with eigenvalues $\{|\lambda_j|^2\}$ along the basis $\{e_j\}$), so that T is normal. We see that T is self-adjoint iff all λ_j are real and that T is positive iff $\lambda_j \geq 0$ for all j.

If $\mathfrak{H}$ is finite-dimensional, every normal operator is diagonalizable. This is no longer true in infinite dimensions. As a matter of fact, most of the interesting (normal) operators on infinite-dimensional Hilbert space are nondiagonalizable, and must be handled with a continuous analogue of the concept of basis; see 4.7.12.

3.2.15. A *unitary* operator is an isometric isomorphism of $\mathfrak{H}$ onto itself. (On real Hilbert spaces the term *orthogonal* operator is used.) It follows from the polarization identities in 3.1.2 that unitary operators conserve the inner product, since

$$4(Ux|Uy) = \sum i^k \|U(x + i^k y)\|^2$$
$$= \sum i^k \|x + i^k y\|^2 = 4(x|y)$$

(and similarly in the real case). From this we see that $U^*U = I$ (cf. 3.2.2), and since U is invertible, $U^{-1} = U^*$, so that U is normal with $UU^* = U^*U = I$. Conversely, if $U \in \mathbf{B}(\mathfrak{H})$ such that $U^{-1} = U^*$, then U is unitary, because

$$\|Ux\|^2 = (U^*Ux|x) = \|x\|^2;$$

so that U is an isometry that is surjective because U is invertible.

If $\{e_j|j \in J\}$ is an orthonormal basis for $\mathfrak{H}$ and U is unitary, then $\{Ue_j|j \in J\}$ also is an orthonormal basis for $\mathfrak{H}$. Conversely, we saw in 3.1.14 that every transition between orthonormal bases is given by a unitary operator. Note that the product of unitary operators is again unitary, so that the set $\mathbf{U}(\mathfrak{H})$ of unitary operators on $\mathfrak{H}$ is a group [a subgroup of the general linear group $\mathbf{GL}(\mathfrak{H})$ of invertible operators in $\mathbf{B}(\mathfrak{H})$].

We say that two operators S and T are *unitarily equivalent* if $S = UTU^*$ for some U in $\mathbf{U}(\mathfrak{H})$. Note that unitary equivalence preserves norm, self-adjointness, normality, diagonalizability, and unitarity.

3.2.16. An operator U in $\mathbf{B}(\mathfrak{H})$ is said to be a *partial isometry* if there is a closed subspace $\mathfrak{X}$ of $\mathfrak{H}$ such that $U|\mathfrak{X}$ is isometric and $U|\mathfrak{X}^\perp = 0$. This implies that $\mathfrak{X}^\perp = \ker U$, whence $\mathfrak{X} = (U^*(\mathfrak{H}))^=$ by 3.2.5. Set $P = U^*U$. Then

$$(Px|x) = \|Ux\|^2 = \|x\|^2$$

for every x in $\mathfrak{X}$, whence $Px = x$ by the Cauchy–Schwarz inequality (3.1.3). Moreover, $Px^\perp = 0$ for every $x^\perp$ in $\mathfrak{X}^\perp$, so that P is the projection of $\mathfrak{H}$ on $\mathfrak{X}$.

Conversely, if $U \in \mathbf{B}(\mathfrak{H})$ such that $U^*U = P$ for some projection P, we put

$\mathfrak{X} = P(\mathfrak{H})$ and see from the equation $\|Ux\|^2 = (Px|x)$ that U is isometric on $\mathfrak{X}$ and 0 on $\mathfrak{X}^\perp$. Thus, U is a partial isometry. Now we can also show that U^* is a partial isometry, because $U(I - P) = 0$, whence

$$(UU^*)^2 = UU^*UU^* = UPU^* = UU^*;$$

so that UU^* is a selfadjoint idempotent, i.e. a projection. Note that U^* is the isometry of $U(\mathfrak{H})$ back onto $U^*(\mathfrak{H})$, so that U^* is a "partial inverse" to U.

In an infinite-dimensional Hilbert space $\mathfrak{H}$ one may have isometries that are not unitaries (because they are not surjective). Consider, for example, the case where $\mathfrak{H}$ is separable with orthonormal basis $(e_n | n \in \mathbb{N})$, and define

$$S(\sum \alpha_n e_n) = \sum \alpha_n e_{n+1}, \qquad \sum \alpha_n e_n \in \mathfrak{H}.$$

Then S (the *unilateral shift* operator) is an isometry of $\mathfrak{H}$ onto the subspace $\{e_1\}^\perp$, and, consequently, S^* is a partial isometry of $\{e_1\}^\perp$ onto $\mathfrak{H}$. In particular, $S^*S = I$, whereas SS^* is the projection onto $\{e_1\}^\perp$.

3.2.17. Theorem. *To each operator T in $\mathbf{B}(\mathfrak{H})$ there is a unique positive operator $|T|$ in $\mathbf{B}(\mathfrak{H})$ satisfying*

$$\|Tx\| = \||T|x\|, \quad x \in \mathfrak{H}; \tag{$*$}$$

and we have $|T| = (T^*T)^{1/2}$. Moreover, there is a unique partial isometry U with $\ker U = \ker T$ and $U|T| = T$. In particular, $U^*U|T| = |T|$, $U^*T = |T|$, and $UU^*T = T$.

PROOF. If $S \geq 0$ is metrically identical to T, then

$$(S^2x|x) = \|Sx\|^2 = \|Tx\|^2 = (T^*Tx|x)$$

for every x in $\mathfrak{H}$. From the polarization identities in 3.1.2 (applied to the forms associated with S^2 and T^*T, cf. 3.2.2) it follows that $S^2 = T^*T$, whence $S = (T^*T)^{1/2}$ by 3.2.11. With $|T| = (T^*T)^{1/2}$ we have

$$\ker T = \ker|T| = |T|(\mathfrak{H})^\perp$$

by 3.2.5. For every $y = |T|x$ in $|T|(\mathfrak{H})$ we define

$$U_0 y = U_0|T|x = Tx;$$

and it follows from $(*)$ that U_0 is a well-defined isometry from $|T|(\mathfrak{H})$ onto $T(\mathfrak{H})$. By continuity (2.1.11) U_0 extends to an isometry U of $|T|(\mathfrak{H})^=$ onto $T(\mathfrak{H})^=$. Setting $U = 0$ on $\ker T$ $(= |T|(\mathfrak{H})^\perp)$ we have evidently constructed a partial isometry such that $U|T| = T$.

If V is another partial isometry with $\ker V = \ker T$ and $V|T| = T$, we note that $Vy = Uy$ for every $y = |T|x$, $x \in \mathfrak{H}$, so that $U = V$ on $|T|(\mathfrak{H})^=$. Since both operators are 0 on $|T|(\mathfrak{H})^\perp$, it follows that $U = V$.

Of the last three identities, the first follows from the fact that U^*U is the projection on $(|T|(\mathfrak{H}))^=$, and the others are derived by inserting $T = U|T|$ and multiplying with U^* from the left. $\qquad\square$

3.2.18. Remark. The preceding construction (by von Neumann) is called the *polar decomposition* of the operator T. We may regard the partial isometry U as a generalized "sign" of T and $|T|$ as the "absolute value" of T. From the formula

$$T^* = |T|U^* = U^*(U|T|U^*),$$

and the uniqueness of the polar decomposition, we see that U^* is the sign of T^*, whereas $U|T|U^*$ (somewhat inconvenient) is the absolute value of T^*.

One should not expect the absolute value of a sum or a product of noncommuting operators to have much relation to the sum or the product of the absolute values. Neither should one expect that the sign, as in the finite-dimensional case, always can be chosen to be unitary. The unilateral shift, mentioned in 3.2.16, is an immediate counterexample. In order for T to have the form $U|T|$ for some unitary operator U, it is necessary and sufficient that the closed subspaces $\ker T$ and $\ker T^*$ have the same dimension (finite or infinite). Easy cases of this general theorem are established in the next results.

3.2.19. Proposition. *If T is invertible in $\mathbf{B}(\mathfrak{H})$, the partial isometry in its polar decomposition is unitary.*

PROOF. By (the proof of) 3.2.17 we have $T = U|T|$, where $\ker U = \ker T$ and $U(\mathfrak{H}) = T(\mathfrak{H})^=$. Since T is invertible, it follows that $\ker U = 0$ and $U(\mathfrak{H}) = \mathfrak{H}$, so that U is unitary. □

3.2.20. Proposition. *If T is normal in $\mathbf{B}(\mathfrak{H})$, there is a unitary operator W commuting with T, T^*, and $|T|$ such that $T = W|T|$.*

PROOF. By 3.2.17 we have $T = U|T|$, and the normality of T implies that

$$U|T|U^* = |T^*| = |TT^*|^{1/2} = |T^*T|^{1/2} = |T|,$$

whence

$$U|T| = U|T|U^*U = |T|U.$$

Since $\ker T = \ker T^*$, we have $T(\mathfrak{H})^= = T^*(\mathfrak{H})^=$ by 3.2.5 so that $U^*U = UU^*$. Define $Wx = Ux$ if $x \in |T|(\mathfrak{H})^= (= T^*(\mathfrak{H})^=)$ and $Wx = x$ if $x \in \ker T$ $(= \ker|T|)$. Then W is unitary and $W|T| = U|T| = T$. Moreover, W commutes with $|T|$ (and with W and W^*, of course), so W commutes with T and T^*. □

3.2.21. Lemma. *If $T = T^*$ and $\|T\| \le 1$, the operator $U = T + i(I - T^2)^{1/2}$ is unitary and $T = \frac{1}{2}(U + U^*)$.*

PROOF. Since $I - T^2 \ge 0$, it has a square root by 3.2.11, commuting with T. Direct computation shows that $UU^* = U^*U = I$, and clearly $\frac{1}{2}(U + U^*) = T$. □

3.2.22. Lemma. *If $S \in \mathbf{B}(\mathfrak{H})$ with $\|S\| < 1$, then for every unitary U there are unitaries U_1 and V_1 such that*

$$S + U = U_1 + V_1.$$

PROOF. Replacing, if necessary, S with U^*S we may assume that $U = I$. Since $\|S\| < 1$, both operators $I + S$ and $I + S^*$ are bounded away from zero, so $I + S$ is invertible by 3.2.6. Consequently, $I + S = V|I + S|$ with V unitary by 3.2.19, and since $\|I + S\| \leq 2$, it follows from 3.2.21 that $|I + S| = W + W^*$ for some unitary W. With $U_1 = VW$ and $V_1 = VW^*$ the lemma follows.   □

3.2.23. Proposition. *If $T \in \mathbf{B}(\mathfrak{H})$ with $\|T\| < 1 - 2/n$ for some $n > 2$, there are unitary operators $U_1, U_2, \ldots, U_n$ such that*

$$T = \frac{1}{n}(U_1 + U_2 + \cdots + U_n).$$

PROOF. Put $S = (n-1)^{-1}(nT - I)$, so that $\|S\| < 1$ and $nT = (n-1)S + I$. Applying 3.2.22 $n - 1$ times we obtain unitaries $V_1, \ldots, V_{n-1}$ and $U_1, \ldots, U_n$ (where $U_n = V_{n-1}$) such that

$$
\begin{aligned}
nT &= (n-1)S + I = (n-2)S + (S + I) = (n-2)S + (V_1 + U_1) \\
&= (n-3)S + (S + V_1) + U_1 = (n-3)S + (V_2 + U_2) + U_1 \\
&= (n-4)S + (S + V_2) + U_2 + U_1 \\
&= (n-4)S + (V_3 + U_3) + U_2 + U_1 = \cdots \\
&= (S + V_{n-2}) + U_{n-2} + \cdots + U_1 = U_n + U_{n-1} + U_{n+2} + \cdots + U_1,
\end{aligned}
$$

as desired.   □

3.2.24. Since the self-adjoint elements span $\mathbf{B}(\mathfrak{H})$, we see already from 3.2.21 that every operator in $\mathbf{B}(\mathfrak{H})$ is the linear combination of (four) unitary operators. The estimate in the *Russo–Dye–Gardner theorem* (3.2.23) is much more precise with regard to the norm, and shows that every element in the open unit ball of $\mathbf{B}(\mathfrak{H})$ is a convex combination (indeed, a mean) of unitary operators. On the surface of the ball this need no longer be true; in fact, the unilateral shift (3.2.16) cannot be expressed as a convex combination of unitaries (see E 3.2.8).

3.2.25. Proposition. *If $\mathfrak{H}$ is a complex Hilbert space and $T \in \mathbf{B}(\mathfrak{H})$, we define the numerical radius of T as*

$$\||T\|| = \sup\{|(Tx|x)| \, | \, x \in \mathfrak{H}, \|x\| \leq 1\}.$$

Then $\||\cdot\||$ is a vector space norm on $\mathbf{B}(\mathfrak{H})$ and satisfies

$$\tfrac{1}{2}\|T\| \leq \||T\|| \leq \|T\|, \qquad T \in \mathbf{B}(\mathfrak{H}).$$

PROOF. It is evident from the definition that $\|\|\cdot\|\|$ is homogeneous and sub-additive, and it follows from the Cauchy–Schwarz inequality that $\|\|T\|\| \leq \|T\|$ for every T in $\mathbf{B}(\mathfrak{H})$.

To prove the inequality $\|T\| \leq 2\|\|T\|\|$ we consider unit vectors x and y in $\mathfrak{H}$ and use the polarization and parallellogram identities to get

$$4|(Tx|y)| = \left| \sum_{k=0}^{3} i^k (T(x + i^k y)|x + i^k y) \right|$$

$$\leq \|\|T\|\| (\|x + y\|^2 + \|x - y\|^2 + \|x + iy\|^2 + \|x - iy\|^2)$$

$$= \|\|T\|\| 2(\|x\|^2 + \|y\|^2 + \|x\|^2 + \|iy\|^2) = 8\|\|T\|\|.$$

It follows from 3.2.2 that $\|T\| \leq 2\|\|T\|\|$. $\square$

3.2.26. Proposition. *If we define* $\mathrm{Re}(T) = \frac{1}{2}(T + T^*)$, *then for each T in* $\mathbf{B}(\mathfrak{H})$ *we have*

$$\|\|T\|\| = \max\{\|\mathrm{Re}(\theta T)\| \,|\, \theta \in \mathbb{C}, |\theta| = 1\}.$$

PROOF. Note first that if $H = H^*$ in $\mathbf{B}(\mathfrak{H})$ then by the last statement in 3.2.9 we have $\|\|H\|\| = \|H\|$. Next, for T in $\mathbf{B}(\mathfrak{H})$ write $T = H + iK$ with H and K self-adjoint operators (so that $H = \mathrm{Re}(T)$ and $K = -\mathrm{Re}(iT)$). Then if x is a unit vector in $\mathfrak{H}$ we get $(Tx|x) = (Hx|x) + i(Kx|x)$, whence

$$|(Tx|x)|^2 = (Hx|x)^2 + (Kx|x)^2. \tag{$*$}$$

It follows that

$$\|H\|^2 = \|\|H\|\|^2 \leq \|\|T\|\|^2,$$

and, replacing T with θT, where $|\theta| = 1$, this means that

$$\sup \|\mathrm{Re}(\theta T)\| \leq \|\|T\|\|.$$

Now, assuming that $\gamma = |(Tx|x)| \neq 0$, put

$$\alpha = \gamma^{-1}(Hx|x), \qquad \beta = \gamma^{-1}(Kx|x).$$

Then with $\theta = \alpha - i\beta$ we have $|\theta| = 1$ by $(*)$ and

$$\mathrm{Re}(\theta T) = \mathrm{Re}((\alpha - i\beta)(H + iK)) = \alpha H + \beta K.$$

Consequently

$$(\mathrm{Re}(\theta T)x|x) = \alpha^2 \gamma + \beta^2 \gamma = \gamma,$$

whence $|(Tx|x)| \leq \|\mathrm{Re}(\theta T)\|$. Since this holds for every x we see that

$$\|\|T\|\| \leq \sup \|\mathrm{Re}(\theta T)\|,$$

and thus equality. As the function $\theta \to \|\mathrm{Re}(\theta T)\|$ is continuous, it attains its maximum on the compact set $\{\theta \in \mathbb{C} \,|\, |\theta| = 1\}$, completing the proof. $\square$

3.2.27. Proposition. *For each T in* $\mathbf{B}(\mathfrak{H})$ *we have* $\|\|T^2\|\| \leq \|\|T\|\|^2$. *Moreover, if T is normal,* $\|\|T\|\| = \|T\|$.

PROOF. Write $T = H + iK$ with H and K self-adjoint. Then compute

$$\mathrm{Re}(T^2) = \mathrm{Re}((H + iK)^2) = H^2 - K^2.$$

Consequently, if x is any unit vector,

$$(\mathrm{Re}(T^2)x|x) = ((H^2 - K^2)x|x) \leq (H^2x|x)$$
$$= \|Hx\|^2 \leq \|H\|^2 \leq \|\|T\|\|^2,$$

using 3.2.26. If $\theta \in \mathbb{C}$ with $|\theta| = 1$, replace T by $\theta^{1/2}T$ in the estimate above to get

$$(\mathrm{Re}(\theta T^2)x|x) \leq \|\|T\|\|^2;$$

and since this holds for every θ, it follows by 3.2.26 that $\|\|T^2\|\| \leq \|\|T\|\|^2$.

If T is normal, $T^{*n}T^n = (T^*T)^n$. With $n = 2^p$ we get by iterated use of $(**)$ in 3.2.3 that

$$\|T^n\|^2 = \|(T^*T)^n\| = \|T^*T\|^n = \|T\|^{2n}.$$

Using the result above in conjunction with the estimate in 3.2.25 yields

$$\|T\|^n = \|T^n\| \leq 2\|\|T^n\|\| \leq 2\|\|T\|\|^n.$$

Taking n'th roots we see as $n \to \infty$ that $\|T\| \leq \|\|T\|\|$, which implies equality by 3.2.25.   □

EXERCISES

E 3.2.1. If $T \in \mathbf{B}(\mathfrak{H})$ show that $T + T^* \geq 0$ iff $T + I$ is invertible in $\mathbf{B}(\mathfrak{H})$ with $\|(T - I)(T + I)^{-1}\| \leq 1$.

Hint: Show that $T + T^* \geq 0$ iff

$$\|(T + I)x\| \geq \|x\| \quad \text{and} \quad \|(T + I)x\| \geq \|(T - I)x\|$$

for every x in $\mathfrak{H}$, and use the invertibility criteria from 3.2.6.

E 3.2.2. Let $\mathfrak{H}$ be a Hilbert space with inner product $(\cdot|\cdot)$, and let $(\cdot|\cdot)_1$ be another inner product on $\mathfrak{H}$ such that $(x|x)_1 \leq (x|x)$ for every x in $\mathfrak{H}$. Show that there is an injective, positive operator T in $\mathbf{B}(\mathfrak{H})$, with $0 \leq T \leq I$, such that $(Tx|y) = (x|y)_1$ for all x and y in $\mathfrak{H}$.

E 3.2.3. (*Reflection operators.*) Show that the following conditions on an operator R in $\mathbf{B}(\mathfrak{H})$ are equivalent:

(i) There is a closed subspace $\mathfrak{X}$ of $\mathfrak{H}$ such that

$$x + Rx \in \mathfrak{X}, \quad x - Rx \in \mathfrak{X}^{\perp}$$

for every x in $\mathfrak{H}$ (so that R is the reflection of $\mathfrak{H}$ in $\mathfrak{X}$).

(ii) $R = R^*$ and $R^2 = I$.

(iii) $R^2 = I$ and R is normal.

(iv) $\frac{1}{2}(R + I) = P$ for some projection P.

Hint: Use the fact that P is a projection if $P^2 = P$ and P is normal, because $P(\mathfrak{H})^\perp = \ker P^* = \ker P$.

E 3.2.4. Let $T: \mathfrak{H} \to \mathfrak{H}$ be a function on $\mathfrak{H}$ that satisfies $(Tx|Ty) = (x|y)$ for all x and y. Show that T is linear and thus an isometry in $\mathbf{B}(\mathfrak{H})$.

E 3.2.5. Show that an isometry T in $\mathbf{B}(\mathfrak{H})$ is distance preserving, i.e., $\|Tx - Ty\| = \|x - y\|$ for all x and y. Show, conversely, that if $\mathbb{F} = \mathbb{R}$, every distance preserving function f on $\mathfrak{H}$ has the form $f(x) = f(0) + Tx$ for some isometry T in $\mathbf{B}(\mathfrak{H})$.

Hint: Use the real polarization identity in 3.1.2 and the parallelogram law, and then apply E 3.2.4.

E 3.2.6. Let $\mathfrak{C}$ be a closed, convex subset of $\mathfrak{H}$ and denote by $\mathbf{U}(\mathfrak{C})$ the set of unitary operators U on $\mathfrak{H}$ such that $U(\mathfrak{C}) = \mathfrak{C}$. Show that $\mathfrak{C}_{\text{fix}} \neq \emptyset$, where

$$\mathfrak{C}_{\text{fix}} = \{x \in \mathfrak{C} \,|\, Ux = x, U \in \mathbf{U}(\mathfrak{C})\}.$$

Show that if $\mathfrak{C}_{\text{fix}} = \{x_0\}$, then

$$\mathfrak{C} \subset \{x \in \mathfrak{H} \,|\, \operatorname{Re}(x - x_0|x_0) = 0\}. \tag{$*$}$$

Hint: Let x_0 be the point in $\mathfrak{C}$ nearest 0 (cf. 3.1.6) and show that $x_0 \in \mathfrak{C}_{\text{fix}}$. If $\mathfrak{C}_{\text{fix}} = \{x_0\}$, let y_0 be the point in $\mathfrak{C}$ nearest to $2x_0$. Show that $y_0 \in \mathfrak{C}_{\text{fix}}$, whence $x_0 = y_0$. Knowing that x_0 is the best approximation in $\mathfrak{C}$ both to 0 and $2x_0$, the assertion $(*)$ follows from plane geometry.

E 3.2.7. Show that every unit vector in $\mathfrak{H}$ is an extreme point in the closed unit ball of $\mathfrak{H}$ (cf. E 2.5.2).

E 3.2.8. Show that every isometry U in $\mathbf{B}(\mathfrak{H})$ is an extreme point in the closed unit ball of $\mathbf{B}(\mathfrak{H})$.

Hint: If $U = \lambda S + (1 - \lambda)T$ and $x \in \mathfrak{H}$ with $\|x\| = 1$, then $Ux = Sx = Tx$ by E 3.2.7.

E 3.2.9. Show that every projection P in $\mathbf{B}(\mathfrak{H})$ is an extreme point in the convex set

$$\mathbf{B}_+ = \{T \in \mathbf{B}(\mathfrak{H}) \,|\, T \geq 0, \|T\| \leq 1\}.$$

Hint: If $P = \lambda S + (1 - \lambda)T$ with S and T in $\mathbf{B}_+$, then $x = Sx = Tx$ for every x in $P(\mathfrak{H})$ by E 3.2.7. Furthermore, $(Sy|y) = (Ty|y) = 0$ for every y in $P(\mathfrak{H})^\perp$, and because of the positivity and the Cauchy–Schwarz inequality this implies that $Sy = Ty = 0$.

E 3.2.10. Suppose that $\|Tx\| = \|T\|$ for some T in $\mathbf{B}(\mathfrak{H})_{sa}$ and a unit vector x in $\mathfrak{H}$. Show that x is an eigenvector for T^2 corresponding to the eigenvalue $\|T\|^2 \, (= \|T^2\|)$. Show that either $Tx = \|T\|x$ or $Ty = -\|T\|y$, where $y = \|T\|x - Tx$.

E 3.2.11. Take T in $\mathbf{B}(\mathfrak{H})$ with $\|T\| < 1$. Show that the power series

$$\sum_{n=0}^{\infty} \binom{\frac{1}{2}}{n} T^n$$

converges uniformly in $\mathbf{B}(\mathfrak{H})$ to an operator $(I + T)^{1/2}$ satisfying $((I + T)^{1/2})^2 = I + T$. Show that $(I + T)^{1/2} \geq 0$ when $T = T^*$.

Hint: If $T = T^*$ and $\|x\| = 1$ then

$$((I + T)^{1/2}x|x) = 1 - \sum_{n=1}^{\infty} \binom{\frac{1}{2}}{n}(T^n x|x)$$

$$\geq 1 - \sum_{n=1}^{\infty} (-1)^n \binom{\frac{1}{2}}{n} \|T\|^n = (1 - \|T\|)^{1/2}.$$

E 3.2.12. Given $T \geq 0$ in $\mathbf{B}(\mathfrak{H})$, show that $T^2 \leq \|T\| T$.

Hint 1: Use the Cauchy–Schwarz inequality for the positive sesquilinear form $(\cdot|\cdot)_T = (T \cdot|\cdot)$ to estimate

$$(T^2 x|x)^2 = (Tx|x)_T^2 \leq (Tx|Tx)_T(x|x)_T = (T^3 x|x)(Tx|x),$$

and note that $T^3 \leq \|T\| T^2$ by 3.2.9.

Hint 2: Write $T^2 = T^{1/2} T T^{1/2}$ and use 3.2.9.

E 3.2.13. Show that $S \leq T$ implies $S^{1/2} \leq T^{1/2}$ for all positive operators S and T in $\mathbf{B}(\mathfrak{H})$.

Hint: Assume first that S and T are invertible and prove (using 3.2.3, 3.2.11, and 3.2.12) that $S \leq T$ iff $\|S^{1/2} T^{-1/2}\| \leq 1$. Then use the estimates

$$\|S^{1/4} T^{-1/4}\|^{2^{n+1}} = \|T^{-1/4} S^{1/2} T^{-1/4}\|^{2^n} = \|(T^{-1/4} S^{1/2} T^{-1/4})^{2^n}\|$$

$$= \|T^{-1/4}(S^{1/2} T^{-1/2})^{2^{n-1}} S^{1/2} T^{-1/4}\|$$

$$\leq \|T^{-1/2}\| \|S^{1/2}\| \|S^{1/2} T^{-1/2}\|^{2^{n-1}}.$$

In the general case, take $\varepsilon > 0$ to obtain $(\varepsilon I + S)^{1/2} \leq (\varepsilon I + T)^{1/2}$ from the invertible case, and use the estimate $\varepsilon^{1/2} I \leq (\varepsilon I + T)^{1/2}$ and 3.2.12 to show that

$$\|(\varepsilon I + T)^{1/2} - T^{1/2}\| = \|\varepsilon((\varepsilon I + T)^{1/2} + T^{1/2})^{-1}\| \leq \varepsilon^{1/2}.$$

E 3.2.14. (*Unitary dilation.*) For each T in $\mathbf{B}(\mathfrak{H})$ with $\|T\| \leq 1$ define the operator U in $\mathbf{B}(\mathfrak{H} \oplus \mathfrak{H})$ by

$$U = \begin{pmatrix} T & (I - TT^*)^{1/2} \\ (I - T^*T)^{1/2} & -T^* \end{pmatrix}.$$

Show that $T(I - T^*T)^{1/2} = (I - TT^*)^{1/2} T$, and use this to prove that U is unitary.

Hint: Use E 2.2.3 and 3.2.11.

E 3.2.15. Let $\mathfrak{H}$ and $\mathfrak{K}$ be Hilbert spaces and consider an operator T in $\mathbf{B}(\mathfrak{H}, \mathfrak{K})$. Show that there is a unique operator T^* in $\mathbf{B}(\mathfrak{K}, \mathfrak{H})$ satisfying

$$(Tx|y) = (x|T^*y), \quad x \in \mathfrak{H}, y \in \mathfrak{K}.$$

Show that the map $T \to T^*$ is a conjugate linear isometry of $\mathbf{B}(\mathfrak{H}, \mathfrak{K})$
onto $\mathbf{B}(\mathfrak{K}, \mathfrak{H})$, and satisfies

$$\| T^* T \| = \| T \|^2 = \| T T^* \|.$$

Hint: Use 3.2.3 on the operator

$$\begin{pmatrix} 0 & 0 \\ T & 0 \end{pmatrix} \quad \text{in } \mathbf{B}(\mathfrak{H} \oplus \mathfrak{K}).$$

E 3.2.16. Let $\{e_n | n \in \mathbb{N}\}$ be an orthonormal basis for the Hilbert space $\mathfrak{H}$, and
define for each T in $\mathbf{B}(\mathfrak{H})$ the doubly infinite matrix $A = (\alpha_{nm})$, by
letting $\alpha_{nm} = (Te_m | e_n)$. Show that every row and every column in A
is square summable (i.e. belongs to ℓ^2). Use this to prove that the
matrix product $AB = C$, $C = (\gamma_{nm})$, where $\gamma_{nm} = \sum_k \alpha_{nk} \beta_{km}$, is well-
defined when the matrices A and B correspond to operators T and
S in $\mathbf{B}(\mathfrak{H})$. Show that C is the matrix corresponding to the operator
TS. Also find the matrices corresponding to the operators $S + T$
and T^*.

E 3.2.17. (The *Schur test*.) Given a doubly infinite matrix $A = (\alpha_{nm})$ and a
sequence (a_n) in $\mathbb{R}_+$, such that

$$\sum_{n=1}^{\infty} |\alpha_{nm}| a_n \le b a_m \qquad \text{for every } m,$$

$$\sum_{m=1}^{\infty} |\alpha_{nm}| a_m \le c a_n \qquad \text{for every } n,$$

for suitable constants b and c. Show that there is a T in $\mathbf{B}(\mathfrak{H})$ having
A as its matrix (cf. E 3.2.16), and that $\| T \|^2 \le bc$.

  Hint: For each vector $x = \sum \lambda_m e_m$ in $\mathfrak{H}$, where $\| x \|^2 = \sum |\lambda_m|^2$,
define $Tx = \sum \lambda_m \alpha_{nm} e_n$. By the Cauchy–Schwarz inequality we have

$$\| Tx \|^2 = \sum_n \left| \sum_m \lambda_m \alpha_{nm} \right|^2 \le \sum_n \left(\sum_m |\lambda_m| |\alpha_{nm}| \right)^2$$

$$= \sum_n \left(\sum_m |\lambda_m| |\alpha_{nm}|^{1/2} a_m^{-1/2} a_m^{1/2} |\alpha_{nm}|^{1/2} \right)^2$$

$$\le \sum_n \left(\sum_m |\lambda_m|^2 |\alpha_{nm}| a_m^{-1} \right) \left(\sum_m a_m |\alpha_{nm}| \right)$$

$$\le \sum_n \left(\sum_m |\lambda_m|^2 |\alpha_{nm}| a_m^{-1} c a_n \right)$$

$$\le \sum_m |\lambda_m|^2 a_m^{-1} c \sum_n |\alpha_{nm}| a_n \le \sum_m |\lambda_m|^2 a_m^{-1} cba_m = bc \| x \|^2.$$

E 3.2.18. (*The Hilbert matrix.*) Show that there is an operator T in $\mathbf{B}(\mathfrak{H})$ whose
corresponding matrix relative to an orthonormal basis $\{e_n | n \in \mathbb{N}\}$ is
given by $A = (\alpha_{nm})$, with $\alpha_{nm} = (n + m - 1)^{-1}$. Show that $T = T^*$
and that $\| T \| \le \pi$ (in fact $\| T \| = \pi$).

Hint: Use the Schur test (E 3.2.17) with $a_n = (n - \frac{1}{2})^{-1/2}$, and estimate the sums by a majorizing integral.

E 3.2.19. (*Tensor product.*) Let $\{e_n | n \in \mathbb{N}\}$ and $\{f_m | m \in \mathbb{N}\}$ be orthonormal bases for the Hilbert spaces $\mathfrak{H}$ and $\mathfrak{K}$, respectively. Consider the vector space $\mathfrak{X}$ of formal linear combinations $x = \sum \alpha_{nm} e_n \otimes f_m$, where $e_n \otimes f_m$ is just meant as a symbol. Define an inner product on $\mathfrak{X}$ by

$$(x|y) = \sum \alpha_{nm} \bar{\beta}_{nm}$$

if $y = \sum \beta_{nm} e_n \otimes f_m$, and denote by $\mathfrak{H} \otimes \mathfrak{K}$ the completion of the pre-Hilbert space $\mathfrak{X}$. Show that $\{e_n \otimes f_m | (n, m) \in \mathbb{N}^2\}$ is an orthonormal basis for $\mathfrak{H} \otimes \mathfrak{K}$. Assume that $\mathfrak{H} = L^2(\mathbb{R}^p)$ and $\mathfrak{K} = L^2(\mathbb{R}^q)$ (with respect to Lebesgue measure) and show that

$$\mathfrak{H} \otimes \mathfrak{K} = L^2(\mathbb{R}^{p+q}),$$

when $e_n \otimes f_m$ is identified with the function $(s, t) \to e_n(s) f_m(t)$ on $\mathbb{R}^{p+q}$.

E 3.2.20. Consider operators S and T in $B(\mathfrak{H})$ and $B(\mathfrak{K})$, respectively, and construct $\mathfrak{H} \otimes \mathfrak{K}$ as in E 3.2.19. Show that there is precisely one operator $S \otimes T$ in $B(\mathfrak{H} \otimes \mathfrak{K})$ such that

$$(S \otimes T)(e_n \otimes f_m) = \sum_{k,l} (Se_n | e_k)(Tf_m | f_l) e_k \otimes f_l$$

for all n and m. Show that $\|S \otimes T\| = \|S\| \|T\|$.

E 3.2.21. (*The Schur product.*) Let S and T be operators in $B(\mathfrak{H})$ whose corresponding matrices relative to an orthonormal basis $\{e_n | n \in \mathbb{N}\}$ are $A = (\alpha_{nm})$ and $B = (\beta_{nm})$; cf. E 3.2.16. Show that there is an operator R in $B(\mathfrak{H})$ whose matrix is $C = (\gamma_{nm})$, where $\gamma_{nm} = \alpha_{nm} \beta_{nm}$ for all n and m.

Hint 1: Let P denote the projection of $\mathfrak{H} \otimes \mathfrak{H}$ (cf. E 3.2.19) on the closed subspace spanned by the vectors $\{e_n \otimes e_n | n \in \mathbb{N}\}$, and put $R = P(S \otimes T)P$, where $S \otimes T$ is defined as in E 3.2.20 and $P(\mathfrak{H} \otimes \mathfrak{H})$ is identified canonically with $\mathfrak{H}$.

Hint 2: Use the Schur test (E 3.2.17) on the matrix C, with $a_n = 1$ for all n and $b = c = \|S\| \|T\|$, and estimate the product sums with the Cauchy–Schwarz inequality.

E 3.2.22. (*The Toeplitz–Hausdorff theorem.*) For a (complex) Hilbert space $\mathfrak{H}$ and T in $B(\mathfrak{H})$, define the *numerical range* of T as

$$\Delta(T) = \{(Tx|x) | x \in \mathfrak{H}, \|x\| = 1\}.$$

Show that $\Delta(T)$ is a convex subset of $\mathbb{C}$.

Hint: If $\alpha = (Tx|x)$ and $\beta = (Ty|y)$ for unit vector x, y in $\mathfrak{H}$ and $\alpha \neq \beta$, choose a, b in $\mathbb{C}$ such that $a\alpha + b = 1$ and $a\beta + b = 0$. Replacing T by the operator $aT + bI$ we may assume that $\alpha = 1$ and $\beta = 0$.

Note that the vector $z(t) = \theta tx + (1 - t)y$ is nonzero for $0 \le t \le 1$ and θ in $\mathbb{C}$ with $|\theta| = 1$. For fixed θ the numbers

$$\|z(t)\|^{-2}(Tz(t)|z(t)) = \|z(t)\|^{-2}(t^2 + t(1 - t)\gamma)$$

where $\gamma = \theta(Tx|y) + \bar{\theta}(Ty|x)$, form a curve inside $\Delta(T)$ that joins 0 to 1. Choosing θ so that $\gamma \in \mathbb{R}$ means that the curve covers the whole interval $[0, 1]$.

E 3.2.23. *(The unilateral shift.)* Define S in $\mathbf{B}(\ell^2)$ by

$$(Sx)_1 = 0, \qquad (Sx)_{n+1} = x_n, \qquad x = (x_n) \in \ell^2.$$

Find S^*. Show that S has no eigenvalues, but that every λ in $\mathbb{C}$ with $|\lambda| < 1$ is an eigenvalue for S^* with multiplicity 1. Show that none of the eigenvectors for S^* are orthogonal to each other. Find the numerical range of S (cf. E 3.2.22) and the numerical radius (3.2.24).

   Hint: $\Delta(S) = \overline{\Delta(S^*)}$.

E 3.2.24. *(The Hilbert transform.)* Consider $L^2(\mathbb{T})$ with the orthonormal basis $\{e_n | n \in \mathbb{Z}\}$ described in E 3.1.5. Show that the operator W in $\mathbf{B}(L^2(\mathbb{T}))$, determined by

$$We_0 = e_0, \qquad We_n = -ie_n, \qquad We_{-n} = ie_{-n}, \qquad n \in \mathbb{N},$$

is unitary and maps real functions to real functions in $L^2(\mathbb{T})$. Show that $x + iWx \in H^2$ for every x in $L^2(\mathbb{T})$ (cf. E 3.1.6). Solve the Dirichlet problem on the disk: to find a square integrable, holomorphic function f on the unit disk such that $\operatorname{Re} f$ has a prescribed boundary value x in $L^2(\mathbb{T})$, when restricted to the circle. Show that f is unique up to an additive, imaginary constant.

3.3. Compact Operators

Synopsis. Equivalent characterizations of compact operators. The spectral theorem for normal, compact operators. Atkinson's theorem. Fredholm operators and index. Invariance properties of the index. Exercises.

3.3.1. An operator T on an infinite-dimensional Hilbert space $\mathfrak{H}$ has *finite rank* if $T(\mathfrak{H})$ is a finite-dimensional subspace of $\mathfrak{H}$ (hence closed, cf. 2.1.9). The set $\mathbf{B}_f(\mathfrak{H})$ of operators in $\mathbf{B}(\mathfrak{H})$ with finite rank is clearly a subspace, and it is easily verified that $\mathbf{B}_f(\mathfrak{H})$ is not only a subalgebra, but even an ideal in $\mathbf{B}(\mathfrak{H})$ (cf. 4.1.2).

   If $T \in \mathbf{B}_f(\mathfrak{H})$, we may use 3.2.5 to obtain an orthogonal decomposition $\mathfrak{H} = T(\mathfrak{H}) \oplus \ker T^*$, which shows that $T^*(\mathfrak{H}) = T^*T(\mathfrak{H})$, so that T^* has finite rank. Thus, $\mathbf{B}_f(\mathfrak{H})$ is a *self-adjoint* ideal in $\mathbf{B}(\mathfrak{H})$ [i.e. $(\mathbf{B}_f(\mathfrak{H}))^* = \mathbf{B}_f(\mathfrak{H})$ as a set]. The class $\mathbf{B}_f(\mathfrak{H})$ bears much the same relation to $\mathbf{B}(\mathfrak{H})$ as the class $C_c(X)$ in relation to $C_b(X)$ (when X is a locally compact Hausdorff space; see 1.7.6 and 2.1.14). These classes describe local phenomena on $\mathfrak{H}$ and on X. Passing to a limit in norm may destroy the exact "locality," but enough structure is preserved to make these "quasilocal" operators and functions very attractive. We shall study the closure of $\mathbf{B}_f(\mathfrak{H})$ in this section as a noncommutative analogue of $C_0(X)$ in function theory.

3.3.2. Lemma. *There is a net* $(P_\lambda)_{\lambda \in \Lambda}$ *of projections in* $\mathbf{B}_f(\mathfrak{H})$ *such that* $\|P_\lambda x - x\| \to 0$ *for each x in* $\mathfrak{H}$.

PROOF. Take any orthonormal basis $\{e_j | j \in J\}$ for $\mathfrak{H}$ (cf. 3.1.12), and let Λ be the net of finite subsets of J, ordered under inclusion. For each λ in Λ let P_λ denote the projection of $\mathfrak{H}$ on the subspace $\mathrm{span}\{e_j | j \in \lambda\}$, so that $(P_\lambda)_{\lambda \in \Lambda}$ is indeed a net in $\mathbf{B}_f(\mathfrak{H})$. If $x \in \mathfrak{H}$, we have $x = \sum \alpha_j e_j$, whence $\|P_\lambda x - x\|^2 = \sum |\alpha_j|^2$, the summation being over all $j \notin \lambda$; and this tends to zero by Parseval's identity (3.1.11). $\qquad\square$

3.3.3. Theorem. *Let $\mathfrak{B}$ denote the closed unit ball in a Hilbert space $\mathfrak{H}$. Then the following conditions on an operator T in $\mathbf{B}(\mathfrak{H})$ are equivalent:*

  (i) $T \in (\mathbf{B}_f(\mathfrak{H}))^=$.
 (ii) $T | \mathfrak{B}$ *is a weak–norm continuous function from $\mathfrak{B}$ into $\mathfrak{H}$.*
(iii) $T(\mathfrak{B})$ *is compact in $\mathfrak{H}$.*
 (iv) $(T(\mathfrak{B}))^=$ *is compact in $\mathfrak{H}$.*
  (v) *Each net in $\mathfrak{B}$ has a subnet whose image under T converges in $\mathfrak{H}$.*

PROOF. (i) $\Rightarrow$ (ii). Let $(x_\lambda)_{\lambda \in \Lambda}$ be a weakly convergent net in $\mathfrak{B}$ with limit x. Given $\varepsilon > 0$ there is by assumption an S in $\mathbf{B}_f(\mathfrak{H})$ with $\|S - T\| < \varepsilon/3$, whence

$$\|Tx_\lambda - Tx\| \leq 2\|T - S\| + \|Sx_\lambda - Sx\|$$

$$\leq \tfrac{2}{3}\varepsilon + \|Sx_\lambda - Sx\|.$$

Since every operator in $\mathbf{B}(\mathfrak{H})$ is weak–weak continuous (cf. 3.1.10), we have $Sx_\lambda \to Sx$, weakly. However, if $\{e_1, \ldots, e_n\}$ is an orthonormal basis for the finite-dimensional subspace $S(\mathfrak{H})$ we have

$$\|Sx_\lambda - Sx\|^2 = \sum |(S(x_\lambda - x)|e_j)|^2 \to 0.$$

Eventually, therefore, $\|Tx_\lambda - Tx\| < \varepsilon$. Since ε was arbitrary, T is weak–norm continuous by 1.4.3.

(ii) $\Rightarrow$ (iii) Since $\mathfrak{B}$ is weakly compact (3.1.10), the image $T(\mathfrak{B})$ is norm(!) compact by 1.6.7.

(iii) $\Rightarrow$ (iv) is trivial, since $T(\mathfrak{B})$ is closed by 1.6.5.

(iv) $\Rightarrow$ (v) It follows from 1.6.2(v) that if $T(\mathfrak{B})$ is relatively compact, then every net has a convergent subnet.

(v) $\Rightarrow$ (i) Take $(P_\lambda)_{\lambda \in \Lambda}$ as in 3.3.2. Then $P_\lambda T \in \mathbf{B}_f(\mathfrak{H})$ for every λ, and we claim that $P_\lambda T \to T$. If not, there is an $\varepsilon > 0$ and (passing if necessary to a subnet of

Λ) for every λ a unit vector x_λ with $\|(P_\lambda T - T)x_\lambda\| \geq \varepsilon$. By assumption we may assume that the net $(Tx_\lambda)_{\lambda \in \Lambda}$ is norm(!) convergent in $\mathfrak{H}$ with a limit y. But then by 3.3.2

$$\varepsilon \leq \|(I - P_\lambda)Tx_\lambda\| \leq \|(I - P_\lambda)(Tx_\lambda - y)\| + \|(I - P_\lambda)y\|$$

$$\leq \|Tx_\lambda - y\| + \|(I - P_\lambda)y\| \to 0,$$

a contradiction. Thus, $\|P_\lambda T - T\| \to 0$, as desired. $\qquad\square$

3.3.4. The class of operators satisfying 3.3.3 is called the *compact operators* [after conditions (iii) and (iv)] and is denoted by $\mathbf{B}_0(\mathfrak{H})$ to signify that these operators "vanish at infinity." Unfortunately, this notation is not standard, and the reader will more often find the letters $\mathbf{K}$ or $\mathbf{C}$ [sometimes $\mathbf{K}(\mathfrak{H})$ or $\mathbf{C}(\mathfrak{H})$] used. We see from condition (i) that $\mathbf{B}_0(\mathfrak{H})$ is a norm closed, self-adjoint ideal in $\mathbf{B}(\mathfrak{H})$ (and actually the smallest such; for separable Hilbert spaces even the *only* nontrivial closed ideal). Note that $I \notin \mathbf{B}_0(\mathfrak{H})$ when $\mathfrak{H}$ is infinite-dimensional, but that $\mathbf{B}_0(\mathfrak{H})$ has an approximate unit consisting of projections of finite rank [cf. the proof of the implication (v) $\Rightarrow$ (i)].

3.3.5. Lemma. *A diagonalizable operator T in $\mathbf{B}(\mathfrak{H})$ is compact iff its eigenvalues $\{\lambda_j | j \in J\}$ corresponding to an orthonormal basis $\{e_j | j \in J\}$ belongs to $c_0(J)$.*

PROOF. We have $Tx = \sum \lambda_j(x|e_j)e_j$ for every x in $\mathfrak{H}$, cf. (*) in 3.2.14. If $T \in \mathbf{B}_0(\mathfrak{H})$ and $\varepsilon > 0$ is given, we let

$$J_\varepsilon = \{j \in J \,||\lambda_j| \geq \varepsilon\}.$$

If J_ε is infinite, the net $(e_j)_{j \in J_\varepsilon}$ will converge weakly to zero for any well-ordering of J_ε, because $(e_j|x) \to 0$ by Parseval's identity 3.1.11. Since $\|Te_j\| = |\lambda_j| \geq \varepsilon$ for j in J_ε, this contradict condition (ii) in 3.3.3. Thus, J_ε is finite for each $\varepsilon > 0$, which means that the λ_j's vanish at infinity.

Conversely, if J_ε is finite for each $\varepsilon > 0$, we let

$$T_\varepsilon = \sum_{j \in J_\varepsilon} \lambda_j(\cdot|e_j)e_j.$$

Then T_ε has finite rank and

$$\|(T - T_\varepsilon)x\|^2 = \left\|\sum_{j \notin J_\varepsilon} \lambda_j(x|e_j)e_j\right\|^2$$

$$= \sum_{j \notin J_\varepsilon} |\lambda_j|^2 |(x|e_j)|^2 \leq \varepsilon^2 \|x\|^2.$$

Thus $\|T - T_\varepsilon\| \leq \varepsilon$, whence $T \in \mathbf{B}_0(\mathfrak{H})$ by 3.3.3(i). $\qquad\square$

Lemma 3.3.6. *If x is an eigenvector for a normal operator T in $\mathbf{B}(\mathfrak{H})$, corresponding to the eigenvalue λ, then x is an eigenvector for T^*, corresponding to the eigenvalue $\bar\lambda$. Eigenvectors for T corresponding to different eigenvalues are orthogonal.*

PROOF. The operator $T - \lambda I$ is normal and its adjoint is $T^* - \bar{\lambda}I$. Consequently, we have

$$\|(T^* - \bar{\lambda}I)x\| = \|(T - \lambda I)x\| = 0;$$

cf. 3.2.7. If $Ty = \mu y$ with $\lambda \neq \mu$, we may assume that $\lambda \neq 0$, whence

$$(x|y) = \lambda^{-1}(Tx|y) = \lambda^{-1}(x|T^*y) = \lambda^{-1}(x|\bar{\mu}y) = \lambda^{-1}\mu(x|y),$$

so that $(x|y) = 0$.   □

3.3.7. Lemma. *Every normal, compact operator T on a complex Hilbert space $\mathfrak{H}$ has an eigenvalue λ with $|\lambda| = \|T\|$.*

PROOF. With $\mathfrak{B}$ as the unit ball of $\mathfrak{H}$ we know from 3.3.3 that $T: \mathfrak{B} \to \mathfrak{H}$ is weak–norm continuous. If therefore $x_i \to x$ weakly in $\mathfrak{B}$, then

$$|(Tx_i|x_i) - (Tx|x)| = |(T(x_i - x)|x_i) + (Tx|x_i - x)|$$
$$\leq \|T(x_i - x)\| + |(Tx|x_i - x)| \to 0.$$

This shows that the function $x \to |(Tx|x)|$ is weakly continuous on $\mathfrak{B}$. Since $\mathfrak{B}$ is weakly compact, the function attains its maximum (1.6.7), and by 3.2.27 that maximum is $\|T\|$. Thus,

$$|(Tx|x)| = \|T\|$$

for some x in $\mathfrak{B}$. Now

$$\|T\| = |(Tx|x)| \leq \|Tx\| \|x\| \leq \|T\|,$$

so that, in fact, $|(Tx|x)| = \|Tx\| \|x\|$. But as we saw in 3.1.3, equality holds in the Cauchy–Schwarz inequality only when the vectors are proportional, and, therefore, $Tx = \lambda x$ for some λ. Evidently $|\lambda| = \|T\|$.   □

3.3.8. Theorem. *Every normal, compact operator T on a complex Hilbert space $\mathfrak{H}$ is diagonalizable and its eigenvalues (counted with multiplicity) vanish at infinity. Conversely, each such operator is normal and compact.*

PROOF. We need only show that every normal, compact operator T is diagonalizable, since the rest of the statement in 3.3.8 is contained in 3.3.5. Toward this end, consider the family of orthonormal systems in $\mathfrak{H}$ consisting entirely of eigenvectors for T. Clearly this family is inductively ordered under inclusion, so by Zorn's lemma (1.1.3) it has a maximal element $\{e_j | j \in J\}$. Let $\{\lambda_j | j \in J\}$ denote the corresponding system of eigenvalues, and let P denote the projection on the closed subspace spanned by the e_j's [so that these form a basis for $P(\mathfrak{H})$]. For each x in $\mathfrak{H}$ we then have, by 3.3.6, that

$$TPx = T\sum (x|e_j)e_j = \sum (x|e_j)\lambda_j e_j$$
$$= \sum (x|\bar{\lambda}_j e_j)e_j = \sum (x|T^*e_j)e_j$$
$$= \sum (Tx|e_j)e_j = PTx.$$

It follows that T and P commute, so that the operator $(I - P)T$ is normal and compact. If $P \neq I$, we either have $(I - P)T = 0$, and then every unit vector e_0 in $(I - P)\mathfrak{H}$ is an eigenvector for T, or else $(I - P)T \neq 0$, in which case by 3.3.7 there is a unit vector e_0 in $(I - P)\mathfrak{H}$ with $Te_0 = \lambda e_0$ and $|\lambda| = \|(I - P)T\|$. Both cases contradict the maximality of the system $\{e_j | j \in J\}$, and therefore $P = I$.

$\square$

3.3.9. It will be convenient, especially for the next section, to introduce the notation $x \odot y$ for the rank one operator in $\mathbf{B}(\mathfrak{H})$ determined by the vectors x and y in $\mathfrak{H}$ by the formula

$$(x \odot y)z = (z|y)x.$$

Note that the map $x, y \to x \odot y$ is a sesquilinear map of $\mathfrak{H} \times \mathfrak{H}$ into $\mathbf{B}_f(\mathfrak{H})$. If $\|e\| = 1$, then $e \odot e$ is the one-dimensional projection of $\mathfrak{H}$ on $\mathbb{C}e$. Every normal, compact operator on $\mathfrak{H}$ can now by 3.3.8 be written in the form

$$T = \sum \lambda_j e_j \odot e_j$$

for a suitable orthonormal basis $\{e_j | j \in J\}$. The sum converges in norm, because either the set $J_0 = \{j \in J | \lambda_j \neq 0\}$ is finite [so that $T \in \mathbf{B}_f(\mathfrak{H})$] or else is countably infinite, in which case the sequence $\{\lambda_j | j \in J_0\}$ converges to zero. We say that the compact set

$$\mathrm{sp}(T) = \{\lambda_j | j \in J_0\} \cup \{0\}$$

is the *spectrum* of T.

For every continuous function f on $\mathrm{sp}(T)$ we define

$$f(T) = \sum f(\lambda_j) e_j \odot e_j.$$

Then $f(T)$ is compact iff $f(0) = 0$, and the map $f \to f(T)$ is an isometric *-preserving homomorphism of $C(\mathrm{sp}(T))$ into $\mathbf{B}(\mathfrak{H})$. Moreover, if $f(z) = \sum \alpha_{nm} z^n \bar{z}^m$, a polynomial in the two commuting variables z and $\bar{z}$, then $f(T) = \sum \alpha_{nm} T^n T^{*m}$. This result is the *spectral (mapping) theorem* for normal, compact operators. In the next chapter we shall show a generalized version of the spectral mapping theorem, valid for every normal operator.

3.3.10. Since $\mathbf{B}_0(\mathfrak{H})$ is a closed ideal in $\mathbf{B}(\mathfrak{H})$, the quotient $\mathbf{B}(\mathfrak{H})/\mathbf{B}_0(\mathfrak{H})$ is a Banach algebra under the quotient norm (4.1.2). (Actually the quotient is even a C^*-algebra cf. E 4.3.10.) This quotient is called the *Calkin algebra*, and several properties of operators in $\mathbf{B}(\mathfrak{H})$ are conveniently expressed in terms of the Calkin algebra. If S and T are elements in $\mathbf{B}(\mathfrak{H})$ and $S - T \in \mathbf{B}_0(\mathfrak{H})$, we say that S is a *compact perturbation* of T. It just means that S and T have the same image in the Calkin algebra. Such "local" perturbations occur frequently in applications, and properties of an operator that are invariant under compact perturbations are therefore highly valued. We proceed to investigate the best known of these: the index.

3.3.11. Proposition. *The following conditions on an operator T in $\mathbf{B}(\mathfrak{H})$ are equivalent:*

(i) *There is a unique operator S in* $\mathbf{B}(\mathfrak{H})$ *with* $\ker S = \ker T^*$ *and* $\ker S^* = \ker T$ *such that ST and TS are the projections on* $(\ker T)^{\perp}$ *and* $(\ker T^*)^{\perp}$, *respectively, and both projections have finite co-rank.*

(ii) *For some operator S in* $\mathbf{B}(\mathfrak{H})$ *both operators* $ST - I$ *and* $TS - I$ *are compact.*

(iii) *The image of T is invertible in the Calkin algebra* $\mathbf{B}(\mathfrak{H})/\mathbf{B}_0(\mathfrak{H})$.

(iv) *Both* $\ker T$ *and* $\ker T^*$ *are finite-dimensional subspaces and* $T(\mathfrak{H})$ *is closed.*

PROOF. Clearly (i) $\Rightarrow$ (ii) and (ii) $\Leftrightarrow$ (iii).

(ii) $\Rightarrow$ (iv) Suppose that (x_n) were an orthonormal sequence in $\ker T$. Then with $A = ST - I$ in $\mathbf{B}_0(\mathfrak{H})$ we have $Ax_n = -x_n$ for every n. However, $x_n \to 0$ weakly (by Parseval's identity) so $\|Ax_n\| \to 0$ by 3.3.3(ii), a contradiction. Replacing T by T^* in (ii) we see that also $\ker T^*$ is finite-dimensional. For the last assertion in (iv) choose by 3.3.3(i) an operator B in $\mathbf{B}_f(\mathfrak{H})$ such that $\|A - B\| \le \frac{1}{2}$. Then for every x in $\ker B$ we have

$$\|S\| \|Tx\| \ge \|STx\| = \|(I + A)x\|$$
$$\ge \|x\| - \|Ax\| \ge \tfrac{1}{2}\|x\|.$$

Thus, $T|\ker B$ is bounded away from zero, so that the subspace $\mathfrak{X} = T(\ker B)$ is closed. On the other hand,

$$\mathfrak{Y} = T((\ker B)^{\perp}) = T(B^*(\mathfrak{H}))$$

by 3.2.5, which is finite-dimensional. With Q as the projection of $\mathfrak{H}$ on $\mathfrak{X}^{\perp}$ we see that $Q(\mathfrak{Y})$ is finite-dimensional, hence closed, whence

$$\mathfrak{X} + \mathfrak{Y} = Q^{-1}Q(\mathfrak{Y})$$

is a closed subspace of $\mathfrak{H}$. But $T(\mathfrak{H}) = \mathfrak{X} + \mathfrak{Y}$.

(iv) $\Rightarrow$ (i) The restriction $T|(\ker T)^{\perp}$ is an injective bounded operator from one Hilbert space onto another [viz. $T(\mathfrak{H})$], and therefore has a bounded inverse S by 2.2.5. We extend S to an operator in $\mathbf{B}(\mathfrak{H})$ by letting $S = 0$ on $(T(\mathfrak{H}))^{\perp}$. Thus, TS is the projection of $\mathfrak{H}$ on $T(\mathfrak{H}) = (\ker T^*)^{\perp}$ and ST is the projection on $(\ker T)^{\perp}$; and both of these projections have finite co-rank by assumption. Clearly S is unique. $\square$

3.3.12. The operators satisfying the conditions in *Atkinson's theorem* (3.3.11) are called *Fredholm operators*, and the class is denoted by $\mathbf{F}(\mathfrak{H})$. For each T in $\mathbf{F}(\mathfrak{H})$ we define the *index* of T as

$$\text{index } T = \dim \ker T - \dim \ker T^*.$$

If we choose S for T as in 3.3.11(i) and write $ST = I - P$, $TS = I - Q$ with P and Q projections of finite rank, then evidently

$$\text{index } T = \text{rank } P - \text{rank } Q.$$

From condition (iii) in 3.3.11 it follows that the product of Fredholm operators is again a Fredholm operator. In particular, every product $RT \in$

$F(\mathfrak{H})$ if $T \in F(\mathfrak{H})$ and R is invertible in $B(\mathfrak{H})$; and in this case

$$\text{index } RT = \text{index } TR = \text{index } T,$$

because R is bijective. It is also clear that $T^* \in F(\mathfrak{H})$ if $T \in F(\mathfrak{H})$, with index $T^* = -\text{index } T$. Finally, we observe that the partial inverse S to T in 3.3.11(i) is a Fredholm operator with index $S = -\text{index } T$.

For each n in $\mathbb{Z}$ define

$$F_n(\mathfrak{H}) = \{T \in F(\mathfrak{H}) \mid \text{index } T = n\}.$$

None of these subclasses are empty. For if S denotes the unilateral shift (cf. 3.2.16), then $S^n \in F_{-n}(\mathfrak{H})$, whereas $S^{*n} \in F_n(\mathfrak{H})$ for every $n > 0$.

3.3.13. Lemma. *If $A \in B_f(\mathfrak{H})$, then $I + A \in F_0(\mathfrak{H})$.*

PROOF. Clearly $I + A \in F(\mathfrak{H})$ and we let S denote the partial inverse, so that $S(I + A) = I - P$ and $(I + A)S = I - Q$ as described in 3.3.12. Then $P - Q = AS - SA$. Let R denote the projection on the finite-dimensional subspace $P(\mathfrak{H}) + Q(\mathfrak{H}) + A(\mathfrak{H}) + A^*(\mathfrak{H})$. Then R is a unit for P, Q, and A, so that with $S_1 = RSR$ we have

$$P - Q = AS_1 - S_1 A.$$

This is an equation in the matrix algebra $B(R(\mathfrak{H}))$, so with tr as the trace on $B(R(\mathfrak{H}))$ we have

$$\text{rank } P - \text{rank } Q = \text{tr}(P - Q) = \text{tr}(AS_1 - S_1 A) = 0,$$

whence $I + A \in F_0(\mathfrak{H})$. $\qquad\square$

3.3.14. Lemma. *If $T \in F_0(\mathfrak{H})$ there is a partial isometry V in $B_f(\mathfrak{H})$ such that $T + V$ is invertible. Moreover, if $A \in B_0(\mathfrak{H})$ then $T + A \in F_0(\mathfrak{H})$.*

PROOF. By assumption we can construct a partial isometry V in $B_f(\mathfrak{H})$ such that $V^*V(\mathfrak{H}) = \ker T$ and $VV^*(\mathfrak{H}) = \ker T^*$. Observe that $T + V$ is injective, because $(T + V)x = 0$ implies

$$Tx = -Vx \in T(\mathfrak{H}) \cap \ker T^* = \{0\}$$

by 3.2.5. Hence, $Tx = 0$ and $V^*Vx = 0$, so $x \in \ker T \cap (\ker T)^\perp$, i.e. $x = 0$. Furthermore, $T + V$ is surjective, because

$$(T + V)(\mathfrak{H}) = (T + V)((\ker T)^\perp \oplus \ker T)$$

$$= T(\mathfrak{H}) \oplus \ker T^* = \mathfrak{H}.$$

It follows from 3.2.6 that $T + V$ is invertible in $B(\mathfrak{H})$. Now choose by 3.3.3(i) a B in $B_f(\mathfrak{H})$ such that $\|A - B\| < \|(T + V)^{-1}\|^{-1}$. Then set

$$S = T + V + A - B = (T + V)(I + (T + V)^{-1}(A - B)).$$

With $R = (T + V)^{-1}(A - B)$ we have $\|R\| < 1$, and thus $I + R$ is invertible by 4.1.7 (or by observing directly that both $I + R$ and $I + R^*$ are bounded away from zero, and applying 3.2.6.). It follows that S is invertible in $\mathbf{B}(\mathfrak{H})$, and we have

$$\text{index}(T + A) = \text{index}(S + B - V)$$

$$= \text{index}(S(I + S^{-1}(B - V))) = \text{index}(I + S^{-1}(B - V)) = 0$$

by 3.3.13, since $S^{-1}(B - V) \in \mathbf{B}_f(\mathfrak{H})$. $\square$

3.3.15. Corollary. *If* $A \in \mathbf{B}_0(\mathfrak{H})$ *and* $\lambda \in \mathbb{C} \backslash \{0\}$, *then either* $\lambda I - A$ *is invertible in* $\mathbf{B}(\mathfrak{H})$ *or* λ *is an eigenvalue for* A *with finite multiplicity, in which case* $\bar{\lambda}$ *is an eigenvalue for* A^* *with the same multiplicity.*

PROOF. Set $T = I - \lambda^{-1}A$. Since $T \in \mathbf{F}_0(\mathfrak{H})$ by 3.3.14 we know that $T(\mathfrak{H})$ is closed and that

$$\dim \ker T = \dim \ker T^* < \infty.$$

If these dimensions are 0, then $T(\mathfrak{H}) = (\ker T^*)^{\perp} = \mathfrak{H}$ and $\ker T = \{0\}$, whence T is invertible by 3.2.6. $\square$

3.3.16. Remark. The classical result above is known as the *Fredholm alternative*. It implies that the *spectrum* of a compact operator (i.e. those λ such that $\lambda I - T$ is not invertible, see 4.1.10) consists of $\{0\}$ and the eigenvalues for T. In particular, it is a countable subset of $\mathbb{C}$ with 0 as the only possible accumulation point.

3.3.17. Theorem. *For every Fredholm operator* T *and every compact operator* A *we have*

$$\text{index}(T + A) = \text{index } T.$$

PROOF. We may assume that index $T = n > 0$, since the case $n = 0$ is covered by 3.3.14 and the case $n < 0$ can be derived from the positive case by considering T^*.

Define $\mathfrak{K} = \mathfrak{H} \oplus \ell^2$ and observe that for any operator S in $\mathbf{F}(\ell^2)$ we have

$$\text{index}(T \oplus S) = \text{index } T + \text{index } S$$

[where the index is computed in the algebras $\mathbf{B}(\mathfrak{K})$, $\mathbf{B}(\mathfrak{H})$ and $\mathbf{B}(\ell^2)$, respectively]. Taking S to be the unilateral shift on ℓ^2 (cf. 3.2.16) we see that $T \oplus S^n \in \mathbf{F}_0(\mathfrak{K})$ and thus

$$(T + A) \oplus S^n = T \oplus S^n + A \oplus 0 \in \mathbf{F}_0(\mathfrak{K})$$

by 3.3.14, so that $T + A \in \mathbf{F}_n(\mathfrak{H})$ as desired. $\square$

3.3.18. Proposition. *Each Fredholm class* $F_n(\mathfrak{H})$, $n \in \mathbb{Z}$, *is open in* $\mathbf{B}(\mathfrak{H})$.

PROOF. It suffices to prove this for $n = 0$, since, as in the proof of 3.3.17, we have

$$\mathbf{F}_n(\mathfrak{H}) \oplus S^n = \mathbf{F}_0(\mathfrak{H} \oplus \ell^2) \cap (\mathbf{B}(\mathfrak{H}) \oplus S^n),$$

with S as the unilateral shift on ℓ^2 and $n > 0$. Negative n is handled by applying the adjoint operation, which is a homeomorphism on $\mathbf{B}(\mathfrak{H})$.

Assume, therefore, that $T_n \to T$ in $\mathbf{B}(\mathfrak{H})$ and that $T \in \mathbf{F}_0(\mathfrak{H})$. Then, by 3.3.14, $T + V$ is invertible for some V in $\mathbf{B}_f(\mathfrak{H})$, so that $T + V$ and $(T + V)^*$ are bounded away from zero by some $\varepsilon > 0$ by 3.2.6. As soon as $\|T - T_n\| < \varepsilon$ we see that also $T_n + V$ and $(T_n + V)^*$ are bounded away from zero and, hence, invertible by 3.2.6. Consequently,

$$\text{index } T_n = \text{index } T_n + V = 0. \qquad \square$$

3.3.19. Proposition. *If $T_1 \in \mathbf{F}_n(\mathfrak{H})$ and $T_2 \in \mathbf{F}_m(\mathfrak{H})$, then $T_1 T_2 \in \mathbf{F}_{n+m}(\mathfrak{H})$.*

PROOF. If $m = 0$, then $T_2 + V$ is invertible for some partial isometry V in $\mathbf{B}_f(\mathfrak{H})$, as we saw in the proof of 3.3.14. Therefore,

$$\text{index } T_1 = \text{index } T_1(T_2 + V)$$
$$= \text{index } T_1 T_2 + T_1 V = \text{index } T_1 T_2$$

by 3.3.17, since $T_1 V \in \mathbf{B}_f(\mathfrak{H})$.

In the general case we may assume that $m > 0$, and working in $\mathfrak{K} = \mathfrak{H} \oplus \ell^2$ as in the proofs of 3.3.17 and 3.3.18 we see that $T_2 \oplus S^m \in \mathbf{F}_0(\mathfrak{K})$ if S is the unilateral shift on ℓ^2. Thus by our first result

$$n = \text{index}((T_1 \oplus I)(T_2 \oplus S^m)) = \text{index } T_1 T_2 \oplus S^m$$
$$= \text{index } T_1 T_2 + \text{index } S^m = \text{index } T_1 T_2 - m. \qquad \square$$

3.3.20. A beautiful topological fact illuminates many of the properties of the index. If $\mathfrak{G}$ denotes the group of invertible elements in the Calkin algebra $\mathbf{B}(\mathfrak{H})/\mathbf{B}_0(\mathfrak{H})$ and $\mathfrak{G}_0$ is the connected component of the identity [$=$ the group generated by elements of the form $\exp A$, for some A in $\mathbf{B}(\mathfrak{H})/\mathbf{B}_0(\mathfrak{H})$], then $\mathfrak{G}_0$ is an open and closed subgroup of $\mathfrak{G}$ and $\mathfrak{G}/\mathfrak{G}_0$ is a discrete group that labels the connected components in $\mathfrak{G}$. In the case at hand, $\mathfrak{G}/\mathfrak{G}_0$ is isomorphic to $\mathbb{Z}$, and the index of a Fredholm operator T is the image of T under the composed quotient maps from $\mathbf{F}(\mathfrak{H})$ to $\mathfrak{G}$ and from $\mathfrak{G}$ to $\mathfrak{G}/\mathfrak{G}_0$ ($= \mathbb{Z}$).

EXERCISES

E 3.3.1. Show that $\mathbf{B}_f(\mathfrak{H})$ is a minimal ideal in $\mathbf{B}(\mathfrak{H})$.
Hint: Show first that $\mathbf{B}_f(\mathfrak{H})$ intersects every nonzero ideal in $\mathbf{B}(\mathfrak{H})$. Then show that $\mathbf{B}_f(\mathfrak{H})$ itself has no nontrivial ideals.

E 3.3.2. Show that every operator $T: \mathfrak{H} \to \mathfrak{H}$ that is weak–norm continuous (cf. 3.1.10) belongs to $\mathbf{B}_f(\mathfrak{H})$.
Hint: The function $x \to \|Tx\|$ is weakly continuous, so there is a finite set $x_1, \ldots, x_n$ in $\mathfrak{H}$ such that $|(x|x_k)| < 1$ for $1 \le k \le n$ implies $\|Tx\| < 1$ for every x in $\mathfrak{H}$.

E 3.3.3. Show that if $T \in \mathbf{B}_0(\mathfrak{H})$, then $T(\mathfrak{H})$ contains no closed, infinite-dimensional subspaces.

Hint: If $P(\mathfrak{H}) \subset T(\mathfrak{H})$ for some projection P in $\mathbf{B}(\mathfrak{H})$, then $PT: \mathfrak{H} \to P(\mathfrak{H})$ is surjective, hence open (2.2.4). Thus with $\mathfrak{B}$ and $\mathfrak{B}_1$ the unit balls in $\mathfrak{H}$ and $P(\mathfrak{H})$ we have $\varepsilon \mathfrak{B}_1 \subset PT(\mathfrak{B})$ for some $\varepsilon > 0$. Now apply E 2.1.3 and 3.3.3 (or argue directly on an orthonormal basis in $P(\mathfrak{H})$).

E 3.3.4. If $T \in \mathbf{B}(\mathfrak{H})$, show that T is compact iff $|T|$ is compact.

E 3.3.5. (*Weighted shifts.*) Given a bounded sequence (λ_n) in $\mathbb{C}$ define an operator S in $\mathbf{B}(\ell^2)$ by $(Sx)_1 = 0$ and

$$(Sx)_n = \lambda_n x_{n-1}, \quad n > 1, \quad \text{for } x = (x_n) \text{ in } \ell^2.$$

Find the polar decomposition of S, and characterize those sequences (λ_n) in ℓ^∞ for which S is compact.

E 3.3.6. For every T in $\mathbf{B}(\mathfrak{H})$ define

$$\text{index } T = \dim \ker T - \dim \ker T^*,$$

with the convention that $\infty - \infty = 0$. Thus, the index takes values in $\mathbb{Z} \cup \{\pm\infty\}$. Show that index T can be nonzero for T in $\mathbf{B}_0(\mathfrak{H})$.

Hint: Use E 3.3.5.

E 3.3.7. Assume that $\mathfrak{H}$ is separable and prove that an operator T in $\mathbf{B}(\mathfrak{H})$ has the form $U|T|$ for some unitary operator U (and with $|T| = (T^*T)^{1/2}$) iff index $T = 0$ (cf. E 3.3.6.).

Hint: Any two closed subspaces of $\mathfrak{H}$ with the same (finite or infinite) dimension can be mapped isometrically one onto the other by 3.1.13. Add such a partial isometry to the canonical partial isometry in the polar decomposition of T (3.2.16) to obtain the unitary U.

E 3.3.8. (*Right Fredholm operators.*) Prove that the following conditions on T in $\mathbf{B}(\mathfrak{H})$ are equivalent:

(i) There is a unique S in $\mathbf{B}(\mathfrak{H})$ such that TS is the projection on $(\ker T^*)^\perp$ and has finite co-rank.
(ii) For some operator S in $\mathbf{B}(\mathfrak{H})$ the operator $TS - I$ is compact.
(iii) The image of T in the Calkin algebra $\mathbf{B}(\mathfrak{H})/\mathbf{B}_0(\mathfrak{H})$ is right invertible.
(iv) The subspace $T(\mathfrak{H})$ is closed and has finite co-dimension.

Hint: Use the proof of 3.3.11.

E 3.3.9. Show that T satisfies the conditions in E 3.3.8 iff $(TT^*)^{1/2}$ is a Fredholm operator.

Hint: Use the polar decomposition of T^* to write $T = (TT^*)^{1/2}V$.

E 3.3.10. Show that the class $\mathbf{F}^{(r)}(\mathfrak{H})$ of right Fredholm operators defined in E 3.3.8 is open in $\mathbf{B}(\mathfrak{H})$ and satisfies $\mathbf{F}^{(r)}(\mathfrak{H})\mathbf{F}^{(r)}(\mathfrak{H}) \subset \mathbf{F}^{(r)}(\mathfrak{H})$.

E 3.3.11. Apply the index from E 3.3.6 to the class $\mathbf{F}^{(r)}(\mathfrak{H})$ of right Fredholm operators defined in E 3.3.8. Show that $\mathbf{F}_n^{(r)}(\mathfrak{H}) = \mathbf{F}_n(\mathfrak{H})$ if $n < \infty$. Show that the class $\mathbf{F}_\infty^{(r)}(\mathfrak{H})$ is open in $\mathbf{B}(\mathfrak{H})$ and invariant under compact perturbations and that one has $\mathbf{F}_n(\mathfrak{H})\mathbf{F}_\infty^{(r)}(\mathfrak{H}) \subset \mathbf{F}_\infty^{(r)}(\mathfrak{H})$ and $\mathbf{F}_\infty^{(r)}(\mathfrak{H})\mathbf{F}_n(\mathfrak{H}) \subset \mathbf{F}_\infty^{(r)}(\mathfrak{H})$.

Hint: Use E 3.3.10 and 3.3.19.

E 3.3.12. Assume that the group $\mathbf{GL}(\mathfrak{H})$ of invertible elements in $\mathbf{B}(\mathfrak{H})$ has been proved to be arcwise connected (E 4.5.7). Show that each Fredholm class $\mathbf{F}_n(\mathfrak{H})$, $n \in \mathbb{Z}$, is arcwise connected.

Hint: For T in $\mathbf{F}_0(\mathfrak{H})$ choose A in $\mathbf{B}_0(\mathfrak{H})$ such that $T + A \in \mathbf{GL}(\mathfrak{H})$. Connect T to $T + A$ linearly and then take a path from $T + A$ to I. If $T \in \mathbf{F}_n(\mathfrak{H})$ with $n < 0$, let S be the unilateral shift on $\mathfrak{H}$ $(= \ell^2)$ and connect TS^{*n} to I. Then multiply the path with S^n.

E 3.3.13. Show that the image $\mathfrak{G}_n$ of $\mathbf{F}_n(\mathfrak{H})$ in the Calkin algebra $[\mathbf{B}(\mathfrak{H})/\mathbf{B}_0(\mathfrak{H})]$ is an arcwise connected set that is both open and closed in the group $\mathfrak{G}$ of invertible elements in the Calkin algebra. Show that the connected components of $\mathfrak{G}$ are precisely the $\mathfrak{G}_n$'s, $n \in \mathbb{Z}$; cf. 3.3.20.

Hint: The quotient map $\pi\colon \mathbf{B}(\mathfrak{H}) \to \mathbf{B}(\mathfrak{H})/\mathbf{B}_0(\mathfrak{H})$ is open.

E 3.3.14. (*Toeplitz operators.*) Let $\{e_n \mid n \in \mathbb{Z}\}$ be the orthonormal basis for $L^2(\mathbb{T})$ described in E 3.1.5, and consider the Hardy space H^2 defined in E 3.1.6 as a subspace of $L^2(\mathbb{T})$. If P denotes the projection of $L^2(\mathbb{T})$ onto H^2, we define for each f in $L^\infty(\mathbb{T})$ the Toeplitz operator T_f in $\mathbf{B}(H^2)$ by

$$T_f x = P(fx), \quad x \in H^2.$$

Show that the map $f \to T_f$ is linear and normdecreasing from $L^\infty(\mathbb{T})$ into $\mathbf{B}(H^2)$, and that $T_f^* = T_{\bar{f}}$. Show further that

$$T_f e_n = \sum_{m=0}^\infty \hat{f}(m - n)e_m, \quad n \geq 0,$$

and conclude that the matrix for T_f, cf. E 3.2.16, is constant on diagonals. Insert $f = e_k$, $k \in \mathbb{Z}$, and check that the corresponding Toeplitz operators are kth powers of the unilateral shift (3.2.16) or its adjoint.

E 3.3.15. Take f in $L^\infty(\mathbb{T})$, and let T_f be the corresponding Toeplitz operator; cf. E 3.3.14. Show that T_f is a compact operator on H^2 only if $f = 0$.

Hint: If $T_f \in \mathbf{B}_0(H^2)$, then, since $e_n \to 0$ weakly (cf. E 3.1.10). $\|T_f e_n\| \to 0$. In particular, $(T_f e_n | e_{n+k}) \to 0$ for each k in $\mathbb{Z}$. Now apply E 3.3.14 to show that $\hat{f} = 0$, whence $f = 0$.

E 3.3.16. Take f and g in $L^\infty(\mathbb{T})$, and let T_f and T_g denote the corresponding Toeplitz operators; cf. E 3.3.14. Show that $T_f T_g - T_g T_f$ and $T_f T_g - T_{fg}$ are both compact operators on H^2 if either f or g is continuous.

Hint: Show that the two operators have finite rank if $f = e_k$ for some k in $\mathbb{Z}$. If $f \in C(\mathbb{T})$, use the fact that f can be uniformly approximated by trigonometric polynomials, and apply 3.3.3. (P.S. Don't miss E 4.3.11 later on.)

E 3.3.17. If $f \in C(\mathbb{T})$, show that the Toeplitz operator T_f, cf. E 3.3.14, is a Fredholm operator if f is invertible in $C(\mathbb{T})$. Show in this case that index $T_f =$ index T_u, where $u = f|f|^{-1}$.

Hint: Use first E 3.3.16. Then use that a self-adjoint Fredholm operator has zero index, so that 3.3.19 applies.

E 3.3.18. Show that functions f and g in $C(\mathbb{T}, \mathbb{T})$ that are homotopic (E 1.4.19) inside $C(\mathbb{T}, \mathbb{T})$ give Toeplitz operators of Fredholm type with index $T_f =$ index T_g.

Hint: If $f_t \colon \mathbb{T} \to \mathbb{T}$ is a continuous path in $C(\mathbb{T}, \mathbb{T})$ with $f_0 = f$ and $f_1 = g$, then index $T_{f_s} =$ index T_{f_t} when $|s - t|$ is small enough by 3.3.18.

E 3.3.19. Take f invertible in $C(\mathbb{T})$ and consider the Toeplitz operator T_f; cf. E 3.3.14. Show that the winding number of f around 0 equals $-$ index T_f. Compare with E 4.1.19.

Hint: Use E 3.3.17 and E 3.3.18 plus the fact (to be proved or taken at face value) that the homotopy classes in $C(\mathbb{T}, \mathbb{T})$ are labeled by the winding number. Check the formula with $f = e_k$, where $k \in \mathbb{Z}$; cf. E 3.3.14.

3.4. The Trace

Synopsis. Definition and invariance properties of the trace. The trace class operators and the Hilbert–Schmidt operators. The dualities among $\mathbf{B}_0(\mathfrak{H})$, $\mathbf{B}^1(\mathfrak{H})$, and $\mathbf{B}(\mathfrak{H})$. Hilbert–Schmidt operators as integral operators. The Fredholm equation. The Sturm–Liouville problem. Exercises.

3.4.1. In search for analogies between the theory of functions and the theory of operators on a complex(!) Hilbert space $\mathfrak{H}$, we have already (in 3.3.1 and 3.3.4) mentioned that $\mathbf{B}_f(\mathfrak{H})$ corresponds to the continuous functions with compact supports and $\mathbf{B}_0(\mathfrak{H})$ corresponds to the continuous functions vanishing at infinity. The class $\mathbf{B}(\mathfrak{H})$ plays a double role: sometimes it mimics the set of all bounded continuous functions and sometimes it behaves like an L^∞-space. The latter behavior assumes the existence of an analogue on $\mathfrak{H}$ to Lebesgue measure, an analogue we will now exhibit.

3.4.2. Choose an orthonormal basis $\{e_j | j \in J\}$ for the Hilbert space $\mathfrak{H}$ (cf. 3.1.12), and for every positive operator T in $\mathbf{B}(\mathfrak{H})$ define the *trace* of T by

$$\operatorname{tr}(T) = \sum (Te_j | e_j),$$

with values in $[0, \infty]$.

3.4.3. Proposition. *For every T in $\mathbf{B}(\mathfrak{H})$ we have*

$$\operatorname{tr}(T^*T) = \operatorname{tr}(TT^*).$$

PROOF. For each i and j we have

$$(Te_i|e_j)(e_j|Te_i) = (T^*e_j|e_i)(e_i|T^*e_j) \geq 0.$$

Summing the first expression over j we get

$$\sum_j ((Te_i|e_j)e_j|Te_i) = (Te_i|Te_i) = (T^*Te_i|e_i).$$

Summing the second expression over i we similarly have

$$\sum_i ((T^*e_j|e_i)e_i|T^*e_j) = (T^*e_j|T^*e_j) = (TT^*e_j|e_j).$$

Since the elements in the series are positive, the sum over both i and j does not depend on the order of the summation, whence

$$\operatorname{tr}(T^*T) = \sum_i (T^*Te_i|e_i) = \sum_j (TT^*e_j|e_j) = \operatorname{tr}(TT^*). \qquad \square$$

3.4.4. Corollary. *If U is unitary and $T \geq 0$, then*

$$\operatorname{tr}(UTU^*) = \operatorname{tr}(T).$$

In particular, the definition of tr *is independent of the choice of basis, and, therefore, $\|T\| \leq \operatorname{tr} T$.*

PROOF. Since $T = (T^{1/2})^2$ by 3.2.11, we may replace T by $UT^{1/2}$ in 3.4.3. The last assertions follow from 3.1.14 and 3.2.25 (or E 3.2.1). $\qquad \square$

3.4.5. Lemma. *If $T \in \mathbf{B}(\mathfrak{H})$ such that $\operatorname{tr}(|T|^p) < \infty$ for some $p > 0$, then T is compact.*

PROOF. Given an orthonormal basis $\{e_j | j \in J\}$ and $\varepsilon > 0$ there is a finite subset λ of J such that $\sum_{j \notin \lambda}(|T|^p e_j|e_j) < \varepsilon$. If P_λ denotes the projection of $\mathfrak{H}$ onto the span of $\{e_j | j \in \lambda\}$, then by $(**)$ in 3.2.3 and 3.4.4

$$\| |T|^{p/2}(I - P_\lambda)\|^2 = \|(I - P_\lambda)|T|^p(I - P_\lambda)\|$$

$$\leq \operatorname{tr}((I - P_\lambda)|T|^p(I - P_\lambda)) < \varepsilon.$$

Since ε is arbitrary, we conclude from 3.3.3(i) that $|T|^{p/2} \in \mathbf{B}_0(\mathfrak{H})$. Thus, for a suitable orthonormal basis (which we still denote by $\{e_j | j \in J\}$) we have

$$|T|^{p/2} = \sum \lambda_j e_j \odot e_j$$

by 3.3.8 (cf. 3.3.9) and the λ_j's vanish at infinity. For integer values of p it is clear that

$$|T| = \sum \lambda_j^{2/p} e_j \odot e_j. \qquad (*)$$

To establish the validity of the formula (∗) in general one will have to define the symbol $|T|^p$ for all real $p > 0$; and we must postpone this task until we have the spectral theorem (4.4.1) at hand. Assuming (∗) it is clear that $|T| \in \mathbf{B}_0(\mathfrak{H})$, and from the polar decomposition $T = U|T|$, cf. 3.2.17, it follows that T belongs to the ideal $\mathbf{B}_0(\mathfrak{H})$.   □

3.4.6. We define the sets of *trace class* operators and *Hilbert–Schmidt* operators as

$$\mathbf{B}^1(\mathfrak{H}) = \mathrm{span}\{T \in \mathbf{B}_0(\mathfrak{H}) | T \geq 0, \mathrm{tr}(T) < \infty\},$$

$$\mathbf{B}^2(\mathfrak{H}) = \{T \in \mathbf{B}_0(\mathfrak{H}) | \mathrm{tr}(T^*T) < \infty\}.$$

Since, evidently, $\mathrm{tr}(T_1 + T_2) = \mathrm{tr}(T_1) + \mathrm{tr}(T_2)$ and $\mathrm{tr}(\alpha T_1) = \alpha \mathrm{Tr}(T_1)$ for all positive operators T_1 and T_2 and each $\alpha \geq 0$; and since $T = \sum_{k=0}^{3} i^k T_k$, with $T_k \geq 0$, for every T in $\mathbf{B}^1(\mathfrak{H})$, it follows that the definition $\mathrm{tr}(T) = \sum i^k \mathrm{tr}(T_k)$ extends tr to a linear functional on $\mathbf{B}^1(\mathfrak{H})$. From now on we may therefore apply the function tr to any operator in the set $\mathbf{B}(\mathfrak{H})_+ + \mathbf{B}^1(\mathfrak{H})$ (with the convention that $\alpha + \infty = \infty$ for every α in $\mathbb{C}$).

3.4.7. Just as for vectors in $\mathfrak{H}$, there is a *parallellogram law* for operators in $\mathbf{B}(\mathfrak{H})$, viz.,

$$(S + T)^*(S + T) + (S - T)^*(S - T) = 2(S^*S + T^*T), \qquad (*)$$

easily verified by computation. From this one derives the useful estimate

$$(S + T)^*(S + T) \leq 2(S^*S + T^*T). \qquad (**)$$

By direct computation we also verify the following *polarization identity* for operators on a complex Hilbert space:

$$4T^*S = \sum_{k=0}^{3} i^k(S + i^kT)^*(S + i^kT). \qquad (***)$$

3.4.8. Proposition. *The classes* $\mathbf{B}^1(\mathfrak{H})$ *and* $\mathbf{B}^2(\mathfrak{H})$ *are self-adjoint ideals in* $\mathbf{B}(\mathfrak{H})$ *and*

$$\mathbf{B}_f(\mathfrak{H}) \subset \mathbf{B}^1(\mathfrak{H}) \subset \mathbf{B}^2(\mathfrak{H}) \subset \mathbf{B}_0(\mathfrak{H}).$$

PROOF. If $T \geq 0$ with $\mathrm{tr}(T) < \infty$, and $S \in \mathbf{B}(\mathfrak{H})$, then by (∗∗∗) in 3.4.7

$$4TS = 4T^{1/2}T^{1/2}S = \sum i^k(S + i^kI)^*T(S + i^kI).$$

By 3.4.3 and 3.2.11 we further have

$$\mathrm{tr}(V^*TV) = \mathrm{tr}(V^*T^{1/2}T^{1/2}V)$$

$$= \mathrm{tr}(T^{1/2}VV^*T^{1/2}) \leq \|VV^*\| \mathrm{tr}(T);$$

and applied with $V = S + i^kI$ it shows that $TS \in \mathbf{B}^1(\mathfrak{H})$. Thus, $\mathbf{B}^1(\mathfrak{H})$ is a self-adjoint right ideal and therefore a twosided ideal (4.1.2).

We claim that

$$\mathbf{B}^1(\mathfrak{H}) = \{T \in \mathbf{B}(\mathfrak{H}) | \mathrm{tr}(|T|) < \infty\}. \qquad (*)$$

If $|T| \in \mathbf{B}^1(\mathfrak{H})$, then from the polar decomposition $T = U|T|$ (3.2.17) we see from the first part of the proof that $T \in \mathbf{B}^1(\mathfrak{H})$. Conversely, $|T| = U^*T$, so if $T \in \mathbf{B}^1(\mathfrak{H})$, then $|T| \in \mathbf{B}^1(\mathfrak{H})$.

It follows from (**) in 3.4.7 that $\mathbf{B}^2(\mathfrak{H})$ is a linear subspace of $\mathbf{B}_0(\mathfrak{H})$, and 3.4.3 shows that this subspace is self-adjoint. Since $\mathbf{B}^1(\mathfrak{H})$ is an ideal in $\mathbf{B}(\mathfrak{H})$, it follows from the definition of $\mathbf{B}^2(\mathfrak{H})$ that this set is also an ideal.

If $T \in \mathbf{B}_f(\mathfrak{H})$, then $|T|$ is a diagonalizable operator of finite rank, whence $|T|$ (and T) belongs to $\mathbf{B}^1(\mathfrak{H})$. If $T \in \mathbf{B}^1(\mathfrak{H})$, then by 3.2.11

$$T^*T = |T|^2 = |T|^{1/2}|T||T|^{1/2} \le \|T\| \, |T|,$$

which shows that $\operatorname{tr}(T^*T) < \infty$, i.e. $T \in \mathbf{B}^2(\mathfrak{H})$. The last assertion (used freely throughout the proof) is contained in 3.4.5. □

3.4.9. Theorem. *The ideal* $\mathbf{B}^2(\mathfrak{H})$ *of Hilbert–Schmidt operators form a Hilbert space under the inner product*

$$(S|T)_{\mathrm{tr}} = \operatorname{tr}(T^*S), \quad S, T \in \mathbf{B}^2(\mathfrak{H}).$$

Proof. That $T^*S \in \mathbf{B}^1(\mathfrak{H})$ follows from (***) in 3.4.7. Thus the sesquilinear form $(\cdot|\cdot)_{\mathrm{tr}}$ is well-defined, self-adjoint, and positive. Moreover, it gives an inner product on $\mathbf{B}^2(\mathfrak{H})$ because the associated 2-norm satisfies

$$\|T\|_2^2 = \operatorname{tr}(T^*T) \ge \|T^*T\| = \|T\|^2,$$

by 3.4.4. This inequality also implies that every Cauchy sequence (T_n) in $\mathbf{B}^2(\mathfrak{H})$ for the 2-norm will converge in norm to an element T in $\mathbf{B}_0(\mathfrak{H})$. For every projection P on a finite-dimensional subspace of $\mathfrak{H}$ we estimate

$$\|P(T - T_n)\|_2^2 = \operatorname{tr}((T - T_n)^*P(T - T_n)) = \operatorname{tr}(P(T - T_n)(T - T_n)^*P)$$
$$= \lim_m \operatorname{tr}(P(T_m - T_n)(T_m - T_n)^*P)$$
$$= \lim_m \operatorname{tr}((T_m - T_n)^*P(T_m - T_n))$$
$$\le \lim\sup_m \operatorname{tr}((T_m - T_n)^*(T_m - T_n)) = \lim\sup_m \|T_m - T_n\|_2^2,$$

and, since P is arbitrary, we conclude that

$$\|T - T_n\|_2 \le \lim\sup_m \|T_m - T_n\|_2;$$

which implies that $T \in \mathbf{B}^2(\mathfrak{H})$ and that $T_n \to T$ in 2-norm. □

3.4.10. Lemma. *If* $T \in \mathbf{B}^1(\mathfrak{H})$ *and* $S \in \mathbf{B}(\mathfrak{H})$, *then*

$$|\operatorname{tr}(ST)| \le \|S\| \operatorname{tr}(|T|).$$

Proof. Let $T = U|T|$ be the polar decomposition of T (3.2.17). Then $(SU|T|^{1/2})^* \in \mathbf{B}^2(\mathfrak{H})$ [because $|T|^{1/2} \in \mathbf{B}^2(\mathfrak{H})$], so by the Cauchy–Schwarz inequality (for the trace) we have

$$|\mathrm{tr}(ST)|^2 = |\mathrm{tr}(SU|T|^{1/2}|T|^{1/2})|^2 = |(|T|^{1/2}|(SU|T|^{1/2})^*)_{\mathrm{tr}}|^2$$

$$\leq \||T|^{1/2}\|_2^2 \|(SU|T|^{1/2})^*\|_2^2 = \mathrm{tr}(|T|)\,\mathrm{tr}(|T|^{1/2}U^*S^*SU|T|^{1/2})$$

$$\leq \mathrm{tr}(|T|)\,\mathrm{tr}(\|U^*S^*SU\|\,|T|) \leq \|S\|^2(\mathrm{tr}(|T|))^2;$$

using 3.4.3 and 3.2.9 on the way.   □

3.4.11. Lemma. *If S and T belong to $\mathbf{B}^2(\mathfrak{H})$, then*
$$\mathrm{tr}(ST) = \mathrm{tr}(TS).$$
The same formula holds when $S \in \mathbf{B}(\mathfrak{H})$ and $T \in \mathbf{B}^1(\mathfrak{H})$.

PROOF. The polarization identity [$(***)$ in 3.4.7] in conjunction with 3.4.3 gives

$$4\,\mathrm{tr}(T^*S) = \sum i^k \mathrm{tr}((S + i^k T)^*(S + i^k T))$$
$$= \sum i^k \mathrm{tr}((S^* + i^{-k}T^*)^*(S^* + i^{-k}T^*))$$
$$= \sum i^k \mathrm{tr}((T^* + i^k S^*)^*(T^* + i^k S^*)) = 4\,\mathrm{tr}(ST^*),$$

which proves the first assertion. For the second, we may assume that $T \geq 0$ (the equation is linear in T), and then from the first result we have

$$\mathrm{tr}(ST) = \mathrm{tr}((ST^{1/2})T^{1/2}) = \mathrm{tr}(T^{1/2}(ST^{1/2}))$$
$$= \mathrm{tr}((T^{1/2}S)T^{1/2}) = \mathrm{tr}(T^{1/2}(T^{1/2}S)) = \mathrm{tr}(TS). \qquad □$$

3.4.12. Theorem. *The ideal $\mathbf{B}^1(\mathfrak{H})$ of trace class operators form a Banach algebra under the norm*
$$\|T\|_1 = \mathrm{tr}(|T|), \quad T \in \mathbf{B}^1(\mathfrak{H}).$$

PROOF. Clearly $\|\cdot\|_1$ is a homogeneous function on $\mathbf{B}^1(\mathfrak{H})$, which is faithful because $\|\cdot\|_1 \geq \|\cdot\|$; cf. 3.4.4. To prove subadditivity take S and T in $\mathbf{B}^1(\mathfrak{H})$ with polar decomposition $S + T = W|S + T|$. Then by 3.4.10

$$\|S + T\|_1 = \mathrm{tr}(W^*(S + T)) \leq |\mathrm{tr}(W^*S)| + |\mathrm{tr}(W^*T)|$$
$$\leq \|W^*\|(\mathrm{tr}(|S|) + \mathrm{tr}(|T|)) \leq \|S\|_1 + \|T\|_1.$$

The corresponding inequality for the product is obtained from the polar decomposition $ST = V|ST|$, which gives

$$\|ST\|_1 = \mathrm{tr}(V^*ST) \leq \|V^*S\|\,\mathrm{tr}(|T|)$$
$$\leq \|\|S\|\|\,\mathrm{tr}(|T|) \leq \mathrm{tr}(|S|)\,\mathrm{tr}(|T|) = \|S\|_1\|T\|_1.$$

If (T_n) is a Cauchy sequence in $\mathbf{B}^1(\mathfrak{H})$ for the 1-norm it must converge in norm to an element T in $\mathbf{B}_0(\mathfrak{H})$. For each projection P of finite rank and every partial isometry U we have

$$\mathrm{tr}(PU^*(T - T_n)) = \lim_m \mathrm{tr}(PU^*(T_m - T_n)) \leq \limsup \|T_m - T_n\|_1$$

by 3.4.10, since $\|PU^*\| \leq 1$. Applied to the polar decomposition $T - T_n = U|T - T_n|$ it shows that $\text{tr}(P|T - T_n|) \leq \lim \sup \|T_m - T_n\|$. Since P is arbitrary, we conclude that

$$\|T - T_n\|_1 \leq \lim_m \sup \|T_m - T_n\|_1,$$

which shows that $T \in \mathbf{B}^1(\mathfrak{H})$ and that $T_n \to T$ in 1-norm. $\qquad\qquad\square$

3.4.13. Theorem. *The bilinear form*

$$\langle S, T \rangle = \text{tr}(ST)$$

implements the dualities between the pair of Banach spaces $\mathbf{B}_0(\mathfrak{H})$ *and* $\mathbf{B}^1(\mathfrak{H})$ *and the pair* $\mathbf{B}^1(\mathfrak{H})$ *and* $\mathbf{B}(\mathfrak{H})$. *Thus, (with $*$ as in 2.3.1)*

$$(\mathbf{B}_0(\mathfrak{H}))^* = \mathbf{B}^1(\mathfrak{H}) \quad and \quad (\mathbf{B}^1(\mathfrak{H}))^* = \mathbf{B}(\mathfrak{H}).$$

PROOF. Clearly every T in $\mathbf{B}^1(\mathfrak{H})$ gives rise to a bounded functional $\varphi_T = \langle \cdot, T \rangle$ on $\mathbf{B}_0(\mathfrak{H})$, and $\|\varphi_T\| \leq \|T\|_1$ by 3.4.10. Conversely, if $\varphi \in (\mathbf{B}_0(\mathfrak{H}))^*$, we take S in $\mathbf{B}^2(\mathfrak{H})$ and estimate

$$|\varphi(S)| \leq \|\varphi\| \|S\| \leq \|\varphi\| \|S\|_2.$$

Since $\mathbf{B}^2(\mathfrak{H})$ is a Hilbert space (3.4.9), there is by 3.1.9 a unique element T^* in $\mathbf{B}^2(\mathfrak{H})$ such that $\varphi(S) = \text{tr}(TS) = \text{tr}(ST)$ for all S in $\mathbf{B}^2(\mathfrak{H})$. However, for each projection P on $\mathfrak{H}$ of finite rank we have (with $T = U|T|$) that

$$|\text{tr}(P|T|)| = |\text{tr}(PU^*T)| = |\varphi(PU^*)| \leq \|\varphi\|.$$

Since P is arbitrary, this implies that $T \in \mathbf{B}^1(\mathfrak{H})$ with $\|T\|_1 \leq \|\varphi\|$. Evidently, the correspondence $\varphi \leftrightarrow T$ is a bijective isometry, whence $(\mathbf{B}_0(\mathfrak{H}))^* = \mathbf{B}^1(\mathfrak{H})$.

Clearly every S in $\mathbf{B}(\mathfrak{H})$ determines a bounded functional $\psi_S = \langle S, \cdot \rangle$ on $\mathbf{B}^1(\mathfrak{H})$, and $\|\psi_S\| \leq \|S\|$ by 3.4.10. Conversely, if $\psi \in (\mathbf{B}^1(\mathfrak{H}))^*$, we define a sesquilinear form B on $\mathfrak{H}$ by

$$B(x, y) = \psi(x \odot y), \quad x, y \in \mathfrak{H},$$

with $x \odot y$ as the rank one operator defined in 3.3.9. Straightforward computations show that

$$|x \odot y| = ((x \odot y)^*(x \odot y))^{1/2} = ((y \odot x)(x \odot y))^{1/2}$$
$$= (\|x\|^2 y \odot y)^{1/2} = \|x\| \|y\|(\|y\|^{-1} y \odot \|y\|^{-1} y);$$

and, therefore, the form B is bounded, as

$$|B(x, y)| \leq \|\psi\| \|x \odot y\|_1 = \|\psi\| \text{tr}(|x \odot y|) = \|\psi\| \|x\| \|y\|.$$

By 3.2.2 there is then a unique operator S in $\mathbf{B}(\mathfrak{H})$ such that $\|S\| \leq \|\psi\|$ and

$$\psi(x \odot y) = B(x, y) = (Sx|y).$$

Every self-adjoint T in $\mathbf{B}^1(\mathfrak{H})$ has a diagonal form $T = \sum \lambda_j e_j \odot e_j$ for some orthonormal basis $\{e_j | j \in J\}$ and real eigenvalues λ_j with $\sum |\lambda_j| = \|T\|_1$. Thus,

$$\psi(T) = \sum \lambda_j \psi(e_j \odot e_j) = \sum \lambda_j (Se_j | e_j)$$
$$= \sum (STe_j | e_j) = \text{tr}(ST).$$

Since $\mathbf{B}^1(\mathfrak{H})$ is self-adjoint, the formula $\psi(T) = \text{tr}(ST)$ holds for all T; and again we have constructed a bijective isometry $\psi \leftrightarrow S$, so that $(\mathbf{B}^1(\mathfrak{H}))^* = \mathbf{B}(\mathfrak{H})$. $\qquad\square$

3.4.14. Proposition. *For every orthonormal basis* $\{e_j | j \in J\}$ *in* $\mathfrak{H}$, *the set*

$$\{e_i \odot e_j | (i,j) \in J^2\}$$

of rank one operators form an orthonormal basis for $\mathbf{B}^2(\mathfrak{H})$.

PROOF. Since $(e_i \odot e_j)^* = e_j \odot e_i$ and $(e_i \odot e_j)(e_k \odot e_l) = \delta_{jk} e_i \odot e_l$, it is clear that the operators $e_i \odot e_j$ form an orthonormal set in $\mathbf{B}^2(\mathfrak{H})$. However, if $T \in \mathbf{B}^2(\mathfrak{H})$, then

$$(T | e_i \odot e_j)_{\text{tr}} = \text{tr}((e_j \odot e_i)T) = \text{tr}(e_j \odot T^* e_i)$$
$$= \sum_l (e_l | T^* e_i)(e_j | e_l) = (Te_j | e_i).$$

This shows that the orthogonal complement to the span of the $e_i \odot e_j$'s is $\{0\}$ which means that they form a basis. $\qquad\square$

3.4.15. The result above gives a particularly concrete realization of the Hilbert –Schmidt operators in the case where the underlying Hilbert space has the form $L^2(X)$ with respect to some Radon integral $\int$ on a locally compact Hausdorff space X; see 6.1. If namely $\int \otimes \int$ denotes the product integral on X^2 (6.6.3), we consider the Hilbert space $L^2(X^2)$. If $\{e_j | j \in J\}$ is an orthonormal basis for $L^2(X)$, the set of functions $e_i \otimes \bar{e}_j(x, y) = e_i(x)\bar{e}_j(y)$ on X^2 is an orthonormal basis for $L^2(X^2)$. It follows from 3.1.14 that we have an isometry U of $L^2(X^2)$ onto $\mathbf{B}^2(L^2(X))$ determined by $U(e_i \otimes \bar{e}_j) = e_i \odot e_j$. On the level of functions this isometry gives the following result.

3.4.16. Proposition. *For every* k *in* $L^2(X^2)$ *the integral operator* T_k *defined by*

$$T_k f(x) = \int_y k(x, y)f(y), \quad f \in L^2(X), \qquad (*)$$

is a Hilbert–Schmidt operator on $L^2(X)$; *and the map* $k \to T_k$ *is an isometry of* $L^2(X^2)$ *onto* $\mathbf{B}^2(L^2(X))$ *in 2-norms. With* $k^*(x, y) = \overline{k(y, x)}$ *we have* $T_{k^*} = T_k^*$, *so that* T_k *is self-adjoint iff its kernel* k *is conjugate symmetric.*

PROOF. If $k \in L^2(X^2)$, then for every pair f, g in $L^2(X)$ we have

$$\int \int |k(x, y)f(y)\overline{g(x)}| = \int \otimes \int |k||f \otimes \bar{g}| < \infty.$$

Fubini's theorem (6.6.6) therefore applies to show that the formula $(*)$ defines an operator T_k in $\mathbf{B}(L^2(X))$ with $\|T_k\| \leq \|k\|_2$.

If $\{e_j| j \in J\}$ is an orthonormal basis for $L^2(X)$ and $k = \sum \alpha_{ij} e_i \otimes \bar{e}_j$ is a finite sum, then

$$T_k f = \sum \alpha_{ij}(f|e_j)e_i = (\sum \alpha_{ij} e_i \odot e_j)f = (Uk)f,$$

with U as in 3.4.15. Since functions of this form are dense in $L^2(X^2)$, it follows by continuity that $T_k = U(k)$ for every k in $L^2(X^2)$; and therefore the map $k \to T_k (= U)$ is an isometry.

The identity $T_{k^*} = T_k^*$ is easily verified if $k = \sum \alpha_{ij} e_i \otimes \bar{e}_j$, and by continuity it therefore holds in general.　　　　　　　　　　　　　　　　　□

3.4.17. The Fredholm integral equation

$$\int_y k(x, y)f(y) - \lambda f(x) = g(x), \qquad (*)$$

where k, g, and λ are given, can be solved by the result in 3.4.16, provided that $g \in L^2(X)$ and k is a conjugate symmetric function in $L^2(X^2)$. Using 3.3.8 we find an orthonormal basis $\{e_j| j \in J\}$ for $L^2(X)$ such that $T_k = \sum \lambda_j e_j \odot e_j$, where $\lambda_j \in \mathbb{R}$ and $\sum |\lambda_j|^2 = \|k\|_2^2$ (Parseval's identity). If $\lambda \neq \lambda_j$ for all j, the solution to $(*)$ is unique and is given by

$$f = \sum (\lambda_j - \lambda)^{-1}(g|e_j)e_j.$$

3.4.18. As a final application of Hilbert–Schmidt operators we mention (without proof) the main results in the *Sturm–Liouville problem*.

On the interval $I = [a, b]$ we consider the following slightly generalized versions of ordinary linear second-order differential equations:

$$(pf')' + qf = 0; \qquad (*)$$

$$(pf')' + qf = \lambda f; \qquad (**)$$

$$(pf')' + qf = \lambda f + h. \qquad (***)$$

Here, p, q, and h are continuous, real-valued functions with $p > 0$ and $\lambda \in \mathbb{R}$.

The homogeneous equation $(*)$ has a two-dimensional complete solution, spanned by, say, the solutions u and v that satisfy the boundary conditions $\alpha u(a) + \beta p(a)u'(a) = 0$ and $\gamma v(b) + \delta p(b)v'(b) = 0$ for given numbers α, β, γ, and δ. Moreover, the Wronskian $p(uv' - u'v) = c$ for some constant $c \neq 0$. Define Green's function by

$$g(x, y) = \begin{cases} c^{-1}u(x)v(y) & \text{for } a \leq x \leq y \leq b \\ c^{-1}u(y)v(x) & \text{for } a \leq y \leq x \leq b. \end{cases}$$

Then with T_g the Hilbert–Schmidt operator on $L^2(I)$ as defined in 3.4.16 we have that for each h in $L^2(I)$ [in particular for h in $C(I)$] the function $f = T_g h$ is the unique solution to $(***)$ with $\lambda = 0$ that satisfies the boundary conditions B: $\alpha f(a) + \beta p(a)f'(a) = \gamma f(b) + \delta p(b)f'(b) = 0$.

The discussion above reduces the equations $(*)$, $(**)$, and $(***)$ to questions related to the (unbounded) operator T_g^{-1}. Note that $T_g = T_g^*$ by 3.4.16, so by 3.3.8 there is an orthonormal basis $\{e_n| n \in \mathbb{N}\}$ for $L^2(I)$ such that $T_g =$

$\sum \lambda_n e_n \odot e_n$, where $\{\lambda_n\} \subset \mathbb{R} \setminus \{0\}$ and $\sum |\lambda_n|^2 = \|g\|_2^2$. It follows that a solution to $(**)$ satisfying B exists only when $\lambda = \lambda_n^{-1}$ for some n; and in that case the solution is e_n. A solution to $(***)$ satisfying B, where $\lambda = \lambda_n^{-1}$ for some n is only possible if $h \perp e_n$. A solution to $(***)$ satisfying B, where $\lambda \neq \lambda_n^{-1}$ for all n is always possible. In both cases the solution is

$$f = \sum \lambda_n (1 - \lambda \lambda_n)^{-1} (h|e_n) e_n.$$

EXERCISES

E 3.4.1. Show that the projective tensor product (cf. E 2.3.11) between two copies of the Hilbert space $\mathfrak{H}$ is isometrically isomorphic to $\mathbf{B}^1(\mathfrak{H})$, whereas the Hilbert space tensor product defined in E 3.2.19 gives $\mathbf{B}^2(\mathfrak{H})$.

E 3.4.2. Consider positive, compact operators S and T, such that S^p and T^q are trace class operators, where $1 < q \leq 2 \leq p < \infty$ and $p^{-1} + q^{-1} = 1$. Show that
(i) $$(S^2 e|e)^{p/2} \leq (S^p e|e)$$

for every unit vector e in $\mathfrak{H}$. Choose an orthonormal basis (e_n) for $\mathfrak{H}$, such that $T = \sum \lambda_n e_n \odot e_n$, and prove that

(ii) $$(|ST|e_n|e_n) \leq (|ST|^2 e_n|e_n)^{1/2} \leq \lambda_n (S^p e_n|e_n)^{1/p}$$

for every n. Show finally that $ST \in \mathbf{B}^1(\mathfrak{H})$ and that

(iii) $$\mathrm{tr}(|ST|) \leq (\mathrm{tr}(S^p))^{1/p} (\mathrm{tr}(T^q))^{1/q}.$$

Hints: For (i), choose a diagonal form for S and use the fact that the function $t \to t^{p/2}$ is convex. For (ii), use the Cauchy–Schwarz inequality and (i). For (iii), use (ii) and the Hölder inequality for the sequence spaces ℓ^p and ℓ^q.

E 3.4.3. (*The Schatten p-ideals.*) For $1 \leq p < \infty$ let $\mathbf{B}^p(\mathfrak{H})$ denote the set of operators T in $\mathbf{B}_0(\mathfrak{H})$ such that $|T|^p \in \mathbf{B}^1(\mathfrak{H})$. Define

$$\|T\|_p = (\mathrm{tr}(|T|^p))^{1/p}, \quad T \in \mathbf{B}^p(\mathfrak{H}).$$

Show that $\mathbf{B}^p(\mathfrak{H})$ is a Banach space under the p-norm and a $*$-invariant ideal of $\mathbf{B}(\mathfrak{H})$ contained in $\mathbf{B}_0(\mathfrak{H})$.
Hints: Given S and T in $\mathbf{B}^p(\mathfrak{H})$, take polar decompositions $S = U|S|$, $T = V|T|$, $S + T = W|S + T|$. For any projection Q of finite rank commuting with $|S + T|$ estimate

$$\mathrm{tr}(|S + T|^p Q) = \mathrm{tr}(W^*(S + T)|S + T|^{p-1}Q)$$

$$= \mathrm{tr}(W^* U|S||S + T|^{p-1}Q)$$

$$+ \mathrm{tr}(W^* V|T||S + T|^{p-1}Q).$$

Take $q = (1 - p^{-1})^{-1}$ and use 3.4.10 and E 3.4.2 to obtain

$$\||S + T|Q\|_p^p \leq (\|S\|_p + \|T\|_p) \||S + T|Q\|_p^{p-1}.$$

Deduce that $S + T \in \mathbf{B}^p(\mathfrak{H})$ and that $\|\cdot\|_p$ is subadditive. Prove completeness of $\mathbf{B}^p(\mathfrak{H})$ as is done in 3.4.9 and 3.4.12, using the fact that $\|\cdot\| \leq \|\cdot\|_p$. Note that $UT \in \mathbf{B}^p(\mathfrak{H})$ and $TU \in \mathbf{B}^p(\mathfrak{H})$ for every T in $\mathbf{B}^p(\mathfrak{H})$ and every unitary U in $\mathbf{B}(\mathfrak{H})$, and use 3.2.23 to prove that $\mathbf{B}^p(\mathfrak{H})$ is an ideal in $\mathbf{B}(\mathfrak{H})$ and therefore $*$-invariant.

E 3.4.4. Take $1 < p \leq 2 \leq q < \infty$ such that $p^{-1} + q^{-1} = 1$, and consider the Banach spaces $\mathbf{B}^p(\mathfrak{H})$ and $\mathbf{B}^q(\mathfrak{H})$ introduced in E 3.4.3. Prove the *Hölder–von Neumann inequality*

$$|\mathrm{tr}(ST)| \leq \|S\|_p \|T\|_q, \quad S \in \mathbf{B}^p(\mathfrak{H}), \ T \in \mathbf{B}^q(\mathfrak{H}).$$

Show that the bilinear form $\langle S, T \rangle = \mathrm{tr}(ST)$ implements a duality, such that $\mathbf{B}^p(\mathfrak{H})$ and $\mathbf{B}^q(\mathfrak{H})$ are isometrically isomorphic to the dual space of each other.

Hints: Use E 3.4.2, 3.4.10, and polar decompositions to prove the inequality, providing a normdecreasing injection of $\mathbf{B}^p(\mathfrak{H})$ into $(\mathbf{B}^q(\mathfrak{H}))^*$. If $\varphi \in (\mathbf{B}^q(\mathfrak{H}))^*$, use the inclusion $\mathbf{B}^2(\mathfrak{H}) \subset \mathbf{B}^q(\mathfrak{H})$ and $\|\cdot\|_2 \geq \|\cdot\|_q$ to find S in $\mathbf{B}^2(\mathfrak{H})$ (by 3.4.9) with $\mathrm{tr}(ST) = \varphi(T)$ for all T in $\mathbf{B}^2(\mathfrak{H})$. Prove that $S \in \mathbf{B}^p(\mathfrak{H})$ and that $\|S\|_p \leq \|\varphi\|$, by inserting $T = |S|^{p-1}QU^*$, where $S = U|S|$ is the polar decomposition and Q is an arbitrary projection in $\mathbf{B}_f(\mathfrak{H})$ commuting with $|S|$.

E 3.4.5. (*Volterra operators.*) Let $I = [0, 1]$ and consider functions k in $L^2(I^2)$ such that $k(x, y) = 0$ for $x < y$. Show that if k_1 and k_2 are two such Volterra functions, then for the corresponding integral operators we have $T_{k_1} T_{k_2} = T_k$ for some Volterra function k. Assume that k is a bounded Volterra function, $\|k\|_\infty \leq c$, and let $k^{(n)}$ denote the kernel corresponding to the operator $(T_k)^n$. Show that $|k^{(n+1)}(x, y)| \leq c^{n+1}(n!)^{-1}(x - y)^n$ and deduce that $\|T_k^{n+1}\| \leq (n!)^{-1}c^{n+1}$. Prove that the only eigenvalue for such a Volterra operator is 0.

E 3.4.6. Let $I = [0, 1]$ and consider the Volterra operator (cf. E 3.4.5) given by

$$Tf(x) = \int_0^x f(y)\,dy, \quad f \in L^2(I).$$

Find T^* and show that $T + T^*$ is the projection of rank one on the subspace spanned by the vector 1.

E 3.4.7. Take T as in E 3.4.6 and find the eigenvalues for the operator T^*T on $L^2(I)$. Prove that $\|T\| = 2\pi^{-1}$.

E 3.4.8. Take $[a, b] = [0, 1]$ and consider the differential equations $(*)$, $(**)$, and $(***)$ in 3.4.18 in the case where $p = 1$ and $q = 0$. Find solutions u and v corresponding to the boundary data $(\alpha, \beta, \gamma, \delta) = (1, 0, 1, 0)$, and compute Green's function g. Find (or guess) an orthonormal basis for $L^2([0, 1])$ that diagonalizes T_g, and compute the eigenvalues.

CHAPTER 4

Spectral Theory

A *function calculus* for a Banach algebra $\mathfrak{A}$ is a collection of algebra isomorphisms of the form $\Phi\colon \mathscr{C} \to \mathfrak{A}$, where $\mathscr{C}$ is an algebra of continuous functions on some compact Hausdorff space X. Loosely speaking, a function calculus is deemed the better, the larger the function algebra $\mathscr{C}$ is inside $C(X)$.

Given an isomorphism $\Phi\colon \mathscr{C} \to \mathfrak{A}$ as above, we may take f in $\mathscr{C}$ with image $A = \Phi(f)$ in $\mathfrak{A}$, and restrict Φ to the algebra $\mathscr{C}(A)$ generated by f, to obtain what might be called the function calculus for a single element A. Now $\mathscr{C}(A)$ is naturally isomorphic to a subalgebra of $C(f(X))$, and $f(X)$ may be characterized as the set of complex numbers λ for which $\lambda 1 - f$ is not invertible [in $C(X)$]. We are led to the definition of a function calculus for an element A of $\mathfrak{A}$, to be an isomorphism $\Phi\colon \mathscr{C}(A) \to \mathfrak{A}$, where $\mathscr{C}(A) \subset C(\mathrm{sp}(A))$ and $\mathrm{sp}(A)$ is the set of complex numbers λ for which $\lambda I - A$ is not invertible (in $\mathfrak{A}$). In this setting it is further required that $\mathscr{C}(A)$ contains the constant function 1 and the identical function id [where $\mathrm{id}(\lambda) = \lambda$ for all λ], and that $\Phi(1) = I$ and $\Phi(\mathrm{id}) = A$. It follows that $\Phi(f) = f(A)$ for every polynomial $f(\lambda) = \sum \alpha_n \lambda^n$. The set $\mathrm{sp}(A)$ is known as the *spectrum* of A—a notion explained by Fourier analysis—and the function calculus is the *spectral theory* for A.

For a general Banach algebra $\mathfrak{A}$ one may develop a spectral theory for an element A, where $\mathscr{C}(A)$ is the class of functions holomorphic in a neighborhood of $\mathrm{sp}(A)$. If $\mathbb{C} \setminus \mathrm{sp}(A)$ is connected, the theory is easily established, because each such function can then be approximated by polynomials uniformly on $\mathrm{sp}(A)$ by Runge's theorem. The more general case uses the vector-valued Cauchy integral formula on the resolvent function (cf. E 4.1.14). We shall not pursue this matter very far (cf. 4.1.11), but aim instead at a theory for special Banach algebras (C^*-algebras), which allow a spectral theory with $\mathscr{C}(A) = C(\mathrm{sp}(A))$ [or even $\mathscr{C}(A) = L^\infty(\mathrm{sp}(A))$], when A is a normal element of $\mathfrak{A}$. En route we pass through Gelfand's theory of commutative (semisimple) Banach algebras and the Stone–Weierstrass theorem for function algebras.

The last four sections are devoted to spectral theory for operators on Hilbert space, and we present, in increasing order of complexity, the various forms the spectral theorem may take. The culmination is the spatial version of the theorem, which unfortunately requires a bit of operator algebra theory and some key results from integration theory. Realizing the maximal commutative subalgebras of $\mathbf{B}(\mathfrak{H})$ as a natural generalization of orthonormal bases, the spectral theorem returns in the end to its finite-dimensional origin: given a normal operator, there is a generalized orthonormal basis in which it becomes a multiplication operator.

4.1. Banach Algebras

Synopsis. Ideals and quotients. Unit and approximate units. Invertible elements. C. Neumann series. Spectrum and spectral radius. The spectral radius formula. Mazur's theorem. Exercises.

4.1.1. A *Banach algebra* is (as was already mentioned in 2.1.3) an algebra $\mathfrak{A}$ with a submultiplicative norm, such that the underlying vector space is a Banach space (i.e. complete).

It is straightforward to check that if $\mathfrak{A}$ is only a normed algebra, then the completion defined in 2.1.12 becomes a Banach algebra in a natural way.

4.1.2. An *ideal* (more precisely, a twosided ideal) in an algebra $\mathfrak{A}$ is a subspace $\mathfrak{I}$ such that $\mathfrak{A}\mathfrak{I} \subset \mathfrak{I}$ and $\mathfrak{I}\mathfrak{A} \subset \mathfrak{I}$. If only one of these inclusions are satisfied, we talk about a left or a right ideal. Given an ideal $\mathfrak{I}$ in $\mathfrak{A}$, the quotient space $\mathfrak{A}/\mathfrak{I}$ of cosets (cf. 2.1.5) becomes an algebra with the product defined as $(A + \mathfrak{I})(B + \mathfrak{I}) = AB + \mathfrak{I}$. If, furthermore, $\mathfrak{I}$ is a closed ideal in the Banach algebra $\mathfrak{A}$, then the quotient $\mathfrak{A}/\mathfrak{I}$ is again a Banach algebra. Indeed, by 2.1.5 it suffices to show that the quotient norm is submultiplicative on $\mathfrak{A}/\mathfrak{I}$. But this is evident from the estimate

$$\|A + \mathfrak{I}\|\,\|B + \mathfrak{I}\|$$
$$= \inf_{S \in \mathfrak{I}} \|A + S\| \inf_{T \in \mathfrak{I}} \|B + T\|$$
$$\geq \inf_{S \in \mathfrak{I}} \inf_{T \in \mathfrak{I}} \|AB + (AT + SB + ST)\| \geq \inf_{R \in \mathfrak{I}} \|AB + R\| = \|AB + \mathfrak{I}\|.$$

Thus, every closed ideal is the kernel of a continuous (even norm decreasing) homomorphism $\Phi\colon \mathfrak{A} \to \mathfrak{A}/\mathfrak{I}$. Conversely, if $\Phi\colon \mathfrak{A} \to \mathfrak{B}$ is a continuous homomorphism between Banach algebras $\mathfrak{A}$ and $\mathfrak{B}$, then ker Φ is a closed ideal.

4.1.3. Many Banach algebras are *unital*, i.e. they have an element I such that $IA = AI = A$ for every A in $\mathfrak{A}$. Such a unit element is unique and $\|I\| \geq 1$ (if $\mathfrak{A} \neq \{0\}$).

If a Banach algebra $\mathfrak{A}$ has no unit, we may try to embed it isometrically into a larger unital Banach algebra $\mathfrak{B}$, in such a way that $\mathfrak{A}$ becomes an ideal in $\mathfrak{B}$, so large that every nonzero ideal of $\mathfrak{B}$ has a nonzero intersection with $\mathfrak{A}$ (an essential ideal). This process is the algebraic counterpart of the compactification of a topological space (1.7.2). For the classical Banach algebras there is often a natural way of *adjoining a unit*; but there is always an abstract procedure—the counterpart of the one-point compactification. Take the direct sum $\mathfrak{A} \oplus \mathbb{F}$ (cf. 2.1.18) equipped with the product

$$(A, \alpha)(B, \beta) = (AB + \alpha B + \beta A, \alpha\beta)$$

and the submultiplicative 1-norm. Then $\mathfrak{A} \oplus \mathbb{F}$ is a unital Banach algebra with $I = (0, 1)$, and contains $\mathfrak{A}$ as an ideal of co-dimension 1.

4.1.4. Given a Banach algebra $\mathfrak{A}$ we define the (left) *regular representation* $\rho: \mathfrak{A} \to \mathbf{B}(\mathfrak{A})$ by $\rho(A)B = AB$, for A and B in $\mathfrak{A}$. It is easily verified that ρ is a norm decreasing algebra homomorphism. If $\mathfrak{A}$ is unital, the inequalities

$$\|A\| \leq \|\rho(A)\| \, \|I\| \leq \|A\| \, \|I\|$$

show that ρ is a homeomorphism. Renorming $\mathfrak{A}$ by using the equivalent norm from $\mathbf{B}(\mathfrak{A})$, we may therefore assume that $\|I\| = 1$ (if $\mathfrak{A} \neq \{0\}$), and this will be done from now on.

4.1.5. A net $(E_\lambda)_{\lambda \in \Lambda}$ in the unit ball of a Banach algebra $\mathfrak{A}$ is an *approximate unit* if

$$\lim E_\lambda A = \lim A E_\lambda = A$$

for every A in $\mathfrak{A}$. All the classical nonunital Banach algebras [such as $C_0(X)$ (2.1.14), $\mathbf{B}_0(\mathfrak{H})$ (3.3), and $L^1(\mathbb{R}^n)$ (4.2.8)] have approximate units. Note that the existence of an approximate unit guarantees that the regular representation ρ is an isometry.

4.1.6. In a unital Banach algebra $\mathfrak{A}$ an element A is *invertible* if there are elements B and C in $\mathfrak{A}$ with $BA = AC = I$. Since

$$B = BI = BAC = IC = C,$$

we see that the left and right inverses for A coalesce and are uniquely determined, so we shall just write A^{-1} for the inverse. The set of invertible elements in $\mathfrak{A}$ is denoted by $\mathbf{GL}(\mathfrak{A})$ (cf. the general linear group in the matrix algebra $\mathfrak{A} = \mathbf{M}_n$).

4.1.7. Lemma. *If A is an element in a unital Banach algebra $\mathfrak{A}$ with $\|A\| < 1$, then $I - A \in \mathbf{GL}(\mathfrak{A})$ and*

$$(I - A)^{-1} = \sum_{n=0}^{\infty} A^n.$$

PROOF. Since $\|A^n\| \le \|A\|^n$, the series $\sum A^n$ (the *C. Neumann series*) converges in $\mathfrak{A}$ to an element B. As $AB = BA = B - I$, it follows that $B = (I - A)^{-1}$. $\qquad\square$

4.1.8. Proposition. *In a unital Banach algebra* $\mathfrak{A}$ *the multiplicative group* $\mathbf{GL}(\mathfrak{A})$ *is an open subset of* $\mathfrak{A}$, *and the map* $A \to A^{-1}$ *is a homeomorphism of* $\mathbf{GL}(\mathfrak{A})$.

PROOF. By computation we see that if A and B are invertible, then so is AB, with $(AB)^{-1} = B^{-1}A^{-1}$. Thus $\mathbf{GL}(\mathfrak{A})$ is a group (which is abelian iff $\mathfrak{A}$ is commutative).

If $A \in \mathbf{GL}(\mathfrak{A})$ and $B \in \mathfrak{A}$, then by 4.1.7

$$B = A - (A - B) = A(I - A^{-1}(A - B)) \in \mathbf{GL}(\mathfrak{A}), \qquad (*)$$

provided that $\|A^{-1}(A - B)\| < 1$. In particular, the ball $\mathfrak{B}(A, \varepsilon)$ is contained in $\mathbf{GL}(\mathfrak{A})$ for $\varepsilon < \|A^{-1}\|^{-1}$, which proves that $\mathbf{GL}(\mathfrak{A})$ is open. If $B \to A$, we see from $(*)$ and 4.1.7 that $B^{-1} \to A^{-1}$; thus, inverting is a continuous process and therefore a homeomorphism, since $(A^{-1})^{-1} = A$. $\qquad\square$

4.1.9. From here on all Banach algebras are assumed to be *complex*, i.e. $\mathbb{F} = \mathbb{C}$. The reason for this restriction is that the theory of characteristic values, which we hope to generalize, naturally uses only the complex number field. In the matrix case, $\mathfrak{A} = \mathbf{M}_n$, it is fairly easy to complexify the real algebra, and this step is taken (more or less tacitly) whenever one wishes to determine the roots in the characteristic polynomial. The complexification of a general real Banach algebra is a more delicate matter (see E 2.1.13), which we shall leave aside, because the algebras that appear in applications are (or easily are transformed into) complex algebras.

4.1.10. Let $\mathfrak{A}$ be a (complex!) unital Banach algebra. For every A in $\mathfrak{A}$ we define the *spectrum* of A as the set

$$\mathrm{sp}(A) = \{\lambda \in \mathbb{C} \,|\, \lambda I - A \notin \mathbf{GL}(\mathfrak{A})\}.$$

The smallest number $r \ge 0$ such that $\mathrm{sp}(A) \subset B(0, r)$ is called the *spectral radius* of A, and is denoted by $\mathrm{r}(A)$. Thus,

$$\mathrm{r}(A) = \sup\{|\lambda| \,|\, \lambda \in \mathrm{sp}(A)\}.$$

The complement of $\mathrm{sp}(A)$ is the *resolvent set*. On this set we can define the *resolvent* (function) $R(A, \lambda)$ [or just $R(\lambda)$ for short] by $R(A, \lambda) = (\lambda I - A)^{-1}$.

4.1.11. Lemma. *If* $A \in \mathfrak{A}$ *and* $f(z) = \sum_{n=0}^{\infty} \alpha_n z^n$ *is a holomorphic function in a region that contains the closed disk* $B(0, r)$ *with* $\|A\| \le r$, *then we can define* $f(A) = \sum \alpha_n A^n$ *in* $\mathfrak{A}$. *Moreover, if* $\lambda \in \mathrm{sp}(A)$ *and* $|\lambda| \le r$, *then* $f(\lambda) \in \mathrm{sp}(f(A))$.

PROOF. The series for $f(A)$ is absolutely convergent (i.e. $\sum |\alpha_n| \|A\|^n < \infty$), so $f(A) \in \mathfrak{A}$. Moreover,

$$f(\lambda)I - f(A) = \sum_{n=1}^{\infty} \alpha_n(\lambda^n I - A^n)$$

$$= (\lambda I - A) \sum_{n=1}^{\infty} \alpha_n P_{n-1}(\lambda, A) = (\lambda I - A)B.$$

Here $P_{n-1}(\lambda, A) = \sum_{k=0}^{n-1} \lambda^k A^{n-k-1}$, so that $\|P_{n-1}(\lambda, A)\| \le nr^{n-1}$, which implies that the series of polynomials converges in $\mathfrak{A}$ to an element B commuting with A. We see from this computation that if $f(\lambda)I - f(A) \in \mathbf{GL}(\mathfrak{A})$ with inverse C, then BC will be the inverse of $\lambda I - A$, so that $\lambda I - A \in \mathbf{GL}(\mathfrak{A})$. $\qquad\square$

4.1.12. Lemma. *For every A in $\mathfrak{A}$ we have* $r(A) \le \inf \|A^n\|^{1/n}$.

PROOF. If $|\lambda| > \|A\|$, then by 4.1.7

$$(\lambda I - A)^{-1} = \lambda^{-1}(I - \lambda^{-1}A)^{-1} = \sum_{n=0}^{\infty} \lambda^{-n-1} A^n. \qquad (*)$$

This shows that $r(A) \le \|A\|$.

Now if $\lambda \in \mathrm{sp}(A)$, then $\lambda^n \in \mathrm{sp}(A^n)$ by 4.1.11. Therefore, $|\lambda^n| \le \|A^n\|$ from the above. It follows that $r(A) \le \|A^n\|^{1/n}$ for every n. $\qquad\square$

4.1.13. Theorem. *For every element A in a unital Banach algebra $\mathfrak{A}$, the spectrum of A is a compact, nonempty subset of $\mathbb{C}$, and the spectral radius of A is the limit of the convergent sequence* $(\|A^n\|^{1/n})$.

PROOF. With $R(\lambda)$ as the resolvent for A (cf. 4.1.10) we see from 4.1.7 that if $\lambda \notin \mathrm{sp}(A)$ and $|\zeta| < \|R(\lambda)\|^{-1}$, then $\lambda - \zeta \notin \mathrm{sp}(A)$ and

$$R(\lambda - \zeta) = (\lambda I - A - \zeta I)^{-1}$$

$$= ((\lambda I - A)(I - R(\lambda)\zeta))^{-1} = \sum_{n=0}^{\infty} R(\lambda)^{n+1} \zeta^n.$$

It follows that $\mathbb{C} \setminus \mathrm{sp}(A)$ is open, whence $\mathrm{sp}(A)$ is closed and therefore compact by 4.1.12.

Fix a functional φ in $\mathfrak{A}^*$ and consider the complex function $f(\lambda) = \varphi(R(\lambda))$, which is holomorphic in the region $\mathbb{C} \setminus \mathrm{sp}(A)$, because it has a local power series expansion [viz. $f(\lambda - \zeta) = \sum \alpha_n \zeta^n$, with $\alpha_n = \varphi(R(\lambda)^{n+1})$]. If $|\lambda| > \|A\|$, we see from $(*)$ in the proof of 4.1.12 that

$$f(\lambda) = \sum_{n=0}^{\infty} \lambda^{-n-1} \varphi(A^n). \qquad (**)$$

This means that

$$|f(\lambda)| \le \sum |\lambda|^{-n-1} \|A\|^n \|\varphi\| = |\lambda|^{-1} \|\varphi\| (1 - |\lambda|^{-1} \|A\|)^{-1}$$

$$= \|\varphi\| (|\lambda| - \|A\|)^{-1}.$$

Thus, $|f(\lambda)| \to 0$ as $|\lambda| \to \infty$.

If $sp(A) = \emptyset$, then f is an entire analytic function that belongs to $C_0(\mathbb{C})$. By Liouville's theorem f must be constant and therefore identically zero. Consequently, $\varphi((\lambda I - A)^{-1}) = 0$ for every φ in $\mathfrak{A}^*$, so that $(\lambda I - A)^{-1} = 0$ by 2.3.4, an obvious contradiction. We conclude that $sp(A) \neq \emptyset$.

The power series expansion for f in $(**)$ is valid for $|\lambda| > \|A\|$. But the function is holomorphic for $|\lambda| > r(A)$. From the Cauchy integral formula [used, say, on the function $f(\lambda^{-1})$ for $|\lambda| < r(A)^{-1}$] it follows that the series converges uniformly on every region $\{\lambda \in \mathbb{C} | |\lambda| \geq r\}$, where $r > r(A)$. Taking $\lambda = re^{i\theta}$ and integrating the series for $\lambda^{n+1} f(\lambda)$ term by term with respect to θ we have

$$\int_0^{2\pi} r^{n+1} e^{i(n+1)\theta} f(re^{i\theta}) d\theta = \sum_{m=0}^{\infty} \int_0^{2\pi} r^{n-m} e^{i(n-m)\theta} \varphi(A^m) d\theta$$

$$= 2\pi \varphi(A^n).$$

With $M(r) = \sup_\theta \|R(re^{i\theta})\|$ we get the estimate

$$|\varphi(A^n)| \leq r^{n+1} M(r) \|\varphi\|.$$

Since this holds for every φ in $\mathfrak{A}^*$, we conclude, again from 2.3.4, that $\|A^n\| \leq r^{n+1} M(r)$. Extracting nth roots and passing to the limit gives

$$\limsup \|A^n\|^{1/n} \leq \inf r = r(A).$$

In conjunction with 4.1.12 it proves that $(\|A^n\|^{1/n})$ converges to $r(A)$ from above.   □

4.1.14. Corollary. *If $\mathfrak{A}$ is a division ring [i.e. $\mathbf{GL}(\mathfrak{A}) = \mathfrak{A} \setminus \{0\}$], then $\mathfrak{A} = \mathbb{C}$.*

PROOF. For each A in $\mathfrak{A}$ there is some λ in $sp(A)$ by 4.1.13. Thus $\lambda I - A \notin \mathbf{GL}(\mathfrak{A})$, whence $A = \lambda I$.   □

EXERCISES

E 4.1.1. Let $\{\mathfrak{A}_j | j \in J\}$ be a family of Banach algebras. Show that the direct product (cf. 2.1.16) consisting of the bounded functions $A: J \to \bigcup \mathfrak{A}_j$ such that $A(j) \in \mathfrak{A}_j$ for every j is a Banach algebra under the pointwise product $AB(j) = A(j)B(j)$. Show that the direct sum (cf. 2.1.18) consisting of those functions A for which $j \to \|A(j)\|$ belongs to $C_0(J)$ is a norm closed ideal in the direct product.

Show that if $\{\mathfrak{X}_j | j \in J\}$ is a family of Banach spaces and $\mathfrak{X}$ denotes the completion of $\sum \mathfrak{X}_j$ in some p-norm (for $1 \leq p \leq \infty$), then there is a natural isometric isomorphism of the direct product algebra of the $\mathbf{B}(\mathfrak{X}_j)$'s, $j \in J$, into $\mathbf{B}(\mathfrak{X})$.

E 4.1.2. (*The resolvent equation.*) Let A be an element in a complex, unital Banach algebra, and consider the resolvent function $R(\lambda) =$

$(\lambda I - A)^{-1}$ defined on $\mathbb{C}\backslash\text{sp}(A)$. Show that R satisfies the equation

$$R(\lambda) - R(\mu) = (\mu - \lambda)R(\lambda)R(\mu).$$

Use this to prove that the resolvent function is complex differentiable at every point of definition (it is a holomorphic vector function) and that $R'(\lambda) = -R(\lambda)^2$.

E 4.1.3. Let A and B be elements in a complex, unital Banach algebra. Show that

$$\text{sp}(AB)\backslash\{0\} = \text{sp}(BA)\backslash\{0\}.$$

Hint: If $\lambda \notin \text{sp}(AB) \cup \{0\}$, consider the element $\lambda^{-1}(I + B(\lambda I - AB)^{-1}A)$ as a candidate for $(\lambda I - BA)^{-1}$.

E 4.1.4. Let A and B be elements in a unital algebra. Show that if both AB and BA are invertible, then so is A and B.

E 4.1.5. Let $\mathfrak{A}$ be a complex, nonunital Banach algebra. An element B in $\mathfrak{A}$ is called a *left* (*right*) *adverse* to some A in $\mathfrak{A}$ if $A + B = BA$ ($A + B = AB$). Show that if A has both a left and a right adverse then these coincide. We then say that A is *adversible* with adverse A°. Define the associative (but not distributive) product $A \circ B = A + B - AB$ in $\mathfrak{A}$. Verify that 0 is the neutral element for this product and that the adverse is the inverse in the new product. Show that A is adversible when $\|A\| < 1$, with $A^\circ = -\sum_{n=1}^\infty A^n$. Verify the formula $(A \circ B)^\circ = B^\circ \circ A^\circ$ for adversible elements in $\mathfrak{A}$. Show that if $\widetilde{\mathfrak{A}}$ is any unital Banach algebra containing $\mathfrak{A}$ as an ideal of co-dimension 1 (i.e. $\widetilde{\mathfrak{A}} = \mathfrak{A} \oplus \mathbb{C}I$) and $A \in \mathfrak{A}$, then $I - A$ is invertible (in $\widetilde{\mathfrak{A}}$) iff A is adversible (in $\mathfrak{A}$); and in that case $(I - A)^{-1} = I - A^\circ$. Show that $\text{sp}(A)$ (in $\widetilde{\mathfrak{A}}$) consists of 0 together with those λ for which $\lambda^{-1}A$ has no adverse (in $\mathfrak{A}$).

E 4.1.6. Let A be an element in a unital, complex Banach algebra $\mathfrak{A}$, and let Ω be an open subset of $\mathbb{C}$ containing $\text{sp}(A)$. Show that there is an $\varepsilon > 0$ such that $\|A - B\| < \varepsilon$ implies that $\text{sp}(B) \subset \Omega$ for all B in $\mathfrak{A}$.

E 4.1.7. Let $\mathfrak{A}$ be a unital Banach algebra and (A_n) be a sequence in $\text{GL}(\mathfrak{A})$ converging to some A in $\mathfrak{A}$. Show that if $(\|A_n^{-1}\|)$ is a bounded sequence, then $A \in \text{GL}(\mathfrak{A})$.

Hint: For large n we will then have

$$\|I - A_n^{-1}A\| = \|A_n^{-1}(A_n - A)\| < 1.$$

E 4.1.8. (*Topological zerodivisors.*) A *topological zerodivisor* in a unital Banach algebra $\mathfrak{A}$ is an element A for which there is a sequence (B_n) with $\|B_n\| = 1$ for all n, such that $\lim \|B_n A\| = \lim \|AB_n\| = 0$. Show that no topological zerodivisor belongs to $\text{GL}(\mathfrak{A})$. Show that every element on the boundary of $\text{GL}(\mathfrak{A})$ (cf. 1.2.7) is a topological zerodivisor. Conclude that if $\mathbb{F} = \mathbb{C}$ and $\mathfrak{A} \neq \mathbb{C}$, then $\mathfrak{A}$ contains topological zerodivisors different from 0.

Hint: If $A_n \to A$ and $(A_n) \subset \mathbf{GL}(\mathfrak{A})$, define $B_n = \|A_n^{-1}\|^{-1} A_n^{-1}$ and use E 4.1.7.

E 4.1.9. Let $\mathfrak{X} = C([0, 1])$ and define T in $\mathbf{B}(\mathfrak{X})$ by

$$Tf(x) = \int_0^x f(y)dy, \quad f \in \mathfrak{X}.$$

Show that T is injective. Show further that for $n \geq 1$ we have

$$(T^{n+1}f)(x) = (n!)^{-1} \int_0^x (x - y)^n f(y)dy,$$

and conclude that the spectral radius of T is zero. Compare with the Volterra operators in E 3.4.4 and E 3.4.5.

E 4.1.10. Let $\mathfrak{X}$ be a Banach space and T be an element in the Banach algebra $\mathbf{B}(\mathfrak{X})$. The set of *eigenvalues* or the *point spectrum*, $\mathrm{sp}(T)_a$, consists of those λ for which $\lambda I - T$ is not injective. The *continuous spectrum*, $\mathrm{sp}(T)_c$, consists of those λ for which $\lambda I - T$ is injective with dense range, but not surjective. The *residual spectrum*, $\mathrm{sp}(T)_r$, consists of those λ for which $\lambda I - T$ is injective, but $((\lambda I - T)(\mathfrak{X}))^= \neq \mathfrak{X}$. Show that these definitions give a disjoint decomposition

$$\mathrm{sp}(T) = \mathrm{sp}(T)_a \cup \mathrm{sp}(T)_c \cup \mathrm{sp}(T)_r.$$

Hint: Use the open mapping theorem (2.2.5) to show that the three partial spectra exhaust $\mathrm{sp}(T)$.

E 4.1.11. Take T in $\mathbf{B}(\mathfrak{X})$ as in E 4.1.10 and consider T^* in $\mathbf{B}(\mathfrak{X}^*)$; cf. 2.3.9. Show that $\mathrm{sp}(T) = \mathrm{sp}(T^*)$. Show further that

$$\mathrm{sp}(T)_r \subset \mathrm{sp}(T^*)_a \subset \mathrm{sp}(T)_a \cup \mathrm{sp}(T)_r.$$

Hint: If $S \in \mathbf{B}(\mathfrak{X})$, then $\ker S^* = S(\mathfrak{X})^\perp$; cf. 3.2.5.

E 4.1.12. Let $\mathfrak{A}$ be a complex, unital, commutative Banach algebra. Show that the spectral radius r is a submultiplicative seminorm on $\mathfrak{A}$. Show that the *radical*

$$R(\mathfrak{A}) = \{A \in \mathfrak{A} | r(A) = 0\}$$

is a closed ideal in $\mathfrak{A}$ containing all nilpotent elements.

Hint 1: To show subadditivity take A and B and $\varepsilon > 0$, and find m such that $\|A^p\| \leq (r(A) + \varepsilon)^p$ and $\|B^p\| \leq (r(B) + \varepsilon)^p$ for all $p \geq m$. Take $n > 2m$ and estimate

$$\|(A + B)^n\| \leq \left\| \sum_{p=0}^m \binom{n}{p} A^p B^{n-p} \right\|$$

$$+ \left\| \sum_{p=m+1}^{n-m-1} \binom{n}{p} A^p B^{n-p} \right\| + \left\| \sum_{p=0}^m \binom{n}{p} A^{n-p} B^p \right\|$$

$$\leq \sum_{p=0}^{m} \binom{n}{p} (\|A\|^p (\mathrm{r}(B) + \varepsilon)^{n-p} + (\mathrm{r}(A) + \varepsilon)^{n-p} \|B\|^p)$$

$$+ \sum_{p=m+1}^{n-m-1} \binom{n}{p} (\mathrm{r}(A) + \varepsilon)^p (\mathrm{r}(B) + \varepsilon)^{n-p}$$

$$\leq n^m (\mathrm{r}(B) + \varepsilon)^n c(A, B, m) + n^m (\mathrm{r}(A) + \varepsilon)^n c(B, A, m)$$

$$+ (r(A) + \mathrm{r}(B) + 2\varepsilon)^n,$$

where $c(A, B, m)$ is a constant independent of n.
Hint 2: Use 4.2.3.

E 4.1.13. Let $\mathfrak{A}$ be a unital Banach algebra, and for each A in $\mathfrak{A}$ define

$$\exp A = \sum_{n=0}^{\infty} (n!)^{-1} A^n,$$

noting that the power series converges uniformly in $\mathfrak{A}$. Show that

$$\exp(A + B) = (\exp A)(\exp B)$$

for every pair of commuting elements A and B in $\mathfrak{A}$. Show that
$\exp A \in \mathbf{GL}(\mathfrak{A})$ for every A in $\mathfrak{A}$.

E 4.1.14. Let A be an element in a unital Banach algebra $\mathfrak{A}$, and let f be a
holomorphic function in a region that contains the closed disk
$B = B(z, r)$. Show that if $\mathrm{sp}(A) \subset B^o$, then the power series

$$\sum_{n=0}^{\infty} (n!)^{-1} f^{(n)}(z)(A - zI)^n$$

converges uniformly in $\mathfrak{A}$ to an element $f(A)$; cf. 4.1.11. Show that if
g is another such holomorphic function, then $fg(A) = f(A)g(A)$; cf.
E 4.1.13. Conclude that

$$f(\mathrm{sp}(A)) \subset \mathrm{sp}(f(A)) \subset f(B^o).$$

Show that we have the equation

$$f(A) = (2\pi \mathrm{i})^{-1} \int_{\partial B} f(\lambda) R(\lambda, A) d\lambda \qquad (*)$$

$$(=) \quad (2\pi)^{-1} \int_0^{2\pi} f(r \exp \mathrm{i}\theta - z)((r \exp \mathrm{i}\theta - z)I - A)^{-1} r d\theta,$$

where we use the vector integral defined in 2.5.15 (or just the poor
man's version from E 2.5.8).

    Hints: Multiply the power series for f and g. Then use $g(\mu) = (\lambda - f(\mu))^{-1}$ for $\lambda \notin B^o$. To prove $(*)$, recall that $2\pi \mathrm{i} f^{(n)}(z) = n! \int_{\partial B} f(\lambda)(\lambda - z)^{-n-1} d\lambda$.

E 4.1.15. Let f and g be holomorphic functions in regions that contain the
disks $B_1 = B(z_1, r_1)$ and $B_2 = B(z_2, r_2)$, respectively, and let A be an

element in a unital Banach algebra $\mathfrak{A}$, such that $\mathrm{sp}(A) \subset B_1^o$ and $f(B_1^o) \subset B_2^o$. Show that

$$g(f(A)) = g \circ f(A).$$

Hint: Use E 4.1.14, especially the equation (∗), together with the fact that $R(\lambda, f(A)) = (\lambda - f)^{-1}(A)$ for each λ in ∂B_2.

E 4.1.16. If $A \in \mathfrak{A}$ and $\mathrm{sp}(A) \subset B(1,1)^o$, define

$$\log A = -\sum_{n=1}^\infty n^{-1}(I - A)^n = (2\pi i)^{-1} \int_{\partial B} \log \lambda R(\lambda, A) d\lambda,$$

where $B = B(1,r)$ with $r < 1$, but $\mathrm{sp}(A) \subset B^o$; cf. E 4.1.14. Show that $\exp \log A = A$.

Hint: Use E 4.1.15.

E 4.1.17. For a unital Banach algebra $\mathfrak{A}$, let $\mathfrak{G}$ denote the arcwise connected component of $\mathbf{GL}(\mathfrak{A})$ containing I, i.e. $\mathfrak{G}$ is the union of all arcwise connected subsets (E 1.4.14) that contain I. Show that $\mathfrak{G}$ is both open and closed and is a normal subgroup of $\mathbf{GL}(\mathfrak{A})$, generated by elements of the form $\exp A$, $A \in \mathfrak{A}$; cf. E 4.1.13.

Hints: Use 4.1.8 and the fact that balls are arcwise connected to show that $\mathfrak{G}$ is open. Show that $I \in A^{-1}\mathfrak{G}$ if $A \in \mathfrak{G}$, and that $I \in B\mathfrak{G}B^{-1}$ for any B in $\mathbf{GL}(\mathfrak{A})$ and deduce from connectivity that $A^{-1}\mathfrak{G} \subset \mathfrak{G}$ and $B\mathfrak{G}B^{-1} \subset \mathfrak{G}$. Conclude that $\mathfrak{G}$ is a normal subgroup of $\mathbf{GL}(\mathfrak{A})$. Write $\mathbf{GL}(\mathfrak{A}) \backslash \mathfrak{G} = \bigcup B\mathfrak{G}$, $B \in \mathbf{GL}(\mathfrak{A}) \backslash \mathfrak{G}$, and deduce that every open subgroup has an open complement. If $B \in \mathfrak{A}$ and $A = \exp B$, define $A_t = \exp tB$, $0 \le t \le 1$, to verify that $A \in \mathfrak{G}$. Use E 4.1.16 to show that $\exp \mathfrak{A}$ contains a neighborhood of I. Deduce that the group $\mathfrak{G}_0$ generated by $\exp \mathfrak{A}$ is open, hence also closed and therefore equals $\mathfrak{G}$ by connectivity.

E 4.1.18. Let $\mathfrak{A}$ be a complex, commutative, unital Banach algebra. Show that the group $\mathfrak{G}$ defined in E 4.1.17 is equal to $\exp \mathfrak{A}$, and that the quotient group $\mathbf{GL}(\mathfrak{A})/\mathfrak{G}$ is torsion free.

Hint: If $A \in \mathbf{GL}(\mathfrak{A})$ such that $A^n = \exp B$ for some number n and some B in $\mathfrak{A}$ [i.e. if $A\mathfrak{G}$ has order n in $\mathbf{GL}(\mathfrak{A})/\mathfrak{G}$], put $S = A \exp(-n^{-1}B)$ and note that $S^n = I$. Write $S_\lambda = \lambda S + (I - \lambda)I$ and

$$\Omega = \{\lambda \in \mathbb{C} \,|\, S_\lambda \in \mathbf{GL}(\mathfrak{A})\}.$$

Show that the complement of Ω is finite as $\lambda \notin \Omega$ iff $1 - \lambda^{-1} \in \mathrm{sp}(S)$ and deduce S can be connected by an arc to I. Conclude that S, $\exp(n^{-1}B)$, and A belong to $\mathfrak{G}$.

E 4.1.19. Let $\mathfrak{A}$ be the Banach algebra $C(\mathbb{T})$ of complex, continuous functions on the circle $\mathbb{T}$. With notations as in E 4.1.18, show that two functions belong to the same coset in $\mathbf{GL}(\mathfrak{A})$ modulo $\mathfrak{G}$ iff they have the same winding number. Deduce that $\mathbf{GL}(\mathfrak{A})/\mathfrak{G} = \mathbb{Z}$.

4.2. The Gelfand Transform

Synopsis. Characters and maximal ideals. The Gelfand transform. Examples, including Fourier transforms. Exercises.

4.2.1. As was mentioned in the introduction to this chapter we aim at the construction of isomorphisms of function algebras into a given Banach algebra $\mathfrak{A}$. Since the range of such an isomorphism is commutative, there is no loss of generality in developing spectral theory only for commutative Banach algebras. Gelfand's theory establishes for any such algebra $\mathfrak{A}$ a homomorphism $\Gamma\colon \mathfrak{A} \to C(\hat{\mathfrak{A}})$, where $\hat{\mathfrak{A}}$ is the compact Hausdorff space of characters of $\mathfrak{A}$. This would seem to be a map in the opposite direction of what we want; but, as we will learn in the next section, there are important cases where the Gelfand transform is an isometry and therefore allows us to use the inverse map.

In this section we assume that $\mathfrak{A}$ is a complex, unital, commutative Banach algebra. Recall that an ideal (i.e. a twosided ideal) $\mathfrak{I}$ of $\mathfrak{A}$ is *maximal* if $\mathfrak{I} \neq \mathfrak{A}$ and $\mathfrak{A}$ is the only ideal containing $\mathfrak{I}$ as a proper subset. Recall further that a *character* of $\mathfrak{A}$ is a surjective (i.e. nonzero) homomorphism $\gamma\colon \mathfrak{A} \to \mathbb{C}$.

4.2.2. Proposition. *In a commutative, unital Banach algebra $\mathfrak{A}$ there is a bijective correspondence, given by $\gamma \leftrightarrow \ker \gamma$, between the set $\hat{\mathfrak{A}}$ of characters of $\mathfrak{A}$ and the set $\mathcal{M}(\mathfrak{A})$ of maximal ideals in $\mathfrak{A}$. Every γ in $\hat{\mathfrak{A}}$ is automatically continuous, and every $\mathfrak{I}$ in $\mathcal{M}(\mathfrak{A})$ is closed. Finally, we have for each A in $\mathfrak{A}$ that*

$$\mathrm{sp}(A) = \{\langle A, \gamma \rangle | \gamma \in \hat{\mathfrak{A}}\}.$$

PROOF. For every ideal $\mathfrak{I} \neq \mathfrak{A}$ we have $\mathfrak{I} \cap \mathbf{GL}(\mathfrak{A}) = \emptyset$. Therefore, $\|I - A\| \geq 1$ for every A in $\mathfrak{I}$ by 4.1.7. Thus $\mathfrak{I}^= \neq \mathfrak{A}$; and since the closure of an ideal is again an ideal, we conclude that maximal ideals are closed.

Suppose now that $A \in \mathfrak{A} \setminus \mathbf{GL}(\mathfrak{A})$. Then $I \notin \mathfrak{A}A$, so that A is contained in a proper ideal (viz. $\mathfrak{A}A$). The set of ideals that contain A but not I is inductively ordered by inclusion (because a union of an increasing family of ideals is an ideal); and a maximal element in this ordering is clearly a maximal ideal. From Zorn's lemma (1.1.3) we see that every A in $\mathfrak{A} \setminus \mathbf{GL}(\mathfrak{A})$ is contained in a maximal ideal.

Take $\mathfrak{I}$ in $\mathcal{M}(\mathfrak{A})$, and consider the quotient Banach algebra $\mathfrak{A}/\mathfrak{I}$ (cf. 4.1.2), which has no proper ideals. If, therefore, $A \in \mathfrak{A}/\mathfrak{I}$ and $A \neq 0$, then $A \in \mathbf{GL}(\mathfrak{A}/\mathfrak{I})$ (since otherwise $A\mathfrak{A}/\mathfrak{I}$ would be a proper ideal). It follows from 4.1.14 that $\mathfrak{A}/\mathfrak{I} = \mathbb{C}$, so that the quotient map $\gamma\colon \mathfrak{A} \to \mathfrak{A}/\mathfrak{I}$ belongs to $\hat{\mathfrak{A}}$ and is continuous. Conversely, we see that if $\gamma \in \hat{\mathfrak{A}}$, then $\ker \gamma$ is an ideal of $\mathfrak{A}$ of co-dimension 1 and therefore maximal. In particular, $\ker \gamma$ is closed, so that γ is continuous. This establishes the bijective correspondence between $\mathcal{M}(\mathfrak{A})$ and $\hat{\mathfrak{A}}$.

If $A \in \mathfrak{A}$ and $\lambda \in \mathrm{sp}(A)$, then $\lambda I - A \notin \mathbf{GL}(\mathfrak{A})$. From the preceding we conclude that for some γ in $\hat{\mathfrak{A}}$ we have $\langle \lambda I - A, \gamma \rangle = 0$, i.e. $\lambda = \langle A, \gamma \rangle$. Conversely, if $\langle A, \gamma \rangle = \lambda$ for some γ in $\hat{\mathfrak{A}}$, then $\lambda I - A \in \ker \gamma$, whence $\lambda I - A \notin \mathbf{GL}(\mathfrak{A})$, i.e. $\lambda \in \mathrm{sp}(A)$. $\square$

4.2.3. Theorem. *Given a commutative, unital Banach algebra* $\mathfrak{A}$, *the set* $\hat{\mathfrak{A}}$ *of characters has a compact Hausdorff topology, such that the map* Γ *[where we write* $\Gamma(A) = \hat{A}$*] defined by*

$$\Gamma(A)(\gamma) = \hat{A}(\gamma) = \langle A, \gamma \rangle, \quad A \in \mathfrak{A}, \gamma \in \hat{\mathfrak{A}},$$

is a norm decreasing homomorphism of $\mathfrak{A}$ *onto a subalgebra of* $C(\hat{\mathfrak{A}})$ *that separates points in* $\hat{\mathfrak{A}}$. *For every* A *in* $\mathfrak{A}$ *we have*

$$\hat{A}(\hat{\mathfrak{A}}) = \mathrm{sp}(A), \quad \|\hat{A}\|_\infty = \mathrm{r}(A).$$

PROOF. For each A in $\mathfrak{A}$ and γ in $\hat{\mathfrak{A}}$ we have by 4.2.2

$$|\langle A, \gamma \rangle| \leq \mathrm{r}(A) \leq \|A\|.$$

This means that $\|\gamma\| \leq 1$, regarding γ as an element in the dual $\mathfrak{A}^*$. Embedding $\hat{\mathfrak{A}}$ as a subset of the unit ball $\mathfrak{B}^*$ of the dual space $\mathfrak{A}^*$ of $\mathfrak{A}$ equipped with the w^*-topology (2.4.8), it inherits a Hausdorff topology. Moreover, if $(\gamma_\lambda)_{\lambda \in \Lambda}$ is a net in $\hat{\mathfrak{A}}$ that is w^*-convergent to some γ in $\mathfrak{B}^*$, then for every A and B in $\mathfrak{A}$ we have

$$\langle AB, \gamma \rangle = \lim \langle AB, \gamma_\lambda \rangle$$
$$= \lim \langle A, \gamma_\lambda \rangle \langle B, \gamma_\lambda \rangle = \langle A, \gamma \rangle \langle B, \gamma \rangle.$$

Thus, $\gamma \in \hat{\mathfrak{A}}$, so that $\hat{\mathfrak{A}}$ is a w^*-closed subset of the compact set $\mathfrak{B}^*$ (cf. 2.5.2), and thus itself compact.

Since the w^*-topology is the "pointwise convergence" topology, it follows that each function $\hat{A}$ on $\hat{\mathfrak{A}}$, where $A \in \mathfrak{A}$ and $\hat{A}(\gamma) = \langle A, \gamma \rangle$, is continuous. Moreover, $\hat{A}(\hat{\mathfrak{A}}) = \mathrm{sp}(A)$ by 4.2.2, whence $\|\hat{A}\|_\infty = \mathrm{r}(A)$.

Finally, the map $\Gamma: A \to \hat{A}$ is a homomorphism of $\mathfrak{A}$ into $C(\hat{\mathfrak{A}})$, because every γ in $\hat{\mathfrak{A}}$ is multiplicative; and if $\gamma_1 \neq \gamma_2$ in $\hat{\mathfrak{A}}$, then certainly $\langle A, \gamma_1 \rangle \neq \langle A, \gamma_2 \rangle$ for some A in $\mathfrak{A}$, so that the function algebra $\Gamma(\mathfrak{A})$ separates points in $\hat{\mathfrak{A}}$. $\square$

4.2.4. The map $\Gamma: A \to \hat{A}$ is the *Gelfand transform* (Gelfand 1941). At first sight it seems to reduce the theory of commutative Banach algebras to the study of function algebras. The reality is not quite that simple though. For one thing we observe that the kernel of the Gelfand transform is the ideal $R(\mathfrak{A}) = \bigcap \mathfrak{I}$, $\mathfrak{I} \in \mathscr{M}(\mathfrak{A})$, (the *radical* of $\mathfrak{A}$) consisting of those elements A in $\mathfrak{A}$ with $\mathrm{r}(A) = 0$. In the case where $\mathfrak{A}$ is a radical algebra, i.e. $R(\mathfrak{R}) = \mathfrak{R}$, or (if we insist on the unital case) in the case where $R(\mathfrak{A})$ is a maximal ideal in $\mathfrak{A}$, the Gelfand transform is trivial. However, the classical Banach algebras are *semisimple* [i.e. $R(\mathfrak{A}) = \{0\}$], so this objection can be overruled. Worse is the fact that the image of $\mathfrak{A}$ under the Gelfand transform may be exceedingly difficult to characterize inside $C(\hat{\mathfrak{A}})$. What has been gained by a concrete realization of $\mathfrak{A}$ as functions may easily be lost, when one is unable to decide whether a given function in $C(\hat{\mathfrak{A}})$ belongs to $\Gamma(\mathfrak{A})$. A few examples will illuminate the problems.

4.2.5. Example. Let X be a compact Hausdorff space, and put $\mathfrak{A} = C(X)$, regarded as a Banach algebra with pointwise sum and product of functions and ∞-norm. We have an injective map $\iota: X \to \hat{\mathfrak{A}}$ given by

$$\langle f, \iota(x) \rangle = f(x), \quad x \in X, f \in C(X).$$

Since the topology on $\hat{\mathfrak{A}}$ is the w^*-topology, we see that ι is continuous and therefore a homeomorphism on its image (1.6.8). If $\gamma \in \hat{\mathfrak{A}} \setminus \iota(X)$, then for each x in X there is an f in $\mathfrak{A}$ such that $\langle f, \gamma \rangle = 0$, but $f(x) \neq 0$. A standard compactness argument produces a finite set $\{f_1, \ldots, f_n\}$ in $\mathfrak{A}$, contained in $\ker \gamma$, such that the co-zero sets $\{x \in X | f_k(x) \neq 0\}$, $1 \leq k \leq n$, cover X. The element $f = \sum \bar{f}_k f_k$ belongs to the proper ideal $\ker \gamma$; but $f(x) > 0$ for every x in X, which means that f is invertible in $\mathfrak{A}$, a contradiction since $\mathbf{GL}(\mathfrak{A}) \cap \ker \gamma = \emptyset$. Thus $\hat{\mathfrak{A}} = \iota(X)$, and the Gelfand transform of $C(X)$ is simply the identity map. The example is therefore not particularly enlightening, but certainly reassuring.

4.2.6. Example. Let $\mathfrak{A}$ be the Banach space $\ell^1(\mathbb{Z})$ of doubly infinite summable sequences (2.1.18). Equipped with the convolution product

$$(AB)_n = \sum_{-\infty}^{\infty} A_k B_{n-k}$$

this is a unital, commutative Banach algebra. The element E in $\mathfrak{A}$, with $E_1 = 1$ and $E_n = 0$ for $n \neq 1$, is a generator for $\mathfrak{A}$, since each element has the form

$$A = \sum_{-\infty}^{\infty} A_n E^n \quad \text{(uniform convergence)}.$$

A character γ is therefore determined by its value on E. Since $E \in \mathbf{GL}(\mathfrak{A})$, we have

$$|\langle E, \gamma \rangle| \leq 1, \quad |\langle E^{-1}, \gamma \rangle| \leq 1,$$

as $\|\gamma\| \leq 1$. Consequently, $\gamma(E) \in \mathbb{T} = \{\lambda \in \mathbb{C} \, | \, |\lambda| = 1\}$. Conversely, we see that for each λ in $\mathbb{T}$ we can define a γ in $\hat{\mathfrak{A}}$ by

$$\langle A, \gamma \rangle = \sum_{-\infty}^{\infty} A_n \lambda^n, \quad A \in \mathfrak{A}.$$

This establishes a continuous map, thus a homeomorphism, from $\hat{\mathfrak{A}}$ onto $\mathbb{T}$. The Gelfand transform is therefore the map of $\ell^1(\mathbb{Z})$ into $C(\mathbb{T})$ given by

$$\hat{A}(\lambda) = \sum_{-\infty}^{\infty} A_n \lambda^n, \quad A \in \ell^1(\mathbb{Z}).$$

Identifying $\mathbb{T}$ with $\mathbb{R}/2\pi\mathbb{N}$ we see that the set $\Gamma(\mathfrak{A})$ of Gelfand transforms in $C(\mathbb{T})$ is the set of continuous, periodic functions on $[0, 2\pi]$ whose Fourier transforms are absolutely convergent.

Suppose that f is a continuous, periodic function on $[0, 2\pi]$ whose Fourier series is absolutely convergent, and suppose that $f(x) \neq 0$ for every x in $[0, 2\pi]$. Then Wiener showed that the reciprocal function f^{-1} also has an absolutely convergent Fourier series. Gelfand theory makes this result quite obvious (a fact that helped considerably in making the theory acceptable to the mathematical community). Indeed, $f = \hat{A}$ for some A in $\ell^1(\mathbb{Z})$, and since $0 \notin \text{sp}(A)$ by assumption, it follows that $A^{-1} \in \ell^1(\mathbb{Z})$ with $(A^{-1})\hat{\,} = \hat{A}^{-1} = f^{-1}$.

4.2.7. Example. Let $\mathfrak{A}$ be the subalgebra ℓ^1 [i.e. $\ell^1(\mathbb{N}_0)$] of $\ell^1(\mathbb{Z})$ consisting of those elements A for which $A_n = 0$ for $n < 0$. The element E is no longer invertible (in ℓ^1) but its positive powers will still generate ℓ^1. Thus, for each λ in $\Delta = \{\lambda \in \mathbb{C} \,|\, |\lambda| \leq 1\}$ we can define a character γ by

$$\langle A, \gamma \rangle = \sum_0^\infty A_n \lambda^n, \quad A \in \ell^1.$$

Reasoning as in 4.2.6 we find that $\hat{\mathfrak{A}} = \Delta$, and that the Gelfand transform is the map of ℓ^1 into $C(\Delta)$ given by

$$\hat{A}(\lambda) = \sum_0^\infty A_n \lambda^n, \quad A \in \ell^1.$$

The image of ℓ^1 in $C(\Delta)$ consists of those continuous functions on Δ that are holomorphic in the interior of Δ and whose Taylor series coefficients are absolutely summable.

4.2.8. Example. The Banach space $L^1(\mathbb{R})$ is a Banach algebra under the convolution product

$$(f \times g)(x) = \int f(y)g(x - y)dy, \quad f, g \in L^1(\mathbb{R}),$$

whose existence (almost everywhere) is guaranteed by Fubini's theorem (6.6.6). We let δ denote the Dirac point measure at 0, and note that δ is a unit for $L^1(\mathbb{R})$ under convolution. Thus $\mathfrak{A} = L^1(\mathbb{R}) + \mathbb{C}\delta$ is a unital Banach algebra.

For each t in $\mathbb{R}$ we define $\gamma(t)$ in $\hat{\mathfrak{A}}$ by

$$\langle f + \lambda\delta, \gamma(t) \rangle = \int f(x)\exp(-ixt)dx + \lambda,$$

for $f + \lambda\delta$ in $\mathfrak{A}$. Furthermore, we define $\gamma(\infty)$ in $\hat{\mathfrak{A}}$ by

$$\langle f + \lambda\delta, \gamma(\infty) \rangle = \lambda.$$

This gives an injective map $\gamma: \mathbb{R} \cup \{\infty\} \to \hat{\mathfrak{A}}$. If $t_n \to t$ in $\mathbb{R}$, then $\gamma(t_n) \to \gamma(t)$ in w^*-topology by Lebesgue's dominated convergence theorem (6.1.15). If $t_n \to \infty$, then $\gamma(t_n) \to \gamma(\infty)$ in w^*-topology by Riemann–Lebesgue's lemma [or by approximation with Schwartz functions; see E 3.1.14(c)]. Regarding $\mathbb{R} \cup \{\infty\}$ as the one-point compactification of $\mathbb{R}$ (1.7.3), it follows that γ is continuous and therefore a homeomorphism of $\mathbb{R} \cup \{\infty\}$ into $\hat{\mathfrak{A}}$. Suppose that $\gamma_0 \in \hat{\mathfrak{A}}$, $\gamma_0 \neq \gamma(\infty)$, so that $\gamma_0 | L^1(\mathbb{R}) \neq 0$. Since $(L^1(\mathbb{R}))^* = L^\infty(\mathbb{R})$ by 6.5.11, there is an h in $L^\infty(\mathbb{R})$ with $\|h\|_\infty \leq 1$ such that $\langle f, \gamma_0 \rangle = \int f(x)h(x)\,dx$ for every f in $L^1(\mathbb{R})$. Set $_y f(x) = f(x - y)$ and note that the map $y \to {}_y f$ is continuous from $\mathbb{R}$ into $L^1(\mathbb{R})$ (6.6.19). For f and g in $L^1(\mathbb{R})$ we compute

$$\int \langle {}_y f, \gamma_0 \rangle g(y) dy = \int\int f(x - y)h(x)g(y)dxdy$$

$$= \int\int f(x - y)g(y)h(x)dydx = \langle f \times g, \gamma_0 \rangle$$

$$= \langle f, \gamma_0 \rangle \langle g, \gamma_0 \rangle = \langle f, \gamma_0 \rangle \int h(y)g(y)dy.$$

Since this holds for every g in $L^1(\mathbb{R})$, it follows that

$$\langle {}_y f, \gamma_0 \rangle = \langle f, \gamma_0 \rangle h(y) \tag{$*$}$$

for almost all y. Choosing f such that $\langle f, \gamma_0 \rangle \neq 0$ we see that the left-hand side of the equation $(*)$ depends continuously on y and therefore allows a choice of h in $C_b(\mathbb{R})$ that satisfies $(*)$ for every y in $\mathbb{R}$. Repeated use of $(*)$ yields

$$\langle f, \gamma_0 \rangle h(y + z) = \langle {}_{y+z} f, \gamma_0 \rangle = \langle {}_y({}_z f), \gamma_0 \rangle$$

$$= \langle {}_z f, \gamma_0 \rangle h(y) = \langle f, \gamma_0 \rangle h(y)h(z).$$

We conclude that $h(x) = \exp(-itx)$, $x \in \mathbb{R}$, for some t in $\mathbb{R}$, which shows that $\hat{\mathfrak{A}}$ is homeomorphic to $\mathbb{R} \cup \{\infty\}$.

The Gelfand transform on $L^1(\mathbb{R}) + \mathbb{C}\delta$ is determined by $\hat{\delta} = 1$ (the constant function) and

$$\hat{f}(t) = \int f(x)\exp(-itx)dx, \quad f \in L^1(\mathbb{R}).$$

In other words, Gelfand transformation is the classical Fourier transform. The same holds when $\mathbb{R}$ is replaced by any other locally compact abelian group G, e.g. $\mathbb{R}^n$, $\mathbb{T}^n$, or $\mathbb{Z}^n$. In all cases we obtain the Fourier transform of $L^1(G)$ into $C_0(\hat{G})$, where $\hat{G}$ is the dual group of G, consisting of the *characters* of G, i.e. the continuous group homomorphisms $\gamma: G \to \mathbb{T}$.

4.2.9. Remark. From the examples mentioned above we can trace the origin of the notion of spectrum. If we take $\mathfrak{A} = L^1(\mathbb{T})$, then $\hat{\mathfrak{A}} = \mathbb{Z}$, and $\hat{A}(n) = \int A(x)\exp(-inx)dx$ for every A in $L^1(\mathbb{T})$ and n in $\mathbb{Z}$. Thus, the spectrum of A is (by 4.2.3) the set of Fourier coefficients of A, in accordance with any reasonable terminology in harmonic analysis (musical or mathematical).

If $\mathfrak{A} = L^1(\mathbb{R})$, then $\hat{\mathfrak{A}} = \mathbb{R}$, and the spectrum of an element A in $L^1(\mathbb{R})$ is the set

$$\left\{ \int A(x)\exp(-ixy)dx \,\big|\, y \in \mathbb{R} \right\},$$

which again is exactly the information needed to perform a spectral analysis of A.

EXERCISES

E 4.2.1. Let $\mathfrak{A}$ be a nonunital, commutative Banach algebra and $\tilde{\mathfrak{A}}$ be any unital Banach algebra containing $\mathfrak{A}$ as an ideal of co-dimension one. We say that an ideal $\mathfrak{I}$ of $\mathfrak{A}$ is *regular* if $A - AE \in \mathfrak{I}$ for some E in $\mathfrak{A}$ and all A in $\mathfrak{A}$. Show that the map $\mathfrak{I} \to \mathfrak{I} \cap \mathfrak{A}$ induces a bijection between the maximal ideals of $\tilde{\mathfrak{A}}$, and the regular maximal ideals of $\mathfrak{A}$ together with the improper ideal $\mathfrak{A}$. Show that every regular maximal ideal in $\mathfrak{A}$ is closed and that every closed maximal ideal is regular.

E 4.2.2. Let $\mathfrak{A}$ be a nonunital, complex commutative Banach algebra. Show that the set $\hat{\mathfrak{A}}$ of continuous characters of $\mathfrak{A}$ has a locally compact Hausdorff topology such that the map $\Gamma: A \to \hat{A}$ given by $\Gamma(A)(\gamma) = \hat{A}(\gamma) = \langle A, \gamma \rangle$ is a norm decreasing homomorphism of $\mathfrak{A}$ onto a subalgebra of $C_0(\hat{\mathfrak{A}})$ that separates points in $\hat{\mathfrak{A}}$ and does not vanish identically at any point. Show that for every A in $\mathfrak{A}$ we have $\hat{A}(\hat{\mathfrak{A}}) \cup \{0\} = \mathrm{sp}(A)$ and $r(A) = \|\hat{A}\|_\infty$.
  Hint: Use E 4.2.1 to reduce the problem to the unital case (4.2.3).

E 4.2.3. Set $\mathbb{N}_0 = \mathbb{N} \cup \{0\}$ and consider the Banach space $\ell^1 = \ell^1(\mathbb{N}_0)$; cf. 2.1.18. Let (α_n), $n \in \mathbb{N}_0$, be a sequence in $\mathbb{R}_+$ such that $\alpha_n \alpha_m \leq \alpha_{n+m}$ for all n and m and $\alpha_0 = \alpha_1 = 1$. Show that the definition

$$(AB)_n = \alpha_n^{-1} \sum_{p=0}^{n} \alpha_p A_p \alpha_{n-p} B_{n-p}$$

for A and B in ℓ^1 gives a product under which ℓ^1 is a commutative, unital Banach algebra. Show that the sequence $(\alpha_n^{-1/n})$ is convergent with a limit ρ. If $\rho = 0$, show that the set $\mathfrak{I}$ consisting of those A in ℓ^1 for which $A_0 = 0$, is the only maximal ideal in ℓ^1. If $\rho > 0$, show that $(\ell^1)\hat{\,}$ is homeomorphic to the disk $B(0, \rho)$ in $\mathbb{C}$, and that the Gelfand transform is injective and takes ℓ^1 into the set of functions in $C(B(0, \rho))$ that are holomorphic in the interior. Show that

$$\hat{A}(z) = \sum_{n=0}^{\infty} A_n \alpha_n^{-1} z^n, \quad A \in \ell^1.$$

E 4.2.4. (*Laplace transform*.) Consider the Banach space $L^1(\mathbb{R}_+)$ of Lebesgue integrable functions on $[0, \infty[$. Show that the definition

$$f \times g(x) = \int_0^x f(y)g(x - y)dy, \quad f, g \in L^1(\mathbb{R}_+)$$

gives a product under which $L^1(\mathbb{R}_+)$ is a commutative, nonunital, Banach algebra. Show that $(L^1(\mathbb{R}_+))\hat{}$ is homeomorphic with $\mathbb{C}_+ = \{z \in \mathbb{C} | \operatorname{Re} z \geq 0\}$ and that the Gelfand transform is given by

$$\hat{f}(z) = \int_0^\infty f(x)\exp(-xz)dx, \quad f \in L^1(\mathbb{R}_+), z \in \mathbb{C}_+.$$

Show that $\hat{f}$ is holomorphic in the interior of $\mathbb{C}_+$ and belongs to $C_0(\mathbb{C}_+)$.

Hint: Use the nonunital Gelfand transform given in E 4.2.2, or adjoin the Dirac point measure at 0 as a unit. Then mimic the proof in 4.2.8.

E 4.2.5. Let $\mathfrak{A}$ be a commutative, complex, unital Banach algebra, and assume that there are elements (generators) $A_1, \ldots, A_n$ of $\mathfrak{A}$ such that the algebra generated by I and these elements is dense in $\mathfrak{A}$. Show that $\hat{\mathfrak{A}}$ is homeomorphic with a closed subset of the product space $\operatorname{sp}(A_1) \times \cdots \times \operatorname{sp}(A_n)$ in $\mathbb{C}^n$. Show in particular that if $n = 1$, then $\hat{\mathfrak{A}}$ is homeomorphic with $\operatorname{sp}(A_1)$.

Hint: Show that the map $\gamma \to (\gamma(A_1), \ldots, \gamma(A_n))$ is continuous and injective from $\hat{\mathfrak{A}}$ into $\mathbb{C}^n$.

E 4.2.6. Let $\Delta = \{z \in \mathbb{C} | |z| \leq 1\}$, and denote by $H(\Delta)$ the Banach algebra of functions in $C(\Delta)$ that are holomorphic in the interior of Δ. Show that the Gelfand spectrum $(H(\Delta))\hat{}$ is homeomorphic to Δ.

Hint: Use E 4.2.5 and (if needed) the functions f_ε given by $f_\varepsilon(z) = f((1 - \varepsilon)z)$. If $f \in H(\Delta)$, then $\|f - f_\varepsilon\|_\infty \to 0$, and f_ε can surely be approximated by a polynomial in the generator function id.

E 4.2.7. Let $C^n(I)$ denote the space of n-times continuously differentiable functions on the interval $I = [0, 1]$; cf. E 2.1.9. Show that $C^n(I)$ is a Banach algebra under the usual pointwise operations if the norm is defined by

$$\|f\| = \sup\left\{\sum_{k=0}^n (k!)^{-1} |f^{(k)}(x)| \,\Big|\, x \in I\right\}.$$

Show that the identical function id is a generator for $C^n(I)$ and find hereby the Gelfand spectrum of characters of $C^n(I)$.

Hint: Use E 4.2.5.

E 4.2.8. Let $\mathfrak{A}$ be the set of formal polynomials of degree at most n, i.e.

$$P(x) = \sum_{p=0}^n \alpha_p x^p.$$

Define sum and product as for functions, but with the convention that $x^p = 0$ if $p > n$. Show that $\mathfrak{A}$ is a (finite-dimensional) Banach algebra under the norm $\|P\| = \sum |\alpha_p|$. Show that $\mathfrak{A}$ has only one maximal ideal (equal to the radical of $\mathfrak{A}$, cf. 4.2.4), consisting entirely of nilpotent elements.

E 4.2.9. Let $\mathfrak{A}(S)$ denote the subalgebra of $\mathbf{M}_n$—the $n \times n$-matrices—generated by I and the matrix S, where $S_{ij} = 0$ if $j \neq i + 1$ and $S_{i,i+1} = 1$. Show that the assignment $x \to S$ extends to an isomorphism of the algebra $\mathfrak{A}$ defined in E 4.2.8 onto $\mathfrak{A}(S)$.

E 4.2.10. Show that the Gelfand transform $\Gamma: \mathfrak{A} \to C(\widehat{\mathfrak{A}})$ of a complex, commutative, unital Banach algebra $\mathfrak{A}$, is an isometry iff $\|A^2\| = \|A\|^2$ for every A in $\mathfrak{A}$.

E 4.2.11. Let $\mathfrak{A}$ and $\mathfrak{B}$ be complex, commutative, unital Banach algebras and $\Phi: \mathfrak{A} \to \mathfrak{B}$ a unit preserving algebra homomorphism. Show that the definition

$$\langle \Phi(A), \gamma \rangle = \langle A, \Phi_*(\gamma) \rangle, \quad A \in \mathfrak{A}, \gamma \in \widehat{\mathfrak{B}},$$

defines a map $\Phi_*: \widehat{\mathfrak{B}} \to \widehat{\mathfrak{A}}$. Show that Φ is automatically continuous if $\mathfrak{B}$ is semisimple (i.e. if ker $\Gamma = \{0\}$; cf. 4.2.4).
Hint: Use Φ_* to prove that Φ has closed graph.

E 4.2.12. Take $\mathfrak{A}$, $\mathfrak{B}$, Φ, and Φ_* as in E 4.2.11 and show that

(a) The continuity of Φ implies that of Φ_*.
(b) If $\Phi(\mathfrak{A})$ is dense in $\mathfrak{B}$, then Φ_* is injective.
(c) If Φ is continuous, $\Phi(\mathfrak{A})$ is dense in $\mathfrak{B}$, and $\mathbf{GL}(\mathfrak{B}) \cap \Phi(\mathfrak{A}) = \Phi(\mathbf{GL}(\mathfrak{A}))$, then Φ_* is a surjection of $\widehat{\mathfrak{B}}$ onto the set hull ker $\Phi = \{\gamma \in \widehat{\mathfrak{A}} | \text{ker } \Phi \subset \text{ker } \gamma\}$.

Hint: If $\gamma_0 \in$ (hull ker Φ)$\backslash \Phi_*(\widehat{\mathfrak{B}})$ use compactness to find $A_1, \ldots,$ A_n in ker γ_0 such that $\langle A_k, \Phi_*(\gamma) \rangle \neq 0$ for every γ in $\widehat{\mathfrak{B}}$ and some A_k (depending on γ). Deduce that the ideal $\mathfrak{B}\Phi(A_1) + \cdots + \mathfrak{B}\Phi(A_n)$ in $\mathfrak{B}$ is equal to $\mathfrak{B}$ and write $I = \sum B_k \Phi(A_k)$. Approximate the B_k's from $\Phi(\mathfrak{A})$ to find $C_1, \ldots, C_n$ in $\mathfrak{A}$ with $A = \sum C_k A_k \in$ ker γ_0 but $\Phi(A)$ invertible in $\mathfrak{B}$. By assumption, find C in $\mathbf{GL}(\mathfrak{A})$ such that $A - C \in$ ker Φ, and see the contradiction.

E 4.2.13. Show that if $\mathfrak{A}$ is a complex, commutative, unital, semisimple Banach algebra, then any new norm on $\mathfrak{A}$ which makes it into a Banach algebra is equivalent with the old norm. In particular, the Banach algebra norm with $\|I\| = 1$ is unique.
Hint: Use E 4.2.11.

4.3. Function Algebras

Synopsis. The Stone–Weierstrass theorem. Involution in Banach algebras. C^*-algebras. The characterization of commutative C^*-algebras. Stone-Čech compactification of Tychonoff spaces. Exercises.

4.3.1. Gelfand's theorem (4.2.3) gives one pertinent fact about the size of the image of a Banach algebra $\mathfrak{A}$ in $C(\widehat{\mathfrak{A}})$, namely, that the Gelfand transforms separate points in $\widehat{\mathfrak{A}}$. Under certain extra conditions this implies that the algebra $\Gamma(\mathfrak{A})$ of Gelfand transforms is dense in $C(\widehat{\mathfrak{A}})$. The relevant tool here is the *Stone–Weierstrass theorem* (4.3.4). This is at heart a theorem about real-valued functions; but the simple condition below also makes it applicable in the complex case.

We say that a space (usually an algebra) $\mathfrak{A}$ of (complex) functions is *self-adjoint* if $\bar{f} \in \mathfrak{A}$ for every f in $\mathfrak{A}$. Since $f = \frac{1}{2}(f + \bar{f}) + i(f - \bar{f})/2i$, this is equivalent to the condition that $\mathfrak{A} = \mathfrak{A}_{sa} + i\mathfrak{A}_{sa}$, where $\mathfrak{A}_{sa}$ denotes the set of real-valued functions in $\mathfrak{A}$.

4.3.2. Lemma. *Let $\mathfrak{A}$ be a vector space of continuous, real-valued functions on a compact Hausdorff space X. If $f \vee g$ and $f \wedge g$ belong to $\mathfrak{A}$ for all f and g in $\mathfrak{A}$, then every continuous function on X that can be approximated from $\mathfrak{A}$ in every pair of points in X can in fact be approximated uniformly from $\mathfrak{A}$.*

PROOF. Let f be a function that can be approximated as described above. For every $\varepsilon > 0$ and x, y in X there is thus an f_{xy} in $\mathfrak{A}$ such that both x and y are in the sets

$$U_{xy} = \{z \in X \mid f(z) < f_{xy}(z) + \varepsilon\},$$

$$V_{xy} = \{z \in X \mid f_{xy}(z) < f(z) + \varepsilon\}.$$

For fixed x and variable y the open sets U_{xy} cover X. Since X is compact, we can therefore find $y_1, \ldots, y_n$ such that $X = \bigcup U_{xy_k}$. By assumption $f_x = \bigvee f_{xy_k} \in \mathfrak{A}$, and we see that $f(z) < f_x(z) + \varepsilon$ for every z in X. At the same time $f_x(z) < f(z) + \varepsilon$ for every z in $W_x = \bigcap V_{xy_k}$, which is an open neighborhood of x. Varying now x we find $x_1, \ldots, x_m$ such that $X = \bigcup W_{x_k}$, and we have $f_\varepsilon = \bigwedge f_{x_k} \in \mathfrak{A}$ with

$$f_\varepsilon(z) - \varepsilon < f(z) < f_\varepsilon(z) + \varepsilon$$

for every z in X.   □

4.3.3. Lemma. *If $\mathfrak{A}$ is a uniformly closed algebra of continuous, bounded, real-valued functions on a topological space X, then $\mathfrak{A}$ is stable under the lattice operations $f \vee g$ and $f \wedge g$ in $C(X)$.*

PROOF. For $\varepsilon > 0$ the function $t \to (\varepsilon^2 + t)^{1/2}$ has a power series expansion that converges uniformly on $[0, 1]$ (using e.g. $t = \frac{1}{2}$ as expansion point). We can thus find a polynomial p such that $|(\varepsilon^2 + t)^{1/2} - p(t)| < \varepsilon$ for every t in $[0, 1]$, in particular, $p(0) < 2\varepsilon$. With $q(t) = p(t) - p(0)$ we know that $q(f) \in \mathfrak{A}$ for every f in $\mathfrak{A}$. Now take f in $\mathfrak{A}$ with $\|f\|_\infty \leq 1$. Then

$$\|q(f^2) - |f|\|_\infty = \sup |q(f^2(x)) - (f^2(x))^{1/2}|$$

$$\leq \sup_{0 \leq t \leq 1} |p(t) - p(0) - t^{1/2}| \leq 3\varepsilon + \sup_{0 \leq t \leq 1} |(t + \varepsilon^2)^{1/2} - t^{1/2}| \leq 4\varepsilon.$$

Since ε is arbitrary and $\mathfrak{A}$ is uniformly closed, it follows that $|f| \in \mathfrak{A}$ for every f in $\mathfrak{A}$. As

$$f \vee g = \tfrac{1}{2}(f + g + |f - g|), \qquad f \wedge g = \tfrac{1}{2}(f + g - |f - g|),$$

we immediately have the desired conclusions. $\square$

4.3.4. Theorem. *Let X be a compact Hausdorff space and $\mathfrak{A}$ be a self-adjoint subalgebra of $C(X)$ containing the constants and separating points in X. Then $\mathfrak{A}$ is uniformly dense in $C(X)$.*

PROOF. The uniform closure $\mathfrak{A}^=$ of $\mathfrak{A}$ is still a self-adjoint algebra, so the set $\mathfrak{A}_{sa}^=$ of real-valued functions from $\mathfrak{A}^=$ is a uniformly closed, real algebra. By 4.3.3 it is therefore a function lattice. Given x and y in X there is a g in $\mathfrak{A}$ such that $g(x) \neq g(y)$. Since both $\operatorname{Re} g$ and $\operatorname{Im} g$ belong to $\mathfrak{A}_{sa}$, we see that $\mathfrak{A}_{sa}^=$ separates points in X and contains the constants. Given a continuous, real-valued function f we can therefore choose h in $\mathfrak{A}_{sa}^=$ such that $h(x) \neq h(y)$, and then find a suitable linear combination f_{xy} of h and 1, such that $f_{xy}(x) = f(x)$ and $f_{xy}(y) = f(y)$. Thus $\mathfrak{A}_{sa}^=$ fulfills the assumptions in 4.3.2, whence $f \in \mathfrak{A}_{sa}^=$. Consequently,

$$C(X) = \mathfrak{A}_{sa}^= + i\mathfrak{A}_{sa}^= = \mathfrak{A}^=. \qquad \square$$

4.3.5. Corollary. *Let X be a locally compact Hausdorff space and $\mathfrak{A}$ be a self-adjoint subalgebra of $C_0(X)$ that separates points in X and does not vanish identically at any point of X. Then $\mathfrak{A}$ is uniformly dense in $C_0(X)$.*

PROOF. We compactify X (cf. 1.7.3), and embed $\mathfrak{A}$ and $C_0(X)$ into $C(X \cup \{\infty\})$. Then $\mathfrak{A} + \mathbb{C}1$ is a self-adjoint subalgebra of $C(X \cup \{\infty\})$ and separates points not only in X but in $X \cup \{\infty\}$. Indeed, if $x \in X$, there is by assumption a g in $\mathfrak{A}$ with $g(x) \neq 0$, whereas $g(\infty) = 0$ since $g \in C_0(X)$. By 4.3.4 there is to each f in $C_0(X)$ and $\varepsilon > 0$ a g in $\mathfrak{A}$ and λ in $\mathbb{C}$ such that $\|f - (g + \lambda 1)\|_\infty < \varepsilon$. As $f(\infty) = g(\infty) = 0$, we see that $|\lambda| < \varepsilon$, whence $\|f - g\|_\infty < 2\varepsilon$. $\square$

4.3.6. Remark. Stone's theorem (published, by request, in 1948) yields a number of classical approximation theorems as corollaries. Most important, of course, the seminal theorem of Weierstrass (1895) that every continuous, real-valued function on a closed, bounded interval can be approximated uniformly by polynomials. Note also that every continuous, periodic function on $[0, 2\pi]$ can be approximated uniformly by trigonometric polynomials (despite the fact that the Fourier series for the function need not be uniformly convergent). Finally, it should be emphasized that the demand in 4.3.4 of self-adjointness is necessary. The example $H(\Delta)$ of continuous functions on the closed unit disk Δ that are holomorphic in the interior of Δ (see 4.2.7) gives a proper, closed subalgebra of $C(\Delta)$ that separates points and contains the constants.

4.3.7. An *involution* on an algebra $\mathfrak{A}$ is a map $A \to A^*$ of $\mathfrak{A}$ onto itself of period two that is conjugate linear and antimultiplicative. As examples of involutions we mention complex conjugation of functions and the adjoint operation on $\mathbf{B}(\mathfrak{H})$ (3.2.3). We shall borrow the terminology developed in 3.2.4 and 3.2.7 wholesale. Likewise we shall talk about projections $(P = P^* = P^2)$, unitary elements $(U^*U = UU^* = I)$, and partial isometries $(U^*U = P)$ in the algebra; cf. 3.2.13, 3.2.15, and 3.2.16.

The involutions that naturally occur on Banach algebras are isometric, i.e. $\|A^*\| = \|A\|$ for every A in $\mathfrak{A}$. A considerably stronger demand on the involution is that it satisfies the equation

$$\|A^*A\| = \|A\|^2, \quad A \in \mathfrak{A}. \tag{$*$}$$

A Banach algebra $\mathfrak{A}$ with an involution satisfying $(*)$ is called a *C*-algebra*. Note that $(*)$ implies that $\|A\|^2 \leq \|A^*\|\,\|A\|$, whence $\|A\| \leq \|A^*\|$ and thus, by symmetry, $\|A\| = \|A^*\|$. We see immediately that $C_0(X)$ is a C^*-algebra (with complex conjugation as involution) for every locally compact Hausdorff space X. Moreover, it follows from 3.2.3 that $\mathbf{B}(\mathfrak{H})$ (with adjoint operation) is a C^*-algebra. Consequently, every closed, self-adjoint subalgebra of $\mathbf{B}(\mathfrak{H})$ [e.g. $\mathbf{B}_0(\mathfrak{H})$, see 3.3] is a C^*-algebra; and, as shown by Gelfand and Naimark (1943), every C^*-algebra is isometrically $*$-isomorphic to one of these.

Returning to the general theory, we say that an involution on $\mathfrak{A}$ is *symmetric* if $\mathrm{sp}(A) \subset \mathbb{R}$ for every self-adjoint A in $\mathfrak{A}$. This condition is fairly elusive and hard to express directly in topological and algebraical terms. The same can be said of the demand that $\mathrm{sp}(A^*A) \subset \mathbb{R}_+$ for every A in $\mathfrak{A}$—a condition that seems necessary if "positivity" is going to be a useful concept. Both of these conditions are, however, satisfied in a C^*-algebra.

4.3.8. Example. Let G be a locally compact, abelian group, for example, $\mathbb{R}^n$, $\mathbb{T}^n$, or $\mathbb{Z}^n$, $n \in \mathbb{N}$. Consider the Banach algebra $L^1(G)$ with the convolution product

$$f \times g(x) = \int f(y)g(x - y)dy;$$

cf. 4.2.6 and 4.2.8. On this algebra we have an involution, given by

$$f^*(x) = \bar{f}(-x), \quad f \in L^1(G).$$

The involution is isometric, but does not satisfy the C^*-condition $(*)$ in 4.3.7 (as will be evident from 4.3.13). However, the involution is symmetric, because every γ in $(L^1(G))\hat{\ }$ has the form

$$\langle f, \gamma \rangle = \int f(x)\exp(-itx)dx,$$

where t belongs to the dual (or character) group $\hat{G}$ of G (in our situation $\mathbb{R}^n$, $\mathbb{Z}^n$, or $\mathbb{T}^n$, $n \in \mathbb{N}$). Therefore,

$$\langle f^*, \gamma \rangle = \int \bar{f}(-x)\exp(-itx)dx$$

$$= \int \overline{f(x)\exp(-itx)}dx = \overline{\langle f, \gamma \rangle}.$$

If $f = f^*$, then $\langle f, \gamma \rangle \in \mathbb{R}$, whence $\mathrm{sp}(f) \subset \mathbb{R}$ by 4.2.2.

It follows from this that the algebra $A(G)$ of Fourier transforms is a self-adjoint subalgebra of $C_0(\hat{G})$ [identifying $(L^1(G))\hat{}$ with $\hat{G}$, cf. 4.2.8], and thus $A(G)$ is uniformly dense in $C_0(\hat{G})$ by 4.3.5. A similar density result is of course valid for any commutative Banach algebra with a symmetric involution.

4.3.9. Lemma. *To every nonunital C^*-algebra $\mathfrak{A}$ there is a unital C^*-algebra $\tilde{\mathfrak{A}}(=\mathfrak{A} + \mathbb{C}I)$, containing $\mathfrak{A}$ as a maximal ideal of co-dimension one.*

PROOF. The regular representation $\rho: \mathfrak{A} \to \mathbf{B}(\mathfrak{A})$ defined by $\rho(A)B = AB$, cf. 4.1.4, is always a norm decreasing homomorphism. But by (*) in 4.3.7 we have

$$\|A\| = \|AA^*\| \, \|A^*\|^{-1} = \|\rho(A)(A^*\|A^*\|^{-1})\| \leq \|\rho(A)\|,$$

so that ρ is actually an isometry. We define $\tilde{\mathfrak{A}} = \rho(\mathfrak{A}) + \mathbb{C}I$, equipped with the norm from $\mathbf{B}(\mathfrak{A})$ and the involution $(\rho(A) + \lambda I)^* = \rho(A^*) + \bar{\lambda}I$.

Given $\tilde{A} = \rho(A) + \lambda I$ in $\tilde{\mathfrak{A}}$ there is for each $\varepsilon > 0$ a B in $\mathfrak{A}$ with $\|B\| = 1$ such that

$$\|\tilde{A}\|^2 - \varepsilon \leq \|\tilde{A}B\|^2 = \|AB + \lambda B\|^2$$

$$= \|(AB + \lambda B)^*(AB + \lambda B)\| = \|(B^*A^* + \bar{\lambda}B^*)(AB + \lambda B)\|$$

$$= \|\rho(B^*)\tilde{A}^*\tilde{A}B\| \leq \|\tilde{A}^*\tilde{A}B\| \leq \|\tilde{A}^*\tilde{A}\|.$$

It follows that $\|\tilde{A}\|^2 \leq \|\tilde{A}^*\tilde{A}\|$. Since we always have $\|\tilde{A}^*\tilde{A}\| \leq \|\tilde{A}^*\| \, \|\tilde{A}\|$, we deduce first that $\|\tilde{A}\| \leq \|\tilde{A}^*\|$, whence by symmetry $\|\tilde{A}\| = \|\tilde{A}^*\|$, and then $\|\tilde{A}\|^2 = \|\tilde{A}^*\tilde{A}\|$. Thus $\tilde{\mathfrak{A}}$ is a C^*-algebra and [identifying $\mathfrak{A}$ with its image $\rho(\mathfrak{A})$ in $\tilde{\mathfrak{A}}$] it contains $\mathfrak{A}$ as an ideal of co-dimension one, therefore a maximal ideal. $\qquad\square$

4.3.10. We define the spectrum of an element A in a nonunital C^*-algebra $\mathfrak{A}$ to be the (usual) spectrum of A computed as an element of the algebra $\tilde{\mathfrak{A}}$ defined in 4.3.9. Note that this has the effect that $0 \in \mathrm{sp}(A)$ for every A in $\mathfrak{A}$, because $\mathfrak{A}$ is an ideal in $\tilde{\mathfrak{A}}$.

4.3.11. Lemma. *If A is a normal element in a C^*-algebra $\mathfrak{A}$, then $\mathrm{r}(A) = \|A\|$.*

PROOF. Assume first that $A = A^*$. Repeated applications of (*) in 4.3.7 show that $\|A^{2^n}\| = \|A\|^{2^n}$ for every n. In the general (normal) case we apply this identity to A^*A, and note that $(A^*A)^m = A^{*m}A^m$ by normality. Thus,

$$\|A\|^{2^n} = \|A^*A\|^{2^{n-1}} = \|A^{*2^n}A^{2^n}\|^{1/2}$$
$$= \|A^{2^n *}A^{2^n}\|^{1/2} = \|A^{2^n}\|.$$

Since $\|A^{2^n}\|^{2^{-n}} \to r(A)$ by 4.1.13, it follows that $\|A\| = r(A)$. □

4.3.12. Lemma. *If A is a self-adjoint element in a C^*-algebra $\mathfrak{A}$, then* $\mathrm{sp}(A) \subset \mathbb{R}$. *If $\mathfrak{A}$ is unital and U is unitary in $\mathfrak{A}$, then* $\mathrm{sp}(U) \subset \mathbb{T}$.

PROOF. Since $(T^{-1})^* = (T^*)^{-1}$ for every T in $\mathbf{GL}(\mathfrak{A})$, we see that $\lambda \in \mathrm{sp}(S)$ implies $\bar{\lambda} \in \mathrm{sp}(S^*)$ for every S in $\mathfrak{A}$. Furthermore, we see from the formula

$$\lambda^{-1}(\lambda I - T)T^{-1} = -(\lambda^{-1}I - T^{-1})$$

that if $\lambda \in \mathrm{sp}(T)$, then $\lambda^{-1} \in \mathrm{sp}(T^{-1})$. Now if U is unitary in $\mathfrak{A}$, then $U^* = U^{-1}$. This means that $\bar{\lambda}^{-1} \in \mathrm{sp}(U)$ for every λ in $\mathrm{sp}(U)$. Since $\|U\| = 1$, we conclude that $|\lambda| \leq 1$ and $|\lambda^{-1}| \leq 1$, whence $\lambda \in \mathbb{T}$.

If $A = A^* \in \mathfrak{A}$, we form the element

$$\exp iA = \sum_{0}^{\infty} (n!)^{-1}(iA)^n$$

(in $\tilde{\mathfrak{A}}$ if $I \notin \mathfrak{A}$). Multiplying the series (or appealing to the holomorphic spectral theory, cf. E 4.1.14–15) it follows that

$$(\exp iA)^* = \exp - iA = (\exp iA)^{-1},$$

so that $\exp iA$ is unitary. Now if $\lambda \in \mathrm{sp}(A)$, then $\exp i\lambda \in \mathrm{sp}(\exp iA)$ by 4.1.11, whence $|\exp i\lambda| = 1$ from the result above, and, consequently, $\lambda \in \mathbb{R}$. □

4.3.13. Theorem. *Every commutative, unital C^*-algebra $\mathfrak{A}$ is isometrically $*$-isomorphic to $C(\hat{\mathfrak{A}})$, where $\hat{\mathfrak{A}}$ is the compact Hausdorff space of characters of $\mathfrak{A}$.*

PROOF. Since every element in $\mathfrak{A}$ is normal, the Gelfand transform is an isometry by 4.2.3 and 4.3.11. If $A = A^*$, then for each γ in $\hat{\mathfrak{A}}$ we have

$$\hat{A}(\gamma) = \langle A, \gamma \rangle \in \mathrm{sp}(A) \subset \mathbb{R}$$

by 4.3.12, so that $\hat{A} \in C(\hat{\mathfrak{A}})_{sa}$. Since every T in $\mathfrak{A}$ can be written as $T = A + iB$ with A and B self-adjoint [take $A = \frac{1}{2}(T + T^*)$ and $B = \frac{1}{2}i(T^* - T)$], it follows that

$$\Gamma(T^*) = \Gamma(A - iB) = \Gamma(A) - i\Gamma(B)$$
$$= \overline{\Gamma(A) + i\Gamma(B)} = \overline{\Gamma(A + iB)} = \overline{\Gamma(T)},$$

so that the Gelfand transform is $*$-preserving. In particular, the image $\Gamma(\mathfrak{A})$ is a self-adjoint algebra of functions, whence $\Gamma(\mathfrak{A}) = C(\hat{\mathfrak{A}})$ by the Stone–Weierstrass theorem (4.3.4). □

4.3.14. Corollary. *Every commutative, nonunital C*-algebra* $\mathfrak{A}$ *is isometrically* **-isomorphic to* $C_0(\hat{\mathfrak{A}})$, *where* $\hat{\mathfrak{A}}$ *is the locally compact Hausdorff space of characters of* $\mathfrak{A}$.

PROOF. We consider $\mathfrak{A}$ as a maximal ideal of $\tilde{\mathfrak{A}}$, as described in 4.3.9, and we let γ_∞ denote the unique character of $\tilde{\mathfrak{A}}$ given by the quotient map $\tilde{\mathfrak{A}} \to \tilde{\mathfrak{A}}/\mathfrak{A}$. Every other character γ of $\tilde{\mathfrak{A}}$ must have a nonzero restriction to $\mathfrak{A}$ (otherwise ker $\gamma = \mathfrak{A}$ and $\gamma = \gamma_\infty$ by 4.2.2). Conversely, every character γ of $\mathfrak{A}$ extends uniquely to a character $\tilde{\gamma}$ of $\tilde{\mathfrak{A}}$. Identifying $\hat{\mathfrak{A}}$ with the locally compact Hausdorff space $(\tilde{\mathfrak{A}})\hat{~}\setminus\{\gamma_\infty\}$, we see from 4.3.13 that the Gelfand transform of $\tilde{\mathfrak{A}}$ takes $\mathfrak{A}$ isometrically and *-isomorphically into a subalgebra of $C_0(\hat{\mathfrak{A}})$ that separates points and that does not vanish identically at any point. Thus $\Gamma(\mathfrak{A}) = C_0(\hat{\mathfrak{A}})$ by 4.3.5. □

4.3.15. Proposition. *Let* T *be a normal element in a unital C*-algebra* $\mathfrak{A}$ *and denote by* $C^*(T)$ *the smallest C*-subalgebra of* $\mathfrak{A}$ *that contains* T *and* I. *There is then an isometric *-isomorphism* Φ *of* $C(\mathrm{sp}(T))$ *onto* $C^*(T)$, *such that* $\Phi(1) = I$ *and* $\Phi(id) = T$.

PROOF. Let $Pol(I, T, T^*)$ denote the *-subalgebra of $\mathfrak{A}$ consisting of polynomials in the three commuting variables I, T, and T^*. Evidently, then, $C^*(T) = (Pol(I, T, T^*))^=$; in particular, $C^*(T)$ is a commutative, unital C*-algebra. With $X = (C^*(T))\hat{~}$ it follows from 4.3.13 that the Gelfand transform Γ is an isometric *-isomorphism of $C^*(T)$ onto $C(X)$.

Denote by sp(T) the spectrum of T in $\mathfrak{A}$ and by sp*(T) the spectrum of T in $C^*(T)$. Since $C^*(T) \subset \mathfrak{A}$ we see that sp$(T) \subset$ sp*(T). By 4.2.3 the evaluation map $\gamma \to \langle T, \gamma \rangle$ is a surjection of X onto sp*(T), which is continuous since X has the w^*-topology as a subset of the dual of $C^*(T)$. Now note that if $\langle T, \gamma_1 \rangle = \langle T, \gamma_2 \rangle$, then

$$\langle T^*, \gamma_1 \rangle = \langle \overline{T, \gamma_1} \rangle = \langle \overline{T, \gamma_2} \rangle = \langle T^*, \gamma_2 \rangle,$$

and clearly $\langle I, \gamma_1 \rangle = 1 = \langle I, \gamma_2 \rangle$. Thus, γ_1 and γ_2 agree on $Pol(I, T, T^*)$ and, being continuous, also on its closure $C^*(T)$, whence $\gamma_1 = \gamma_2$. The evaluation map is therefore injective and thus a homeomorphism. Let Ψ denote the transposed map of $C(\mathrm{sp}^*(T))$ onto $C(X)$ given by

$$\Psi(f)(\gamma) = f(\langle T, \gamma \rangle), \quad f \in C(\mathrm{sp}^*(T)), \gamma \in X,$$

which is clearly an isometric *-isomorphism. Finally, put $\Phi = \Gamma^{-1} \circ \Psi$, which is an isometric *-isomorphism of $C(\mathrm{sp}^*(T))$ onto $C^*(T)$. For each γ in X we have

$$\Gamma(T)(\gamma) = \langle T, \gamma \rangle = id(\langle T, \gamma \rangle) = \Psi(id)(\gamma),$$

whence $\Gamma(T) = \Psi(id)$ and thus $T = \Phi(id)$. Likewise, $I = \Phi(1)$.

If $\lambda \in \mathrm{sp}^*(T)$, we can, for each $\varepsilon > 0$, find f in $C(\mathrm{sp}^*(T))$ with $\|f\|_\infty = 1$, such that $f(\lambda) = 1$ but $f(\mu) = 0$ whenever $|\lambda - \mu| \geq \varepsilon$. Take $A = \Phi(f)$ [in

$C^*(T)$] and compute

$$\|(T - \lambda I)A\| = \|\Phi^{-1}((T - \lambda I)A)\|_\infty = \|(id - \lambda)f\|_\infty \leq \varepsilon.$$

It follows that $T - \lambda I$ cannot be invertible in $\mathfrak{A}$ (the inverse would have to have a norm greater than ε^{-1}), so that $\lambda \in \text{sp}(T)$. This shows that $\text{sp}(T) = \text{sp}^*(T)$ and completes the proof. $\qquad\square$

4.3.16. Corollary. *If T is a normal element in a unital C^*-algebra $\mathfrak{A}$ and $\mathfrak{B}$ is any C^*-subalgebra of $\mathfrak{A}$ containing T and I, then the spectrum of T in $\mathfrak{B}$ is the same as the spectrum of T in $\mathfrak{A}$.*

PROOF. Because $C^*(T) \subset \mathfrak{B} \subset \mathfrak{A}$ gives

$$\text{sp}^*(T) \supset \text{sp}_\mathfrak{B}(T) \supset \text{sp}(T) = \text{sp}^*(T). \qquad\square$$

4.3.17. One way of interpreting Gelfand's theorem (4.3.13 + 4.3.14) is that it establishes a bijective correspondence between locally compact Hausdorff spaces (modulo homeomorphism) and commutative C^*-algebras (modulo ∗-isomorphism). Every topological phenomenon must therefore have an algebraic counterpart. As a striking application of this philosophy we mention the *Stone–Čech compactification* (4.3.18), which arises by attaching the largest possible unital C^*-algebra to a given topological space.

A *Tychonoff space* is a Hausdorff space X such that for every x in X and every open neighborhood A of x there is a continuous function $f: X \to [0,1]$ such that $f(x) = 1$ but $f|X \setminus A = 0$. Tychonoff spaces are also called *completely regular* spaces. They appear in 1.5.13 and (burdened with the second axiom of countability) in 1.6.14. Varying the proof of 1.6.14 or of 4.3.18 a little, it is not hard to see that X is a Tychonoff space iff it is a subset of the cube $[0,1]^\alpha$ for a suitable cardinal α (not necessarily countable). Of more immediate interest to us are the facts that normal spaces and locally compact Hausdorff spaces are Tychonoff spaces; cf. 1.5.6 and 1.7.5.

4.3.18. Proposition. *To each Tychonoff space X there is a Hausdorff compactification $\beta(X)$, with the property that every continuous function $\Phi: X \to Y$, where Y is a compact Hausdorff space, extends to a continuous function $\beta\Phi: \beta(X) \to Y$.*

PROOF. Let $C_b(X)$ denote the set of bounded, continuous (complex) functions on X, equipped with the pointwise algebraic operations and ∞-norm. Then $C_b(X)$ is a commutative, unital C^*-algebra and therefore (4.3.13) isometrically ∗-isomorphic to an algebra $C(\beta(X))$, where $\beta(X)$ is a compact Hausdorff space. The map $\iota: X \to \beta(X)$, given by

$$\langle f, \iota(x) \rangle = f(x), \quad x \in X, f \in C_b(X)$$

is evidently continuous. If A is an open subset of X, there is by assumption for every x in A an continuous function $f: X \to [0,1]$ that is 1 at x and 0 on

$X \setminus A$. Consequently,

$$\iota(x) \in \{\gamma \in \beta(X) | \langle f, \gamma \rangle > \tfrac{1}{2}\} \cap \iota(X) \subset \iota(A),$$

which shows that $\iota(A)$ is relatively open in $\iota(X)$. Thus ι is an open map; and with $A = X \setminus \{y\}$ we see from the above that $\iota(x) \neq \iota(y)$ if $x \neq y$, so that ι is a homeomorphism. If $\iota(X)$ was not dense in $\beta(X)$, there would be (1.5.6 + 1.6.6) a nonzero continuous function f on $\beta(X)$ vanishing on $\iota(X)$. Since $C(\beta(X)) = C_b(X)$, this is impossible, and thus $\beta(X)$ is a compactification of X; cf. 1.7.2.

Given a map $\Phi: X \to Y$ we define $\Phi^*: C(Y) \to C_b(X)$ by

$$(\Phi^* f)(x) = f(\Phi(x)), \quad x \in X, f \in C(Y).$$

It is easy to verify that Φ^* is a normdecreasing $*$-homomorphism. We now define $\beta\Phi$ by

$$\langle f, \beta\Phi(\gamma) \rangle = \langle \Phi^* f, \gamma \rangle, \quad \gamma \in \beta(X), f \in C(Y),$$

recalling that $\beta(X)$ and Y are the character spaces for the algebras $C_b(X)$ and $C(Y)$. Since the topology on these spaces is the w^*-topology, it follows that $\beta\Phi$ is continuous. Note finally that if $x \in X$, then for every f in $C(Y)$ we have

$$\langle f, \Phi(x) \rangle = f(\Phi(x)) = (\Phi^* f)(x)$$
$$= \langle \Phi^* f, \iota(x) \rangle = \langle f, \beta\Phi\iota(x) \rangle,$$

so that $\beta\Phi$ is indeed an extension of Φ. $\square$

4.3.19. Corollary. *To every Hausdorff compactification Y of X there is a surjective, continuous map $\Phi: \beta X \to Y$ such that $\Phi | X$ is the identical map.*

EXERCISES

E 4.3.1. Let X and Y be compact Hausdorff spaces. Show that the set of linear combinations of functions $f \otimes g$, where $f \in C(X)$, $g \in C(Y)$, and $f \otimes g(x, y) = f(x)g(y)$, is dense in $C(X \times Y)$.

E 4.3.2. Show that the set of piecewise linear continuous functions on an interval $I = [a, b]$ is uniformly dense in $C(I)$.
   Hint: Use 4.3.2.

E 4.3.3. Let $\mathfrak{A}$ denote the set of linear combinations of functions f in $C_0(\mathbb{R})$ of the form

$$f(x) = (x - z)^{-n}, \quad z \in \mathbb{C} \setminus \mathbb{R}, n \in \mathbb{N}.$$

Show that $\mathfrak{A}$ is dense in $C_0(\mathbb{R})$. Then show that the subset $\mathfrak{A}_1$, spanned by those functions f for which $n = 1$, is still dense in $C_0(\mathbb{R})$.
   Hints: Note that $\mathfrak{A}$ is an algebra and use 4.3.5. Show that for each z in $\mathbb{C} \setminus \mathbb{R}$ and $\varepsilon > 0$ there are pairwise distinct numbers $z_1, \ldots, z_n$ in $\mathbb{C} \setminus \mathbb{R}$ such that

$$\left| (x - z)^{-n} - \prod_{n=1}^{n} (x - z_k)^{-1} \right| \leq \varepsilon$$

for all x in $\mathbb{R}$, and use the fact that the product $\prod(x - z_k)^{-1}$ belongs to $\mathfrak{A}_1$.

E 4.3.4. Consider the Tychonoff cube $T = [0,1]^{\mathbb{N}}$, defined in 1.6.13. Define a multiindex to be an element p in $(\mathbb{N}_0)^{\mathbb{N}}$, such that $p(n) = 0$ except for finitely many n in $\mathbb{N}$, and for each $x = (x_n)$ in T define the monomial $x^p = \prod x_n^{p(n)}$. Show that the set of polynomials $\sum \alpha_p x^p$, where $\alpha_p = 0$ except for finitely many multiindices p, is uniformly dense in $C(T)$.

E 4.3.5. Show that every involution on a commutative, complex, semisimple, unital Banach algebra $\mathfrak{A}$ is continuous.
　　Hint: For every γ in $\hat{\mathfrak{A}}$ define γ^* in $\hat{\mathfrak{A}}$ by $\langle A, \gamma^* \rangle = \overline{\langle A^*, \gamma \rangle}$. Use this to show that the map $A \to A^*$ (regarded as a real linear map of the real Banach space $\mathfrak{A}$) has a closed graph.

E 4.3.6. Show that the following conditions on a self-adjoint element A in a unital C^*-algebra $\mathfrak{A}$ are equivalent:

　(i) $\mathrm{sp}(A) \subset \mathbb{R}_+$.
　(ii) $\|tI - A\| \leq t$ for every $t \geq \|A\|$.
　(iii) $\|tI - A\| \leq t$ for some $t \geq \|A\|$.

　　Hint: Use that $\mathrm{sp}(tI - A) = t - \mathrm{sp}(A)$ and that $\|tI - A\| = r(tI - A)$.

E 4.3.7. The self-adjoint elements A in a unital C^*-algebra $\mathfrak{A}$ satisfying the conditions in E 4.3.6 are called *positive*, in symbols $A \geq 0$. Show that the set of positive elements in $\mathfrak{A}$ form a closed cone.
　　Hint: If $A \geq 0$ and $B \geq 0$, use E 4.3.6 (ii) + (iii) to prove that $A + B \geq 0$, computing

$$\|(\|A\| + \|B\|)I - (A + B)\| = \|(\|A\|I - A) + (\|B\|I - B)\|$$
$$\leq \|\, \|A\|I - A\| + \|\, \|B\|I - B\|$$
$$\leq \|A\| + \|B\|.$$

E 4.3.8. Show that an element A in a unital C^*-algebra $\mathfrak{A}$ is positive (cf. E 4.3.7) iff $A = B^*B$ for some B in $\mathfrak{A}$.
　　Hints: If $A \geq 0$, use the function calculus in $C^*(A)$ $[= C(\mathrm{sp}(A))]$ to write $A = (A^{1/2})^2$. If $A = B^*B$, use the functions $t \to t \vee 0$ and $t \to -(t \wedge 0)$ on $\mathrm{sp}(A)$ $(\subset \mathbb{R})$ to write $A = A_+ - A_-$, a difference between two positive, orthogonal elements. Put $T = BA_-$ and show that $T^*T = -(A_-)^3 \leq 0$. Show further that if $T = H + iK$ with $H = H^*$ and $K = K^*$, then

$$TT^* = 2H^2 + 2k^2 + (-T^*T) \geq 0,$$

by E 4.3.7. Now use E 4.1.3 to prove that $\mathrm{sp}(T^*T) = \mathrm{sp}(TT^*) = \{0\}$, whence $T^*T = 0 = A_-$, i.e. $A = A_+ \geq 0$.

E 4.3.9. Let $\Phi: \mathfrak{A} \to \mathfrak{B}$ be a *-homomorphism between unital C^*-algebras such that $\Phi(I) = I$. Show that Φ is norm decreasing. Show further that Φ is an isometry if it is injective.

Hints: Use the algebraic fact that $\mathrm{sp}(\Phi(A)) \subset \mathrm{sp}(A)$ to estimate

$$\|\Phi(A)\|^2 = \|\Phi(A)^*\Phi(A)\| = \|\Phi(A^*A)\| = \mathrm{r}(\Phi(A^*A))$$
$$\leq \mathrm{r}(A^*A) = \|A^*A\| = \|A\|^2.$$

To prove isometry when $\ker \Phi = \{0\}$ use the idea above to reduce the problem to positive (hence normal) elements. Using 4.3.15, reduce the isometry problem to the case where $\Phi: C(X) \to C(Y)$ is an injective *-homomorphism (with dense range) and X and Y are compact Hausdorff spaces. Show in this case that the induced dual map $\Phi_*: Y \to X$ (cf. E 4.2.11) is surjective (actually a homeomorphism), whence Φ is isometric.

E 4.3.10. For a separable Hilbert space $\mathfrak{H}$, consider the Calkin algebra $\mathbf{C}(\mathfrak{H}) = \mathbf{B}(\mathfrak{H})/\mathbf{B}_0(\mathfrak{H})$, cf. 3.3.10, and the quotient map $\Phi: \mathbf{B}(\mathfrak{H}) \to \mathbf{C}(\mathfrak{H})$. Show that $\mathbf{C}(\mathfrak{H})$ is a C^*-algebra with the quotient norm and the involution defined by $\Phi(T)^* = \Phi(T^*)$, $T \in \mathbf{B}(\mathfrak{H})$.

Hint: Show that $\|\Phi(T)\| = \lim \|T(I - P_n)\|$, where P_n is the projection on the span of the first n vectors in an orthonormal basis for $\mathfrak{H}$. Recall from the proof of 3.3.3 that $\|A - AP_n\| \to 0$ for every A in $\mathbf{B}_0(\mathfrak{H})$.

E 4.3.11. For each f in $C(\mathbb{T})$, let T_f denote the Toeplitz operator in $\mathbf{B}(H^2)$; cf. E 3.3.14. Show that the map $f \to T_f$ is an isometry from $C(\mathbb{T})$ into $\mathbf{B}(H^2)$, and that $\mathrm{r}(T_f) = \|T_f\|$ for every f.

Hint: Let $\Phi: \mathbf{B}(H^2) \to \mathbf{C}(H^2)$ be the quotient map as in E 4.3.10. Use E 3.3.15 and E 3.3.16 to show that the map $f \to \Phi(T_f)$ is an injective *-homomorphism from one C^*-algebra into another (cf. E 4.3.10), and deduce from E 4.3.9 that $\|\Phi(T_f)\| = \|f\|$ for every f in $C(\mathbb{T})$ and that $\Phi(T_f)$ is a normal element in $\mathbf{C}(H^2)$.

E 4.3.12. (*Positive functionals.*) A functional $\varphi: \mathfrak{A} \to \mathbb{C}$ on a unital C^*-algebra $\mathfrak{A}$ is *positive*, in symbols $\varphi \geq 0$, if $\varphi(A) \geq 0$ for every $A \geq 0$ in $\mathfrak{A}$. Show in this case that φ satisfies the Cauchy–Schwarz inequality

$$|\varphi(B^*A)|^2 \leq \varphi(B^*B)\varphi(A^*A), \quad A, B \text{ in } \mathfrak{A},$$

and that φ is bounded with $\|\varphi\| = \varphi(I)$.

Hint: Take λ in $\mathbb{C}$ and use the fact that

$$|\lambda|^2 \varphi(A^*A) + 2\mathrm{Re}\,\lambda\varphi(B^*A) + \varphi(B^*B) = \varphi((\lambda A + B)^*(\lambda A + B)) \geq 0.$$

E 4.3.13. Show that a continuous functional φ on a unital C^*-algebra is positive (cf. E 4.3.12) if $\|\varphi\| = \varphi(I)$.

Hint: Show that if $A = A^*$, then $\|A + inI\|^2 = \|A\|^2 + n^2$ for every n. Use this to prove that if $\varphi(A) = \alpha + i\beta$, then $|\varphi(A + inI)|^2 = \alpha^2 + (\beta + n\|\varphi\|)^2$, which forces $\beta = 0$. Knowing that φ is self-adjoint, take $A \geq 0$ and use E 4.3.6 (ii) to get $t\|\varphi\| - \varphi(A) = \varphi(tI - A) \leq t\|\varphi\|$, i.e. $\varphi(A) \geq 0$.

E 4.3.14. Let $\mathfrak{B}$ be a C^*-subalgebra of a unital C^*-algebra $\mathfrak{A}$ such that $I \in \mathfrak{B}$. Show that every positive functional φ on $\mathfrak{B}$ extends to a positive functional $\tilde{\varphi}$ on $\mathfrak{A}$ such that $\|\tilde{\varphi}\| = \|\varphi\|$.
   Hint: Use 2.3.3 and E 4.3.13.

E 4.3.15. Let A be a normal element in a unital C^*-algebra $\mathfrak{A}$. Show that for each λ in $\mathrm{sp}(A)$ there is a positive functional φ in $\mathfrak{A}$ with $\|\varphi\| = 1$ such that $\varphi(A) = \lambda$.
   Hint: Define φ_λ on $C^*(A)$ by $\varphi_\lambda(B) = \hat{B}(\lambda)$, $B \in C^*(A)$; cf. 4.3.15. Then use E 4.3.14 to get $\varphi = \tilde{\varphi}_\lambda$.

E 4.3.16. (*The Gelfand–Naimark–Segal construction.*) Let φ be a positive functional (cf. E 4.3.12) on a unital C^*-algebra $\mathfrak{A}$. Define

$$\mathfrak{L} = \{A \in \mathfrak{A} \,|\, \varphi(A^*A) = 0\}.$$

Use the Cauchy–Schwarz inequality (E 4.3.12) to show that $A \in \mathfrak{L}$ iff $\mathfrak{A}A \subset \ker \varphi$, and deduce that $\mathfrak{L}$ is a closed left ideal in $\mathfrak{A}$. Show that the quotient space $\mathfrak{A}/\mathfrak{L}$ is a pre-Hilbert space (3.1.4) with the inner product

$$(A + \mathfrak{L} | B + \mathfrak{L}) = \varphi(B^*A), \quad A, B \in \mathfrak{A}.$$

If $\mathfrak{H}_\varphi$ denotes the completion of $\mathfrak{A}/\mathfrak{L}$, show that the definition

$$\Phi(A)(B + \mathfrak{L}) = AB + \mathfrak{L}, \quad A, B \in \mathfrak{A},$$

defines a norm decreasing $*$-homomorphism Φ of $\mathfrak{A}$ as operators on $\mathfrak{A}/\mathfrak{L}$, thus by continuity a $*$-homomorphism $\Phi \colon \mathfrak{A} \to \mathbf{B}(\mathfrak{H}_\varphi)$. Finally show that

$$\ker \Phi = \{A \in \mathfrak{A} \,|\, \mathfrak{A}A\mathfrak{A} \subset \ker \varphi\}.$$

E 4.3.17. Let $\mathfrak{A}$ be a unital C^*-algebra and $\{\mathfrak{H}_j | j \in J\}$ a family of Hilbert spaces. Assume that for every j in J there is a $*$-homomorphism (normdecreasing by E 4.3.9) $\Phi_j \colon \mathfrak{A} \to \mathbf{B}(\mathfrak{H}_j)$. Let $\mathfrak{H}$ denote the orthogonal sum of the $\mathfrak{H}_j$'s (3.1.5) and show that there is a (unique) $*$-homomorphism $\Phi \colon \mathfrak{A} \to \mathbf{B}(\mathfrak{H})$ such that $P_j(\Phi(A)x) = \Phi_j(A)P_j(x)$, $A \in \mathfrak{A}$, $x \in \mathfrak{H}$, where (as usual) P_j is the projection of $\mathfrak{H}$ onto $\mathfrak{H}_j$. Show that

$$\ker \Phi = \bigcap \ker \Phi_j.$$

E 4.3.18. (Abstract versus concrete C^*-algebras.) Show that for every unital C^*-algebra $\mathfrak{A}$ there is a Hilbert space $\mathfrak{H}$ and an isometric $*$-isomorphism Φ of $\mathfrak{A}$ into $\mathbf{B}(\mathfrak{H})$.

Hints: Use E4.3.16 to obtain some ∗-homomorphisms. Add
them up, using E4.3.17. Show that you can get an injective
∗-homomorphism, using E4.3.15, and conclude that it is an isometry by E4.3.9.

4.4. The Spectral Theorem, I

Synopsis. Spectral theory with continuous function calculus. Spectrum versus
eigenvalues. Square root of a positive operator. The absolute value of an
operator. Positive and negative parts of a self-adjoint operator. Fuglede's
theorem. Regular equivalence of normal operators. Exercises.

4.4.1. Theorem. *Let $\mathfrak{H}$ be a complex Hilbert space and T be a normal operator
in $\mathbf{B}(\mathfrak{H})$ with spectrum $\mathrm{sp}(T)$. If $C^*(T)$ denotes the smallest norm closed,
∗-invariant subalgebra of $\mathbf{B}(\mathfrak{H})$ containing T and I, there is an isometric
∗-isomorphism $f \to f(T)$ of $C(\mathrm{sp}(T))$ onto $C^*(T)$ such that $\mathrm{id}(T) = T$ and
$1(T) = I$. Moreover, $\mathrm{sp}(f(T)) = f(\mathrm{sp}(T))$ for every function f in $C(\mathrm{sp}(T))$.*

PROOF. Since $\mathbf{B}(\mathfrak{H})$ is a C^*-algebra, the theorem is just a reformulation of 4.3.15.
□

4.4.2. Remark. The *spectral theorem* above was proved by Hilbert (around
1906) for a self-adjoint operator (disguised as a bounded quadratic form in
infinitely many variables), whereas the complete result for a normal (and
possibly unbounded) operator is due to von Neumann. The proof given here,
where the spectral theorem appears as a corollary to Gelfand's characterization of commutative C^*-algebras (4.3.13), is not the shortest, but it provides
a deeper insight in operator theory. For example, we immediately see how the
result would generalize to an arbitrary commuting family $\{T_j | j \in J\}$ of normal
operators: There is a compact Hausdorff space X [homeomorphic to a closed
subset of $\prod \mathrm{sp}(T_j)$], a family $\{f_j | j \in J\}$ in $C(X)$ separating the points in X, and
an isometric ∗-isomorphism of $C(X)$ onto $C^*\{T_j | j \in J\}$ that sends f_j to T_j for
every j in J. Even though it is not, in general, possible to describe the set X
in any detail, this generalization is sometimes valuable, because computations
with operators are replaced by manipulations with functions.

4.4.3. Comparing the spectral theorem in 4.4.1 with the spectral theorem in
3.3.9 for normal compact operators, one may well ask what sort of subsets of
$\mathbb{C}$ can appear as spectra for operators, and to what extent we may replace
spectral points with eigenvalues.

Let X be an arbitrary compact subset of $\mathbb{C}$. Choosing a dense sequence (λ_n)
in X we define a diagonal operator M on ℓ^2, equipped with the usual orthonormal basis (e_n), by

$$M(\sum \alpha_n e_n) = \sum \lambda_n \alpha_n e_n, \quad (\alpha_n) \in \ell^2.$$

Every eigenvalue is in the spectrum by definition (4.1.10), and since the spectrum is also closed, we have $\mathrm{sp}(M) \supset X$. However, if $\lambda \notin X$ we can define

$$(\lambda I - M)^{-1} \left(\sum \alpha_n e_n \right) = \sum (\lambda - \lambda_n)^{-1} \alpha_n e_n$$

as a bounded operator in $\mathbf{B}(\ell^2)$, because $|\lambda - \lambda_n| \geq \varepsilon$ for some $\varepsilon > 0$ and all n. Thus $\mathrm{sp}(M) = X$.

An operator completely without eigenvalues is obtained by choosing a compact subset X of $\mathbb{C}$ (or of $\mathbb{R}$) such that $(X^\circ)^- = X$. With $\mathfrak{H} = L^2(X)$ with respect to Lebesgue measure on $\mathbb{R}^2$ (or on $\mathbb{R}$), we define M_{id} in $\mathbf{B}(\mathfrak{H})$ by

$$M_{id}f(\lambda) = \lambda f(\lambda), \quad \lambda \in X, f \in \mathscr{L}^2(X). \tag{$*$}$$

It is immediately verified that $(M_{id}^* f)(\lambda) = \bar{\lambda} f(\lambda)$, so that M_{id} is normal (and even self-adjoint if $X \subset \mathbb{R}$). If $\lambda_0 \notin X$, we can define $(\lambda_0 I - M_{id})^{-1}$ in $\mathbf{B}(\mathfrak{H})$ by

$$((\lambda_0 I - M_{id})^{-1} f)(\lambda) = (\lambda_0 - \lambda)^{-1} f(\lambda).$$

If $\lambda_0 \in X$, then $(\lambda_0 I - M_{id})^{-1}$ is not bounded: Take f to be the characteristic function for the disk $X_0 = B(\lambda_0, \varepsilon)$ multiplied by $m(X_0)^{-1/2}$ [note that the measure $m(X_0) \neq 0$]. Then we have the contradiction

$$1 = \|f\|_2 \leq \|(\lambda_0 I - M_{id})^{-1}\| \, \|(\lambda_0 I - M_{id})f\|_2 \leq \|(\lambda_0 I - M_{id})^{-1}\| \varepsilon.$$

Thus $\mathrm{sp}(M_{id}) = X$. It is quite obvious from $(*)$ that M_{id} has no eigenvalues, because the Lebesgue measure of any point is zero.

Employing other continuous (diffuse) measures (see 6.4.4) it is possible in the same manner to find an operator M_{id} without eigenvalues, whose spectrum is any given compact subset of $\mathbb{C}$ without isolated points. As the next results show, these examples are worst possible.

4.4.4. Lemma. *If $T \in \mathbf{B}(\mathfrak{H})$ and $\lambda \in \mathrm{sp}(T)$, there is a sequence (x_n) of unit vectors in $\mathfrak{H}$ such that either $\|Tx_n - \lambda x_n\| \to 0$ or $\|T^*x_n - \bar{\lambda} x_n\| \to 0$.*

PROOF. If not, then both $T - \lambda I$ and $T^* - \bar{\lambda} I$ are bounded away from zero, whence $T - \lambda I$ is invertible by 3.2.6, a contradiction. $\qquad \square$

4.4.5. Proposition. *If T is a normal operator in $\mathbf{B}(\mathfrak{H})$ and $\lambda \in \mathrm{sp}(T)$, then for every $\varepsilon > 0$ there is a unit vector x in $\mathfrak{H}$ with $\|Tx - \lambda x\| < \varepsilon$. If λ is an isolated point in $\mathrm{sp}(T)$, it is an eigenvalue.*

PROOF. The first statement follows from 4.4.4, since $T - \lambda I$ and $T^* - \bar{\lambda} I$ are metrically identical. To prove the second, define a continuous function f on $\mathrm{sp}(T)$ by setting $f(\lambda) = 1$ and $f(\mu) = 0$ for all μ in $\mathrm{sp}(T) \backslash \{\lambda\}$. Then by 4.4.1,

$$\|(\lambda I - T)f(T)\| = \|(\lambda 1 - id)f\|_\infty = 0,$$

because the ∞-norm is computed over the set $\mathrm{sp}(T)$. Since $\lambda \in \mathrm{sp}(T)$, we have

$f(T) \neq 0$, so $x = f(T)y \neq 0$ for some y in $\mathfrak{H}$. Consequently,

$$(\lambda I - T)x = (\lambda I - T)f(T)y = 0. \qquad \square$$

4.4.6. Remark. The function f in the proof above takes only the values 0 and 1 on sp(T); so $f(T)$ is a projection. Thus we may forget about the Hilbert space setting and observe that if λ is an isolated point in the spectrum of a normal element T in a C^*-algebra $\mathfrak{A}$, there is a self-adjoint idempotent (a projection) P in $\mathfrak{A}$ such that $PT = TP = \lambda P$.

In the same manner, the following results are valid in a general C^*-algebra, and just for convenience stated here for the algebra $\mathbf{B}(\mathfrak{H})$. Since every abstract C^*-algebra has a concrete representation as a C^*-subalgebra of some $\mathbf{B}(\mathfrak{H})$ (see E 4.3.18), we have not really lost any information by this procedure.

4.4.7. Proposition. *A normal operator T in $\mathbf{B}(\mathfrak{H})$ is self-adjoint (respectively, positive) iff* sp(T) $\subset \mathbb{R}$ *(respectively,* sp(T) $\subset \mathbb{R}_+$).

PROOF. That $T = T^*$ implies sp(T) $\subset \mathbb{R}$ and $T \geq 0$ implies sp(T) $\subset \mathbb{R}_+$ follows immediately from 4.4.5. Conversely, if T is normal and sp(T) $\subset \mathbb{R}$, then $id = \overline{id}$ on sp(T), whence $T = T^*$ by 4.4.1. If, moreover, sp(T) $\subset \mathbb{R}_+$, then $f(\lambda) = \lambda^{1/2}$ defines a continuous function on sp(T), hence an element $f(T)$ in $\mathbf{B}(\mathfrak{H})$. We have $f(T) = f(T)^*$ and $f(T)^2 = T$, so for every x in $\mathfrak{H}$,

$$(Tx|x) = (f(T)^2 x|x) = \|f(T)x\|^2 \geq 0. \qquad \square$$

4.4.8. Proposition. *To each positive operator T in $\mathbf{B}(\mathfrak{H})$ there is a unique positive operator, denoted by $T^{1/2}$, satisfying $(T^{1/2})^2 = T$. Moreover, $T^{1/2}$ commutes with every operator commuting with T.*

PROOF. Since sp(T) $\subset \mathbb{R}_+$, we can define $T^{1/2} = f(T)$, where $f(\lambda) = \lambda^{1/2}$ as in the proof of 4.4.7. Since $f \geq 0$, we have $T^{1/2} \geq 0$ and clearly $(T^{1/2})^2 = T$. Since f is a uniform limit of polynomials by 4.3.4 (or just by Weierstrass' original result), it follows that $T^{1/2}$ is a limit of polynomials in T; so $T^{1/2}$ commutes with every element commuting with T.

If $S \in \mathbf{B}(\mathfrak{H})$, $S \geq 0$, such that $S^2 = T$, then $T \in C^*(S)$, whence $T^{1/2} \in C^*(S)$ by construction. On sp(S) the element $T^{1/2}$ is represented by a positive, continuous function h such that $h^2(\lambda) = \lambda^2$ for each λ in sp(S) [because $(T^{1/2})^2 = S^2$]. Since sp(S) $\subset \mathbb{R}_+$, it follows that $h = id$, whence $T^{1/2} = S$. $\qquad \square$

4.4.9. Proposition. *To each self-adjoint operator T in $\mathbf{B}(\mathfrak{H})$ there is a unique pair of positive operators, denoted by T_+ and T_-, such that $T_+ T_- = 0$ and $T = T_+ - T_-$. Moreover, T_+ and T_- commute with every operator commuting with T.*

PROOF. Since sp(T) $\subset \mathbb{R}$, we can define T_+ and T_- as the elements in $C^*(T)$ corresponding to the positive functions $id_+: \lambda \to \lambda \vee 0$ and $id_-: \lambda \to -(\lambda \wedge 0)$.

Since $id_+ id_- = 0$ and $id_+ - id_- = id$, the elements T_+ and T_- have the required properties; and since they belong to $C^*(T)$, they commute with every operator commuting with T.

If $A \geq 0$, $B \geq 0$, $AB = 0$, and $A - B = T$, then A and B commute with T, hence with T_+ and T_-. Put $S = T_+ - A = T_- - B$. Then $S = S^*$ and

$$0 \leq S^2 = (T_+ - A)(T_- - B) = -(T_+ B + T_- A) \leq 0.$$

Thus $S = 0$, so $T_+ = A$ and $T_- = B$.   □

4.4.10. Recall from 3.2.17 that we defined $|T| = (T^*T)^{1/2}$ for any T in $\mathbf{B}(\mathfrak{H})$. If T is normal, we see from (the proof of) 4.4.8 that $|T| \in C^*(T)$, and that (very properly) it corresponds to the function $|id|: \lambda \to |\lambda|$ on $\mathrm{sp}(T)$. In particular, if $T = T^*$, we have $|T| = T_+ + T_-$.

4.4.11. In the algebra $\mathbf{M}_n$ $[= \mathbf{B}(\mathbb{C}^n)]$ an operator S commutes with a given normal operator T iff S leaves invariant all the eigenspaces for T. Since T and T^* have the same eigenspaces, it therefore follows that S commutes with T^* whenever it commutes with T. The corresponding result—*Fuglede's theorem* (1950)—for an infinite-dimensional Hilbert space is much less obvious, since the spectrum no longer (necessarily) can be described by eigenvalues. The following elegant proof (by Rosenblum) of Fuglede's theorem may convince the reader that complex function theory is a very sharp tool—also in very abstract settings.

4.4.12. Proposition. *If S and T are operators in $\mathbf{B}(\mathfrak{H})$ and T is normal, then $ST = TS$ implies $ST^* = T^*S$.*

PROOF. For each λ in $\mathbb{C}$ we define

$$\exp(\lambda T) = \sum_{n=0}^{\infty} (n!)^{-1}(\lambda T)^n, \qquad (*)$$

and we note that $\exp(\lambda T) \in C^*(T)$ by 4.4.1. Likewise $\exp(\lambda T^*) \in C^*(T)$, and by the function calculus

$$\exp(\lambda T^*) = \exp(\lambda T^* - \bar{\lambda} T)\exp(\bar{\lambda} T).$$

Note now that $\lambda T^* - \bar{\lambda} T = iR$ with $R = R^*$. Therefore, with $U(\lambda) = \exp(\lambda T^* - \bar{\lambda} T)$ we see that $U(\lambda)$ is a unitary operator in $C^*(T)$ with $U(\lambda)^* = U(-\lambda)$.

From $(*)$ it follows that S commutes with $\exp(\lambda T)$ for every λ, so that

$$\exp(-\lambda T^*)S\exp(\lambda T^*) = U(-\lambda)SU(\lambda). \qquad (**)$$

In particular, the operators in $(**)$ are uniformly bounded in norm by $\|S\|$. Fixing x and y in $\mathfrak{H}$ we define a function f on $\mathbb{C}$ by

$$f(\lambda) = (\exp(-\lambda T^*)S\exp(\lambda T^*)x | y).$$

It follows from (∗) (with T replaced by T^*) that f is an entire analytic function. On the other hand, $|f(\lambda)| \le \|S\| \|x\| \|y\|$, so f is constant by Liouville's theorem. Thus,

$$((\exp(-\lambda T^*)S\exp(\lambda T^*) - S)x|y) = f(\lambda) - f(0) = 0.$$

Since this holds for arbitrary x and y, we conclude that

$$\exp(-\lambda T^*)S\exp(\lambda T^*) - S = 0.$$

From the power series expansion in (∗) we deduce that $ST^* - T^*S = 0$, as desired.   □

4.4.13. Proposition. *If two normal operators T_1 and T_2 in $\mathbf{B}(\mathfrak{H})$ are regular equivalent, they are also unitarily equivalent.*

PROOF. By assumption there is an invertible operator S such that $ST_1 = T_2 S$. Defining the operators $\tilde{S}$ and $\tilde{T}$ on $\mathfrak{H} \oplus \mathfrak{H}$ by the operator matrices

$$\tilde{S} = \begin{pmatrix} 0 & 0 \\ S & 0 \end{pmatrix}, \qquad \tilde{T} = \begin{pmatrix} T_1 & 0 \\ 0 & T_2 \end{pmatrix},$$

the assumption shows that $\tilde{S}\tilde{T} = \tilde{T}\tilde{S}$. Since $\tilde{T}$ is normal, we have $\tilde{S}\tilde{T}^* = \tilde{T}^*\tilde{S}$ by 4.4.12, which means that $ST_1^* = T_2^*S$ (Putnam's theorem). Combining the adjoint of this equation with the original, we obtain $S^*ST_1 = T_1 S^*S$, so that S^*S, hence also $|S|\ [=(S^*S)^{1/2}]$, commutes with T_1. By 3.2.19 we have a polar decomposition $S = U|S|$ with U unitary in $\mathbf{B}(\mathfrak{H})$. Consequently,

$$UT_1 U^* = U|S|T_1|S|^{-1}U^{-1} = ST_1 S^{-1} = T_2.$$ □

EXERCISES

E 4.4.1. Let (T_n) be a sequence of normal operators in $\mathbf{B}(\mathfrak{H})$ converging to an operator T (necessarily normal). If X is a compact subset of $\mathbb{C}$ containing $\mathrm{sp}(T_n)$ for all n and $f \in C(X)$, show that $\mathrm{sp}(T) \subset X$ and that $f(T_n) \to f(T)$.
   Hint: If $\mathrm{dist}(\lambda, X) = \varepsilon > 0$, show that $\|(\lambda I - T_n)^{-1}\| \le \varepsilon^{-1}$ and use E 4.1.7 to prove that $(\lambda I - T_n)^{-1} \to (\lambda I - T)^{-1}$.

E 4.4.2. (Bilateral shift.) Let $\{f_n | n \in \mathbb{Z}\}$ be an orthonormal basis for the Hilbert space $\mathfrak{H}$, and define the *bilateral shift* operator U on $\mathfrak{H}$ by

$$U\left(\sum \alpha_n f_n\right) = \sum \alpha_n f_{n+1}, \quad (\alpha_n) \in \ell^2(\mathbb{Z}).$$

Find U^* and determine $\mathrm{sp}(U)$. Show that U has no eigenvalues [i.e. $\mathrm{sp}(U)_a = \emptyset$] and conclude that U is not diagonalizable. Identify $\mathfrak{H}$ with $L^2(\mathbb{T})$ via the map $f_n \to e_n$, where $e_n(z) = (2\pi)^{-1/2}z^n$, $z \in \mathbb{T}$, cf. E 3.1.5, and describe U as a multiplication operator on $L^2(\mathbb{T})$.

E 4.4.3. Let S and T be normal operators in $\mathbf{B}(\mathfrak{H})$ with $\mathrm{sp}(T) = \mathrm{sp}(S)\,(=X)$. Assume that there are vectors x and y in $\mathfrak{H}$ such that both subspaces

$\{f(S)x | f \in C(X)\}$ and $\{f(T)y | f \in C(X)\}$ are dense in $\mathfrak{H}$ and such that $(f(S)x|x) = (f(T)y|y)$ for every f in $C(X)$. Show that S and T are unitarily equivalent.

Hint: Define $U_0 f(S)x = f(T)y$ and extend it by continuity to an element U in $U(\mathfrak{H})$, where $USU^* = T$. Or use 4.7.11.

E 4.4.4. For T in $B(\mathfrak{H})$ define the numerical range $\Delta(T)$ as in E 3.2.22. Show that $\mathrm{sp}(T) \subset (\Delta(T))^-$.

Hint: Use 4.4.4.

E 4.4.5. Show that if T is normal then with $\Delta(T)$ denoting the numerical range of T (cf. E 3.2.22) we have

$$\mathrm{conv}\, \mathrm{sp}(T) = (\Delta(T))^-.$$

Hint: Given $\varepsilon > 0$ take a finite covering of $\mathrm{sp}(T)$ with (relatively) open sets E_n of diameter $\leq \varepsilon$. Let $\{f_n\}$ be a partition of unit relative to the covering $\{E_n\}$; see 1.7.12. Choose for each n a λ_n in E_n and show that $\|T - \sum \lambda_n f_n(T)\| \leq \varepsilon$. Now note that if $x \in \mathfrak{H}$, $\|x\| = 1$, then

$$\sum \lambda_n (f_n(T)x|x) \in \mathrm{conv}\, \mathrm{sp}(T).$$

E 4.4.6. A continuous, real-valued function f on an interval I is *operator monotone* if whenever A and B are selfadjoint operators in $B(\mathfrak{H})$ with spectra contained in I, such that $A \leq B$, then we have $f(A) \leq f(B)$. Show that the function f_α, given by $f_\alpha(t) = t(1 + \alpha t)^{-1}$, is operator monotone on $\mathbb{R}_+$ if $\alpha \geq 0$ and operator monotone on $[0,1]$ if $-1 < \alpha \leq 0$.

Hint: Write $f_\alpha(t) = \alpha^{-1}(1 - (1 + \alpha t)^{-1})$ and use 3.2.12.

E 4.4.7. Let X be a locally compact Hausdorff space and T be a normal operator in $B(\mathfrak{H})$. Suppose that $\int$ is a finite Radon integral on X and that $h \in C_b(X \times \mathrm{sp}(T))$. Define $f(t) = \int h(\cdot, t)$, so that $f \in C(\mathrm{sp}(T))$. Show that $f(T) = \int h(\cdot, T)$, in the (weak) sense that

$$(f(T)x|y) = \int (h(\cdot, T)x|y) \qquad (*)$$

for all x and y in $\mathfrak{H}$.

Hints: Use polarization to reduce $(*)$ to the case $x = y$. Then note that the map $\int_x : g \to (g(T)x|x)$, $g \in C(\mathrm{sp}(T))$, is a Radon integral on $\mathrm{sp}(T)$, and use the Fubinito theorem (6.6.4) on the product integral $\int \otimes \int_x$ and the function h to establish $(*)$.

E 4.4.8. Let $\int$ be a Radon integral on $\mathbb{R}_+$ such that $\alpha \to f_\alpha(t)$ is integrable for each $t \geq 0$, with f_α as in E 4.4.6. Show that $f = \int f_\alpha$ is operator monotone on $\mathbb{R}_+$.

Hint: Use E 4.4.7.

E 4.4.9. Take f_α as in E 4.4.6. Show that

$$\int_0^\infty f_\alpha(t)\alpha^{-\beta}d\alpha = t^\beta \int_0^\infty (1 + \gamma)^{-1}\gamma^{-\beta}d\gamma.$$

Show further that $t \to t^\beta$ is operator monotone on $\mathbb{R}_+$ for $0 < \beta \le 1$.
Compare this with E 3.2.13.

Hint: Use E 4.4.8.

E 4.4.10. Take f_α as in E 4.4.6. Show that

$$\log(1 + t) = \int_0^\infty f_\alpha(t)d\alpha,$$

and conclude as in E 4.4.9 that $t \to \log(1 + t)$ is an operator monotone function on $\mathbb{R}_+$.

E 4.4.11. A continuous, real-valued function f on an interval I is *operator convex*, if for any two self-adjoint operators A, B with spectrum in I and
every λ in $[0, 1]$ we have $f(\lambda A + (1 - \lambda)B) \le \lambda f(A) + (1 - \lambda)f(B)$.
We say that f is *operator concave* if $-f$ is operator convex. Show
that the function $t \to t^{-1}$ is operator convex on $]0, \infty[$.

Hint: Show first convexity in the (commutative) case where A and
B are replaced by I and $C = A^{-1/2}BA^{-1/2}$, and then use 3.2.9 to obtain
the general result.

E 4.4.12. Show that the functions $f_\alpha(t)$, $0 \le \alpha$ (cf. E 4.4.6), t^β, $0 < \beta \le 1$, and
$\log(1 + t)$ are all operator concave on $]0, \infty[$.

Hint: For f_α, use E 4.4.11. For the others use the same integration
technique as in E 4.4.9 and E 4.4.10.

4.5. The Spectral Theorem, II

Synopsis. Spectral theory with Borel function calculus. Spectral measures.
Spectral projections and eigenvalues. Exercises.

4.5.1. We present in this section a generalized version of the spectral theorem
4.4.1, where the function calculus is extended from continuous to Borel functions. The spectral map $f \to f(T)$ obtained is no longer an isomorphism (but
see 4.7.15), and we assign no continuity properties to it here (that will be taken
up in 4.7.16). Instead we emphasize the property that the spectral map shares
with the integral, namely, preservation of monotone sequential limits. For this
we need the fact that $B(\mathfrak{H})_{sa}$, like $\mathbb{R}$, is order complete in the sense described
in the next result.

4.5.2. Proposition. *If* $(T_\lambda)_{\lambda \in \Lambda}$ *is an increasing net of self-adjoint operators
in* $B(\mathfrak{H})$ *(i.e.* $\lambda \le \mu$ *implies* $T_\lambda \le T_\mu$ *for all* λ *and* μ *in* Λ*), which is bounded
(i.e.* $\|T_\lambda\| \le c$ *for all* λ*), then there is a smallest self-adjoint operator* T *in* $B(\mathfrak{H})$
such that $T_\lambda \le T$ *for all* λ *(i.e.* $T = \text{lub } T_\lambda$*). Moreover,* $\|T_\lambda x - Tx\| \to 0$ *for every
x in* $\mathfrak{H}$*.*

PROOF. For each x in $\mathfrak{H}$ the net $(T_\lambda x | x)_{\lambda \in \Lambda}$ in $\mathbb{R}$ is increasing and bounded above (by $c\|x\|^2$), hence convergent to a number $B(x, x)$. Using the polarization identity (3.1.2) it follows that also every net $(T_\lambda x | y)_{\lambda \in \Lambda}$, with x and y in $\mathfrak{H}$, is convergent in $\mathbb{C}$ with a limit $B(x, y)$ that satisfies

$$4B(x, y) = \sum_{k=0}^{3} i^k B(x + i^k y, x + i^k y).$$

From the limiting process it is clear that B is a self-adjoint, bounded, sesquilinear form on $\mathfrak{H}$, whence $B(x, y) = (Tx | y)$ for some self-adjoint operator T in $\mathbf{B}(\mathfrak{H})$ by 3.2.2. Since $(Tx | x) = \lim (T_\lambda x | x)$ for every x in $\mathfrak{H}$, it is evident that T is the least upper bound (lub) for the net $(T_\lambda)_{\lambda \in \Lambda}$.

Since $T_\lambda \le T$, we have $(T - T_\lambda)^2 \le \|T - T_\lambda\| (T - T_\lambda)$ by 3.2.9, and we note that the net $(\|T - T_\lambda\|)_{\lambda \in \Lambda}$ is bounded (by $\|cI - T_{\lambda_0}\|$, for any fixed λ_0 in Λ). We do not assert, however, that the net converges to zero. But for each x in $\mathfrak{H}$ we can estimate

$$\|(T - T_\lambda)x\|^2 = ((T - T_\lambda)^2 x | x) \le \|T - T_\lambda\| ((T - T_\lambda)x | x),$$

and the last expression tends to zero as $\lambda \to \infty$, by our definition of T. $\qquad \square$

4.5.3. If X is a locally compact Hausdorff space, we denote by $\mathscr{B}_b(X)$ the class of bounded (complex) Borel functions on X. This is a commutative C^*-algebra, and if X is second countable, it follows from 6.2.9 that the real part $\mathscr{B}_b(X)_{sa}$ is the smallest class of bounded, real-valued functions on X that contains $C_0(X)_{sa}$ and which also contains with every bounded, monotone (increasing or decreasing) sequence of functions the pointwise limit.

If T is a normal operator in $\mathbf{B}(\mathfrak{H})$, we denote by $W^*(T)$ the set of operators in $\mathbf{B}(\mathfrak{H})$ that commute with every operator commuting with T. By Fuglede's theorem (4.4.12) this is a commutative C^*-algebra containing $C^*(T)$. Moreover, it also contains with every net $(T_\lambda)_{\lambda \in \Lambda}$ converging "pointwise" to some T_0 in $\mathbf{B}(\mathfrak{H})$ (i.e. $\|T_\lambda x - T_0 x\| \to 0$ for each x in $\mathfrak{H}$) the limit T_0. Indeed, if S is an operator commuting with T, then for every x in $\mathfrak{H}$

$$\|(T_0 S - S T_0)x\| = \lim \|(T_\lambda S - S T_\lambda)x\| = 0,$$

whence $T_0 \in W^*(T)$. More about this in 4.6.7.

With these definitions, we can formulate the *advanced version* of the *spectral theorem*.

4.5.4. Theorem. *Given a normal operator T in $\mathbf{B}(\mathfrak{H})$ there is a norm decreasing *-homomorphism $f \to f(T)$ from $\mathscr{B}_b(\mathrm{sp}(T))$ into $W^*(T)$ that extends the isomorphism of $C(\mathrm{sp}(T))$ onto $C^*(T)$. Moreover, if (f_n) is a bounded, increasing sequence in $\mathscr{B}_b(\mathrm{sp}(T))_{sa}$ and $f = \mathrm{lub}\, f_n$, then $f(T) = \mathrm{lub}\, f_n(T)$ as described in 4.5.2.*

PROOF. For any pair of vectors x, y in $\mathfrak{H}$ the map $f \to (f(T)x | y), f \in C(\mathrm{sp}(T))$, defines a functional $\mu(x, y)$ on $C(\mathrm{sp}(T))$ bounded by $\|x\| \|y\|$. Moreover, if $y = x$, then $\mu(x, x)$ is positive and thus by definition a finite Radon integral on the compact Hausdorff space sp(T); cf. 6.1.2. The Daniell extension theorem (6.1.10) allows us to regard $\mu(x, x)$ as a bounded, positive functional on $\mathscr{B}_b(\mathrm{sp}(T))$, because $\mathscr{B}_b(\mathrm{sp}(T)) \subset \mathscr{L}^1(\mathrm{sp}(T))$ by 6.2.16. We now define, for x, y in $\mathfrak{H}$, an extension of $\mu(x, y)$ from $C(\mathrm{sp}(T))$ to $\mathscr{B}_b(\mathrm{sp}(T))$ by

$$4\langle f, \mu(x, y)\rangle = \sum_{k=0}^{3} \mathrm{i}^k \langle f, \mu(x + \mathrm{i}^k y, x + \mathrm{i}^k y)\rangle. \qquad (*)$$

If (f_n) is a sequence in $\mathscr{B}_b(\mathrm{sp}(T))_{sa}$ converging monotone (increasing or decreasing) to a function f in $\mathscr{B}_b(\mathrm{sp}(T))_{sa}$, then $\langle f, \mu(x, x)\rangle = \lim \langle f_n, \mu(x, x)\rangle$ for every x in $\mathfrak{H}$ by Lebesgue's monotone convergence theorem (6.1.13). It follows from $(*)$ that also $\langle f, \mu(x, y)\rangle = \lim \langle f_n, \mu(x, y)\rangle$ for x and y in $\mathfrak{H}$. This means, in particular, that if $\mathscr{B}_1$ denotes the class of functions f in $\mathscr{B}_b(\mathrm{sp}(T))_{sa}$ such that the expression $\langle f, \mu(x, y)\rangle$ defines a bounded, self-adjoint sesqui-linear form on $\mathfrak{H}$, then $\mathscr{B}_1$ is monotone sequentially complete (cf. 6.2.1). Since, evidently, $C(\mathrm{sp}(T))_{sa} \subset \mathscr{B}_1$ [because in that case $\langle f, \mu(x, y)\rangle = (f(T)x|y)$], it follows that $\mathscr{B}_1 = \mathscr{B}_b(\mathrm{sp}(T))_{sa}$; cf. 6.2.9. Combining this information with 3.2.2 we see that for every f in $\mathscr{B}_b(\mathrm{sp}(T))_{sa}$ there is a unique self-adjoint operator $f(T)$ in $\mathbf{B}(\mathfrak{H})$, such that $(f(T)x|y) = \langle f, \mu(x, y)\rangle$ for all x and y in $\mathfrak{H}$. Moreover, we have shown that if (f_n) is a bounded, increasing sequence in $\mathscr{B}_b(\mathrm{sp}(T))_{sa}$ with $\mathrm{lub}\, f_n = f$, then $(f(T)x|x) = \lim(f_n(T)x|x)$; so that $f(T) = \mathrm{lub}\, f_n(T)$ as described in 4.5.2.

The spectral map $f \to f(T)$ is evidently (real) linear. To show that it is multiplicative, let $\mathscr{B}_2$ denote the class of functions f in $\mathscr{B}_b(\mathrm{sp}(T))_{sa}$, for which $f^2(T) = (f(T))^2$. If (f_n) is a sequence of positive functions in $\mathscr{B}_2$, converging monotone increasing to some function f in $\mathscr{B}_b(\mathrm{sp}(T))_{sa}$, then $f^2 = \mathrm{lub}\, f_n^2$. Thus, for every x in $\mathfrak{H}$ we have by 4.5.2 that

$$(f^2(T)x|x) = \lim(f_n^2(T)x|x) = \lim(f_n(T)^2 x|x)$$
$$= \lim \| f_n(T)x\|^2 = \| f(T)x\|^2 = (f(T)^2 x|x).$$

It follows that $f^2(T) = f(T)^2$, so that $f \in \mathscr{B}_2$. If (f_n) is a general sequence in $\mathscr{B}_2$ increasing to some f, we can choose t in $\mathbb{R}$ such that $f_1 + t \geq 0$. Since $f_n + t \in \mathscr{B}_2$, because the spectral map is linear, and $f + t = \mathrm{lub}(f_n + t)$, we conclude from the above that $f + t \in \mathscr{B}_2$, whence $f \in \mathscr{B}_2$. Thus $\mathscr{B}_2$ is monotone sequentially complete, and since $C(\mathrm{sp}(T)) \subset \mathscr{B}_2$, it follows from 6.2.9 that $\mathscr{B}_2 = \mathscr{B}_b(\mathrm{sp}(T))_{sa}$. Consequently, for every pair f, g in $\mathscr{B}_b(\mathrm{sp}(T))_{sa}$ we have

$$(fg)(T) = \tfrac{1}{2}((f + g)^2 - f^2 - g^2)(T)$$
$$= \tfrac{1}{2}((f(T) + g(T))^2 - f(T)^2 - g(T)^2) = f(T)g(T),$$

as desired.

If $\mathscr{B}_3$ denotes the class of functions f in $\mathscr{B}_b(\mathrm{sp}(T))_{sa}$ for which $f(T) \in W^*(T)$, then $C(\mathrm{sp}(T))_{sa} \subset \mathscr{B}_3$, since $C^*(T) \subset W^*(T)$. If (f_n) is a sequence in $\mathscr{B}_3$ converging monotone to a function f in $\mathscr{B}_b(\mathrm{sp}(T))_{sa}$, we proved that $f_n(T)x \to f(T)x$ for every x in $\mathfrak{H}$ (cf. 4.5.2). Since $(f_n(T)) \subset W^*(T)$ and $W^*(T)$ is closed under pointwise limits (cf. 4.5.3), it follows that $f(T) \in W^*(T)$. Thus $\mathscr{B}_3$ is monotone sequentially complete, whence $\mathscr{B}_3 = \mathscr{B}_b(\mathrm{sp}(T))_{sa}$.

Finally, we extend the map $f \to f(T)$ from $\mathscr{B}_b(\mathrm{sp}(T))_{sa}$ to $\mathscr{B}_b(\mathrm{sp}(T))$ $[= \mathscr{B}_b(\mathrm{sp}(T))_{sa} + i\mathscr{B}_b(\mathrm{sp}(T))_{sa}]$ by complexification, to obtain a $*$-homomorphism from $\mathscr{B}_b(\mathrm{sp}(T))$ into $W^*(T)$ as desired.   □

4.5.5. Remarks. The spectral theorem given above is clearly an extension theorem: We know $f(T)$ for every continuous function f, and wish to define $f(T)$ also for a Borel function, in such a manner that the analogue of Lebesgue's monotone convergence theorem holds. (This, by the way, makes the extension unique.) The proof employed above reduces this extension problem to the (scalar) problem of extending the integral, and then doggedly building up the operator from its sesquilinear form. A more sophisticated proof would instead analyze the extension theorem for the integral, and show that an operator-valued form is available when the image space $W^*(T)$ is a commutative von Neumann algebra.

We have chosen to establish the spectral map $f \to f(T)$, for a single normal operator T, but it is clear that the method is capable of considerable generalization. In fact, if X is any (second countable) compact Hausdorff space and Φ is a (normdecreasing) $*$-homomorphism from $C(X)$ into $\mathbf{B}(\mathfrak{H})$, we can construct an extension $\bar{\Phi}$ of Φ that maps $\mathscr{B}_b(X)$ homomorphically to a commutative $*$-algebra of operators in $\mathbf{B}(\mathfrak{H})$ in such a manner that the monotone convergence theorem holds. The proof of 4.5.4 applies verbatim.

4.5.6. It is a (deplorable) fact that most mathematicians prefer to deal with integrals in terms of measures. The spectral map in 4.5.4, visualized as an operator-valued integral, is no exception.

A *spectral measure* on a set X is a map $E: \mathscr{S} \to \mathbf{B}(\mathfrak{H})_p$, where $\mathscr{S}$ is a σ-algebra of subsets of X (cf. 6.2.3), and $\mathbf{B}(\mathfrak{H})_p$ denotes the class of projections in $\mathbf{B}(\mathfrak{H})$. It is further assumed that E satisfies the following conditions:

(i) $E(\emptyset) = 0, \qquad E(X) = I$;
(ii) $E(Y \cap Z) = E(Y)E(Z), \quad Y, Z \in \mathscr{S}$;
(iii) $E(Y \cup Z) = E(Y) + E(Z), \quad Y, Z \in \mathscr{S}, Y \cap Z = \emptyset$;
(iv) $E(\bigcup Y_n) = \bigvee E(Y_n), \quad \{Y_n\} \subset \mathscr{S}$.

Here the symbol $\bigvee E(Y_n)$ denotes the smallest projection majorizing all $E(Y_n)$, viz. the projection on the closed subspace spanned by $\sum E(Y_n)(\mathfrak{H})$. Note that condition (ii) implies that the projections $E(Y)$ and $E(Z)$ commute (otherwise their product would not be self-adjoint), so that the product equals $E(Y) \wedge E(Z)$—the projection onto the subspace $E(Y)(\mathfrak{H}) \cap E(Z)(\mathfrak{H})$. Note also that condition (iii) is redundant; it follows from (iv).

Given a spectral measure E as above, we obtain for each vector x in $\mathfrak{H}$ an ordinary (scalar) measure μ_x on X by

$$\mu_x(Y) = (E(Y)x|x), \quad Y \in \mathscr{S};$$

see 6.3.1. The family of measures so obtained satisfies a coherence relation inherited from the parallellogram law (3.1.3), viz.

$$\mu_{x+y} + \mu_{x-y} = 2\mu_x + 2\mu_y. \tag{$*$}$$

Moreover, we see from the polarization identity (3.1.2) that each pair of vectors x, y in $\mathfrak{H}$ gives rise to a signed measure μ_{xy} on X, where

$$4\mu_{xy} = \sum_{k=0}^{3} i^k \mu_{x+i^k y}, \tag{**}$$

or, directly, $\mu_{xy}(Y) = (E(Y)x|y)$, $Y \in \mathscr{S}$. Using the integrals obtained from the measures μ_x, $x \in \mathfrak{H}$, or, better, the signed measures μ_{xy}, x, $y \in \mathfrak{H}$, it follows from the Cauchy-Schwarz inequality that every bounded, $\mathscr{S}$-measurable function f on X gives rise to a sesquilinear form on $\mathfrak{H}$ bounded by $\|f\|_\infty$, hence defines an operator T_f in $\mathbf{B}(\mathfrak{H})$, where

$$(T_f x|y) = \int_X f(\lambda)d\mu_{xy}(\lambda). \tag{***}$$

This definition explains the suggestive notation

$$T_f = \int_X f(\lambda)dE(\lambda)$$

for the operators obtained by integrating a spectral measure. The spectral map $f \to T_f$, so obtained from the spectral measure E, clearly defines a norm decreasing, $*$-preserving linear map from the commutative C^*-algebra $\mathscr{B}_b(X)$ of bounded $\mathscr{S}$-measurable functions on X into a commuting family of operators in $\mathbf{B}(\mathfrak{H})$. To show that this map is multiplicative it suffices to prove that $T_f T_g = T_{fg}$, when f and g are simple functions, because such functions are uniformly dense in $\mathscr{B}_b(X)$. But if $f = \sum \alpha_n[Y_n]$, $g = \sum \beta_m[Z_m]$, where $[Y]$ as in 6.2.3 denotes the characteristic function for the subset Y of X, then

$$T_f T_g = \left(\sum \alpha_n E(Y_n) \right)\left(\sum \beta_m E(Z_m) \right)$$
$$= \sum \alpha_n \beta_m E(Y_n)E(Z_m) = \sum \alpha_n \beta_m E(Y_n \cap Z_m) = T_{fg},$$

because $fg = \sum \alpha_n \beta_m [Y_n \cap Z_m]$. We summarize our observations in the following proposition.

4.5.7. Proposition. *To every spectral measure E on a set X with a σ-algebra $\mathscr{S}$, there is a normdecreasing $*$-homomorphism $f \to T_f$ from the algebra $\mathscr{B}_b(X)$ of bounded, $\mathscr{S}$-measurable functions on X into an algebra of normal operators in $\mathbf{B}(\mathfrak{H})$, given by*

$$T_f = \int_X f(\lambda)dE(\lambda) \quad \text{or} \quad (T_f x|y) = \int_X f(\lambda)d\mu_{xy},$$

where $\mu_{xy}(Y) = (E(Y)x|y)$ for every x and y in $\mathfrak{H}$. If (f_n) is a bounded, monotone increasing sequence in $\mathscr{B}_b(X)_{sa}$ and $f = \mathrm{lub}\, f_n$, then $T_f = \mathrm{lub}\, T_{f_n}$ as described in 4.5.2.

PROOF. Only the last assertion is not covered by the discussion in 4.5.6. But if $x \in \mathfrak{H}$ and μ_x is the measure given by $\mu_x(Y) = (E(Y)x|x)$ as in 4.5.6, then from Lebesgue's monotone convergence theorem we have

$$(T_f x | x) = \int_X f(\lambda) d\mu_x(\lambda) = \lim \int_X f_n(\lambda) d\mu_x(\lambda) = \lim(T_{f_n} x | x),$$

for every x in $\mathfrak{H}$, whence $T_f = \text{lub } T_{f_n}$. $\square$

4.5.8. Proposition. *To each normal operator T in $\mathbf{B}(\mathfrak{H})$ there is a spectral measure E on the Borel subsets of $\text{sp}(T)$ taking values in the projections in $W^*(T)$, such that*

$$T = \int_{\text{sp}(T)} \lambda \, dE(\lambda).$$

PROOF. Consider the $*$-homomorphism $f \to f(T)$ given in 4.5.4. If $\mathscr{B}$ denotes the class of Borel subsets of $\text{sp}(T)$ and $Y \in \mathscr{B}$, then the characteristic function $[Y]$ belongs to $\mathscr{B}_b(\text{sp}(T))$, and we define

$$E(Y) = [Y](T), \quad Y \in \mathscr{B}.$$

Since the spectral map is a $*$-homomorphism, the operators $E(Y)$ in $W^*(T)$ are projections [being images of self-adjoint idempotents in $\mathscr{B}_b(\text{sp}(T))$], and it is straightforward to check that E satisfies the conditions (i)–(iv) in 4.5.6.
  If f is a simple function, i.e. $f = \sum \alpha_n [Y_n]$, where $Y_n \in \mathscr{B}$, then

$$\int f \, dE(\lambda) = \sum \alpha_n E(Y_n) = \sum \alpha_n [Y_n](T) = f(T).$$

Since every function f in $\mathscr{B}_b(\text{sp}(T))_{sa}$ is the monotone sequential limit of simple functions it follows that $\int f dE(\lambda) = f(T)$ for every f in $\mathscr{B}_b(\text{sp}(T))$, because both spectral maps preserve such limits, cf. 4.5.4 and 4.5.7. In particular,

$$T = id(T) = \int id \, dE = \int \lambda \, dE(\lambda). \qquad \square$$

4.5.9. Remark. From the presentation of spectral measures given in 4.5.6–4.5.8 it is clear that they are just (in the author's opinion) a rather cumbersome way of expressing a perfectly natural $*$-homomorphic extension of the Gelfand C^*-isomorphism. In one instance, however, they give a superior description of a situation, namely, the position of eigenvalues in the spectrum.

4.5.10. Proposition. *Let T be a normal operator in $\mathbf{B}(\mathfrak{H})$ with spectral measure E. If f is a bounded Borel function on $\text{sp}(T)$, then $\lambda \in \text{sp}(f(T))$ iff*

$$E(f^{-1}(B(\lambda, \varepsilon))) \neq 0$$

for every $\varepsilon > 0$. Moreover, λ is an eigenvalue for $f(T)$ iff $E(f^{-1}(\{\lambda\})) \neq 0$, in which case $E(f^{-1}(\{\lambda\}))$ is the projection on $\mathbf{B}(\mathfrak{H})$ on the eigenspace for $f(T)$ corresponding to λ.

PROOF. Put $E_\varepsilon = E(f^{-1}(B(\lambda, \varepsilon)))$. If $E_\varepsilon = 0$ for some $\varepsilon > 0$, define

$$g(\mu) = \begin{cases} (\lambda - f(\mu))^{-1} & \text{if } \mu \notin f^{-1}(B(\lambda, \varepsilon)); \\ 0 & \text{if } \mu \in f^{-1}(B(\lambda, \varepsilon)). \end{cases}$$

Then $g \in \mathscr{B}_b(\mathrm{sp}(T))$ (indeed, $\|g\|_\infty \leq \varepsilon^{-1}$), and with $X = \mathrm{sp}(T)\backslash f^{-1}(B(\lambda, \varepsilon))$ we have

$$g(T)(\lambda I - f(T)) = \int_X g(\mu)(\lambda - f(\mu))dE(\mu) = I.$$

Conversely, if $E_\varepsilon \neq 0$ for every $\varepsilon > 0$, then for each unit vector x in $E_\varepsilon(\mathfrak{H})$ we have

$$\|(\lambda I - f(T))x\| = \|\lambda I - f(T))E_\varepsilon x\|$$

$$\leq \|(\lambda I - f(T))E_\varepsilon\| \leq \|\lambda - f|f^{-1}(B(\lambda, \varepsilon))\|_\infty \leq \varepsilon,$$

whence $\lambda \in \mathrm{sp}(f(T))$.

To prove the second half of the proposition, put $E_0 = E(f^{-1}(\{\lambda\}))$. If $E_0 \neq 0$, then with $Y = f^{-1}(\{\lambda\})$ we have

$$f(T)E_0 = \int_Y f(\mu)dE(\mu) = \lambda E_0.$$

Thus $E_0 \leq P_\lambda$, where P_λ denotes the projection in $\mathbf{B}(\mathfrak{H})$ on the eigenspace for $f(T)$ corresponding to λ, i.e. $P_\lambda(\mathfrak{H}) = \ker(\lambda I - f(T))$. Conversely, if $P_\lambda \neq 0$, let $f_n = (|\lambda - f| \wedge 1)^{1/n}$. Then $f_n \in \mathscr{B}_b(\mathrm{sp}(T))$, and the sequence (f_n) converges pointwise up to the characteristic function for the set $\mathrm{sp}(T)\backslash Y$. Since $f_n(T)P_\lambda = 0$ for every n, it follows from 4.5.4 that

$$0 = E(\mathrm{sp}(T)\backslash Y)P_\lambda = (I - E_0)P_\lambda,$$

whence $P_\lambda \leq E_0$. Consequently, $E_0 = P_\lambda$, as claimed.   □

4.5.11. Corollary. *If $f \in \mathscr{B}_b(\mathrm{sp}(T))$, then*

$$\mathrm{sp}(f(T)) \subset f(\mathrm{sp}(T))^-.$$

4.5.12. Corollary. *For each λ in $\mathrm{sp}(T)$ we have $E(\{\lambda\}) = 0$ unless λ is an eigenvalue for T; in which case $E(\{\lambda\})$ is the projection in $\mathbf{B}(\mathfrak{H})$ on the eigenspace for T corresponding to λ.*

4.5.13. Corollary. *If T is a diagonalizable operator in $\mathbf{B}(\mathfrak{H})$, in particular, if T is compact and normal, then the spectral measure E for T is totally atomic and concentrated on the subset $\mathrm{sp}(T)_a$ of eigenvalues in $\mathrm{sp}(T)$. Moreover, $\mathrm{sp}(T) = (\mathrm{sp}(T)_a)^-$ and for every x in $\mathfrak{H}$,*

$$Tx = \sum_{\lambda \in \mathrm{sp}(T)_a} \lambda E(\{\lambda\})x.$$

EXERCISES

E 4.5.1. Let T be a normal operator in $\mathbf{B}(\mathfrak{H})$, and let g and f be bounded Borel functions on $\mathrm{sp}(T)$ and X, respectively, where X is a compact subset of $\mathbb{C}$ containing $g(\mathrm{sp}(T))$. Show that

$$(f \circ g)(T) = f(g(T)).$$

Hint: By 4.5.4 the formula is true when f is a polynomial in ζ and $\bar{\zeta}$, restricted to X. By 4.3.4 it is therefore also valid if $f \in C(X)$. Show by 4.5.4 that the set $\mathscr{B}_0$ of elements f in $\mathscr{B}_b(X)_{sa}$, for which the formula holds, is monotone sequentially closed; and conclude from $C(X)_{sa} \subset \mathscr{B}_0$ that $\mathscr{B}_0 = \mathscr{B}_b(X)_{sa}$; cf. 6.2.9.

E 4.5.2. Show that for every normal operator T in $\mathbf{B}(\mathfrak{H})$ and every $\varepsilon > 0$ there is a finite set $\{P_n\}$ of pairwise orthogonal projections with sum I, and a corresponding set $\{\lambda_n\}$ in $\mathbb{C}$, such that $\|T - \sum \lambda_n P_n\| \le \varepsilon$.
Hint: Take $P_n = f_n(T)$, where f_n is the characteristic function corresponding to a small "half-open" square in $\mathrm{sp}(T)$.

E 4.5.3. Show that the norm closure of the set of diagonalizable operators in $\mathbf{B}(\mathfrak{H})$ (cf. 3.2.14) is the set of normal operators.
Hint: Use E 4.5.2.

E 4.5.4. If $0 \le T \le I$ in $\mathbf{B}(\mathfrak{H})$, find a sequence (P_n) of pairwise commuting projections, such that $T = \sum 2^{-n} P_n$.
Hint: Let P_1 be the spectral projection of T corresponding to the interval $]\frac{1}{2}, 1]$. Let P_2 correspond to the union $]\frac{1}{4}, \frac{1}{2}] \cup]\frac{3}{4}, 1]$, and let P_3 correspond to the union $]\frac{1}{8}, \frac{1}{4}] \cup]\frac{3}{8}, \frac{1}{2}] \cup]\frac{5}{8}, \frac{3}{4}] \cup]\frac{7}{8}, 1]$, and continue by induction.

E 4.5.5. Show that for every unitary U in $\mathbf{B}(\mathfrak{H})$ there is a selfadjoint operator T in $\mathbf{B}(\mathfrak{H})$, such that $U = \exp iT$.
Hint: Define $\log(\exp i\theta) = i\theta$ if $-\pi < \theta \le \pi$, and note that $\log$ is a Borel function on the circle such that $\exp \log z = z$ for every z in $\mathbb{T}$. Take $T = -i \log U$ and apply E 4.5.1.

E 4.5.6. Show that the group $\mathbf{U}(\mathfrak{H})$ of unitary operators in $\mathbf{B}(\mathfrak{H})$ is arcwise connected in the norm topology.
Hint: If $U \in \mathbf{U}(\mathfrak{H})$, write $U = \exp iT$ by E 4.5.5, and set $U_t = \exp it T$ for $0 \le t \le 1$.

E 4.5.7. Show that the group $\mathbf{GL}(\mathfrak{H})$ of invertible operators in $\mathbf{B}(\mathfrak{H})$ is arcwise connected in norm.
Hint: If $T \in \mathbf{GL}(\mathfrak{H})$, write $T = U|T|$ with U in $\mathbf{U}(\mathfrak{H})$ by 3.2.19. Define $T_t = U(t|T| + (1 - t)I)$ for $0 \le t \le 1$ and then apply E 4.5.6.

E 4.5.8. Show that if $T = T^*$ in $\mathbf{B}(\mathfrak{H})$ and $x \in \mathfrak{H}$, then for each $\varepsilon > 0$ there is a projection P of finite rank with $Px = x$, such that $\|(I - P)TP\|_2 < \varepsilon$, with $\|\cdot\|_2$ as defined in (the proof of) 3.4.9.

Hints: Assume that $\|T\| = \|x\| = 1$. Let P_n be the spectral projection of T corresponding to the interval $](n - 1)m^{-1}, nm^{-1}]$, where $-m \leq n \leq m$ and $m > 3\varepsilon^{-2}$. Show that $\|TP_n - nm^{-1}P_n\| \leq m^{-1}$. Let P be the projection of $\mathfrak{H}$ on the space spanned by the orthogonal vectors $\{P_n x \mid |n| \leq m\}$. Show that

$$\|(I - P)TP\|_2^2 = \sum_n (PT(I - P)TP\, P_n x \mid P_n x) \|P_n x\|^{-2}$$

$$= \sum_n \|(I - P)TP\, P_n x\|^2 \|P_n x\|^{-2}$$

$$\leq \sum_n m^{-2} \|P_n x\|^2 \|P_n x\|^{-2}$$

$$= (2m + 1)m^{-2} < \varepsilon^2.$$

E 4.5.9. (*Weyl–von Neumann's theorem*.) Show that for each self-adjoint operator T on a separable Hilbert space $\mathfrak{H}$ and every $\varepsilon > 0$, there is a self-adjoint, diagonalizable operator D and a self-adjoint Hilbert–Schmidt operator S with $\|S\|_2 < \varepsilon$, such that $T = D + S$.

Hint: Define inductively a sequence (P_n) of pairwise orthogonal projections of finite rank, such that

$$\left\| T - QTQ - \sum_1^n P_k TP_k \right\|_2 < \sum_1^n 2^{-k}\varepsilon.$$

Here $Q = I - P_1 - \cdots - P_n$, and P_{n+1} is obtained from E 4.5.8 applied to QTQ on $Q(\mathfrak{H})$, with $x = Qx_{n+1}$, where (x_n) is a dense sequence in $\mathfrak{H}$. Note that $\sum P_n = I$ and that each $P_n TP_n$ is diagonalizable, whence $D = \sum P_n TP_n$.

E 4.5.10. Let $\mathfrak{H}$ be a separable Hilbert space and $\mathfrak{I}$ be a norm closed ideal of $\mathbf{B}(\mathfrak{H})$. Show that if $\mathfrak{I} \setminus \mathbf{B}_0(\mathfrak{H}) \neq \emptyset$, then $\mathfrak{I} = \mathbf{B}(\mathfrak{H})$ (and compare with E 3.3.1).

Hint: If $\mathfrak{I} \not\subset \mathbf{B}_0(\mathfrak{H})$, there is a noncompact, positive operator T in $\mathfrak{I}$. Show that $P \notin \mathbf{B}_0(\mathfrak{H})$, where P is the spectral projection of T corresponding to an interval $[\varepsilon, \|T\|]$, where $\varepsilon > 0$ is small enough. Show that $P = Pf(T)$, $f \in C(\mathrm{sp}(T))$ with $f(0) = 0$ but $f(t) = 1$ for $t \geq \varepsilon$, and deduce that $P \in \mathfrak{I}$. Find a partial isometry V such that $V^*V = P$ and $VV^* \in I$ (thus $VP = V$) and conclude that $I \in \mathfrak{I}$.

E 4.5.11. Let X be a locally compact Hausdorff space and T be a normal operator in $\mathbf{B}(\mathfrak{H})$. Suppose that $\int$ is a finite Radon integral on X and that h is a bounded Borel function on $X \times \mathrm{sp}(T)$. Define $f(t) = \int h(\cdot, t)$, so that f is a bounded Borel function on $\mathrm{sp}(T)$. Show that $f(T) = \int h(\cdot, T)$, in the (weak) sense that

$$(f(T)x \mid y) = \int (h(\cdot, T)x \mid y) \tag{$*$}$$

for all x and y in $\mathfrak{H}$.

Hint: As hinted in E 4.4.7, but now with the mature Fubini theorem (6.6.6).

4.6. Operator Algebra

Synopsis. Strong and weak topology on $B(\mathfrak{H})$. Characterization of strongly/weakly continuous functionals. The double commutant theorem. Von Neumann algebras. The σ-weak topology. The σ-weakly continuous functionals. The predual of a von Neumann algebra. Exercises.

4.6.1. For an infinite-dimensional Hilbert space $\mathfrak{H}$, the operator algebra $B(\mathfrak{H})$ has many interesting (vector space) topologies. Besides the (operator) norm topology there are at least six weak topologies that are relevant for operator algebra theory. The two most important are the strong and the weak topology.

The *strong topology* on $B(\mathfrak{H})$ is the locally convex vector space topology induced by the family of seminorms of the form $T \to \|Tx\|$ for various x in $\mathfrak{H}$.

The *weak topology* on $B(\mathfrak{H})$ is the locally convex vector space topology induced by the family of seminorms of the form $T \to |(Tx|y)|$ for various x, y in $\mathfrak{H} \times \mathfrak{H}$.

Similar constructions exist, of course, when $\mathfrak{X}$ is a Banach space, and define on the Banach algebra $B(\mathfrak{X})$ the strong topology, with seminorms indexed by $\mathfrak{X}$, and the weak topology with seminorms indexed by $\mathfrak{X} \times \mathfrak{X}^*$.

Since $|(Tx|y)| \leq \|Tx\|\,\|y\| \leq \|T\|\,\|x\|\,\|y\|$, we immediately observe that the weak topology is weaker than the strong topology, which in turn is weaker than the norm topology. Since $|(Tx|y)| = |(T^*y|x)|$, it is moreover clear that the adjoint operation is weakly continuous. The adjoint operation is not strongly continuous (E 4.6.1), except when restricted to the set of normal elements (E 4.6.2), which is unfortunately not a subspace of $B(\mathfrak{H})$. Multiplication is separately continuous in each variable, in both the strong and the weak topology; but multiplication as a function $B(\mathfrak{H}) \times B(\mathfrak{H}) \to B(\mathfrak{H})$ is neither weakly nor strongly continuous. However, as a function $B(0, n) \times B(\mathfrak{H}) \to B(\mathfrak{H})$, i.e. if the first factor remains bounded, the product *is* strongly continuous (and for most situations that will suffice). This is seen from the rewriting

$$\|(ST - S_0 T_0)x\| \leq \|(S - S_0)T_0 x\| + \|S\|\,\|(T - T_0)x\|. \qquad (*)$$

Observe that the unit ball $\mathbf{B}$ of $B(\mathfrak{H})$ is weakly compact. Indeed, if $(T_\lambda)_{\lambda \in \Lambda}$ is a universal net in $\mathbf{B}$, then for x and y in $\mathfrak{H}$ we have $|(T_\lambda x|y)| \leq \|x\|\,\|y\|$, so that $(T_\lambda x|y) \to B(x, y)$, where B is a sesquilinear form on $\mathfrak{H}$ with $\|B\| \leq 1$ and thus corresponds to an element T in $\mathbf{B}$, such that $T_\lambda \to T$ weakly. The compactness of $\mathbf{B}$ now follows from 1.6.2.

4.6.2. If $\mathfrak{H}$ is separable, the unit ball $\mathbf{B}$ of $B(\mathfrak{H})$ is metrizable in both the weak and the strong topology. Indeed, let (x_n) be a dense sequence in the unit ball of $\mathfrak{H}$ and define the metrics

$$d_s(S, T) = \sum 2^{-n}\|(S - T)x_n\| \quad \text{and} \quad d_w(S, T) = \sum 2^{-n}|(S - T)x_n|x_n)|.$$

Routine arguments show that d_s induces the strong topology on $\mathbf{B}$ and d_w the weak. Of course, $\mathbf{B}$ may be replaced by any other bounded subset of $\mathbf{B}(\mathfrak{H})$ in the argument above. Nevertheless, neither the strong nor the weak topology is metrizable when $\mathfrak{H}$ is infinite-dimensional. In fact, they are not even first countable (E 4.6.4).

If $\mathfrak{H}$ is separable, $\mathbf{B}$ is a metrizable, compact Hausdorff space in the weak topology, and therefore second countable. In particular, $\mathbf{B}$ is weakly separable, and therefore $\mathbf{B}(\mathfrak{H})$ ($= \mathbb{R}_+ \mathbf{B}$) is weakly separable. To see that $\mathbf{B}(\mathfrak{H})$ is even strongly separable, we first note that the algebra $\mathbf{B}_f(\mathfrak{H})$ of operators of finite rank (cf. 3.3.1) is strongly dense in $\mathbf{B}(\mathfrak{H})$, because any strong neighborhood only involves a finite number of vectors. Then we appeal to the easily established fact that for a fixed orthonormal basis in $\mathfrak{H}$, every element in $\mathbf{B}_f(\mathfrak{H})$ can be approximated in norm by a finite matrix (in the given basis) with rational coefficients. Thus $\mathbf{B}_f(\mathfrak{H})$ [and $\mathbf{B}_0(\mathfrak{H})$] is norm separable, therefore also strongly separable; and, consequently, $\mathbf{B}(\mathfrak{H}) = (\mathbf{B}_f(\mathfrak{H}))^{-s}$ is strongly separable.

4.6.3. For a Hilbert space $\mathfrak{H}$ and a natural number n we let $\mathfrak{H}^n$ denote the orthogonal sum (3.1.5) of n copies of $\mathfrak{H}$. If P_k denotes the projection of $\mathfrak{H}^n$ onto the kth summand, then $\sum P_k = I$ in $\mathbf{B}(\mathfrak{H}^n)$, and every element T in $\mathbf{B}(\mathfrak{H}^n)$ can be written $T = \sum P_k T P_l, 1 \le k, l \le n$. Identifying $P_k T P_l$ with an operator T_{kl} in $\mathbf{B}(\mathfrak{H})$ we obtain a matrix representation $T = (T_{kl})$ for the elements in $\mathbf{B}(\mathfrak{H}^n)$ that agrees with the computational rules. Thus we identify $\mathbf{B}(\mathfrak{H}^n)$ with $\mathbf{M}_n(\mathbf{B}(\mathfrak{H}))$—the $n \times n$-operator matrices over $\mathbf{B}(\mathfrak{H})$.

The following construction, known as the *amplification* of $\mathbf{B}(\mathfrak{H})$, will be useful in the sequel. Define $\Phi\colon \mathbf{B}(\mathfrak{H}) \to \mathbf{B}(\mathfrak{H}^n)$ by $\Phi(T)_{kk} = T$ and $\Phi(T)_{kl} = 0$ if $k \ne l$. In other words, $\Phi(T)$ is the diagonal matrix with T in the diagonal. Clearly Φ is a $*$-isomorphism of $\mathbf{B}(\mathfrak{H})$ into $\mathbf{B}(\mathfrak{H}^n)$.

4.6.4. Proposition. *For a functional φ on $\mathbf{B}(\mathfrak{H})$ the following conditions are equivalent:*

(i) *There are vectors $x_1, \ldots, x_n$ and $y_1, \ldots, y_n$ in $\mathfrak{H}$ such that $\varphi(T) = \sum (Tx_k|y_k)$ for all T in $\mathbf{B}(\mathfrak{H})$.*
(ii) *φ is weakly continuous.*
(iii) *φ is strongly continuous.*

PROOF. The implications (i) $\Rightarrow$ (ii) $\Rightarrow$ (iii) are obvious.

(iii) $\Rightarrow$ (i). If φ is strongly continuous, there are by $(*)$ in 2.4.1 vectors $x_1, \ldots, x_n$ in $\mathfrak{H}$ such that $\max\|Tx_k\| \le 1$ implies $|\varphi(T)| \le 1$ for all T in $\mathbf{B}(\mathfrak{H})$. This implies that

$$|\varphi(T)|^2 \le \sum \|Tx_k\|^2. \tag{$*$}$$

With notation as in 4.6.3 we define the vector x in $\mathfrak{H}^n$ as the orthogonal sum of the x_k's, and we define a functional ψ on the subspace of $\mathfrak{H}^n$ consisting of vectors $\Phi(T)x$, $T \in \mathbf{B}(\mathfrak{H})$, by setting

$$\psi(\Phi(T)x) = \varphi(T).$$

Now note that $(*)$ exactly expresses the continuity of ψ, viz. $|\psi(\Phi(T)x)|^2 \leq \|\Phi(T)x\|^2$. By 2.3.3 we can extend ψ to all of $\mathfrak{H}^n$, and by 3.1.9 there is therefore a vector $y = (y_1, \ldots, y_n)$ in $\mathfrak{H}^n$ such that $\psi = (\cdot \,|y)$. In particular,

$$\varphi(T) = \psi(\Phi(T)x) = (\Phi(T)x|y) = \sum (Tx_k|y_k). \qquad \square$$

4.6.5. Corollary. *Every strongly closed, convex set in $\mathbf{B}(\mathfrak{H})$ is weakly closed. In particular, every strongly closed subspace of $\mathbf{B}(\mathfrak{H})$ is weakly closed.*

4.6.6. For any subset $\mathfrak{A}$ of $\mathbf{B}(\mathfrak{H})$ we let $\mathfrak{A}'$ denote the *commutant* of $\mathfrak{A}$, i.e.

$$\mathfrak{A}' = \{T \in \mathbf{B}(\mathfrak{H})| TS = ST, \forall S \in \mathfrak{A}\}.$$

It is easily verified that $\mathfrak{A}'$ is weakly closed and is an algebra. If $\mathfrak{A}$ is a self-adjoint subset of $\mathbf{B}(\mathfrak{H})$ (i.e. $S \in \mathfrak{A} \Rightarrow S^* \in \mathfrak{A}$, we therefore have that $\mathfrak{A}'$ is a weakly closed, unital C^*-subalgebra of $\mathbf{B}(\mathfrak{H})$.

We shall need the iterated commutants $(\mathfrak{A}')'$ and $((\mathfrak{A}')')'$. These will just be written as $\mathfrak{A}''$ and $\mathfrak{A}'''$. Note that if $\mathfrak{A}_1 \subset \mathfrak{A}_2$ then $\mathfrak{A}_1' \supset \mathfrak{A}_2'$. On the other hand we always have $\mathfrak{A} \subset \mathfrak{A}''$. It follows that $\mathfrak{A}''' \subset \mathfrak{A}' \subset \mathfrak{A}'''$ for every subset $\mathfrak{A}$, so that the process of taking commutants stabilizes after at most two steps.

The following *double commutant theorem* by von Neumann (1929) is the fundamental result in operator algebra theory.

4.6.7. Theorem. *For a self-adjoint, unital subalgebra $\mathfrak{A}$ of $\mathbf{B}(\mathfrak{H})$ the following conditions are equivalent:*

(i) $\mathfrak{A} = \mathfrak{A}''$.
(ii) $\mathfrak{A}$ *is weakly closed.*
(iii) $\mathfrak{A}$ *is strongly closed.*

PROOF. The implication (i) $\Rightarrow$ (ii) is clear, and (ii) $\Leftrightarrow$ (iii) follows from 4.6.5. To prove that (iii) $\Rightarrow$ (i) we consider for each x in $\mathfrak{H}$ the projection P of $\mathfrak{H}$ onto $(\mathfrak{A}x)^=$. For each T in $\mathfrak{A}$ we then have that $TP(\mathfrak{H}) \subset P(\mathfrak{H})$, whence $(I - P)TP = 0$. Taking $T = T^*$ this implies that TP ($= PTP$) is self-adjoint, whence $TP = PT$. Since $\mathfrak{A}$ is self-adjoint, it follows that $P \in \mathfrak{A}'$. Now take S in $\mathfrak{A}''$. Then $SP = PS$, in particular, $Sx \in P(\mathfrak{H})$, because $Px = x$ (as $I \in \mathfrak{A}$, $x \in \mathfrak{A}x$). For each $\varepsilon > 0$ we can therefore find T in $\mathfrak{A}$ such that $\|Sx - Tx\| < \varepsilon$. The argument shows that S can be approximated from $\mathfrak{A}$ on each single vector x in $\mathfrak{H}$.

Now take $x_1, \ldots, x_n$ in $\mathfrak{H}$ and define (as in the proof of 4.6.4) x in $\mathfrak{H}^n$ as the orthogonal sum of the x_k's. Identifying $\mathbf{B}(\mathfrak{H}^n)$ with $\mathbf{M}_n(\mathbf{B}(\mathfrak{H}))$ as in 4.6.3 we see that

$$(R\Phi(T) - \Phi(T)R)_{kl} = R_{kl}T - TR_{kl}$$

for every $R = (R_{kl})$ in $\mathbf{M}_n(\mathbf{B}(\mathfrak{H}))$ and T in $\mathbf{B}(\mathfrak{H})$. It follows that the commutant of $\Phi(\mathfrak{A})$ in $\mathbf{M}_n(\mathbf{B}(\mathfrak{H}))$ consists of the matrices with entries in $\mathfrak{A}'$, i.e. $\Phi(\mathfrak{A})' = \mathbf{M}_n(\mathfrak{A}')$. Now apply the first part of the proof with $\mathfrak{H}^n$, $\Phi(\mathfrak{A})$, and $\Phi(S)$ in place of $\mathfrak{H}$, $\mathfrak{A}$, and S, to obtain an element T in $\mathfrak{A}$ with

$$\sum \|(S - T)x_k\|^2 = \|(\Phi(S) - \Phi(T))x\|^2 < \varepsilon^2.$$

This shows that every S in $\mathfrak{A}''$ can be approximated arbitrarily well in the strong topology with elements from $\mathfrak{A}$. Since $\mathfrak{A}$ was assumed strongly closed, $S \in \mathfrak{A}$, and $\mathfrak{A}'' = \mathfrak{A}$. □

4.6.8. Corollary. *If $\mathfrak{A}$ is a self-adjoint, unital subalgebra of $\mathbf{B}(\mathfrak{H})$, its strong (= weak) closure in $\mathbf{B}(\mathfrak{H})$ is precisely $\mathfrak{A}''$.*

4.6.9. Remark. An algebra satisfying the conditions in 4.6.7 is called a *von Neumann algebra* (sometimes also a *W*-algebra*; and by J.v.N. a ring of operators). These algebras appear quite naturally in many connections. From the preceding we see that if $\mathfrak{D}$ is any self-adjoint subset of $\mathbf{B}(\mathfrak{H})$, then $\mathfrak{D}'$ is a von Neumann algebra. We see further that von Neumann algebras come in pairs, $\mathfrak{A}$ and $\mathfrak{A}'$ and that $\mathfrak{A} \cap \mathfrak{A}'$ is the common center for the two algebras.

A von Neumann algebra $\mathfrak{A}$ is called a *factor* if $\mathfrak{A} \cap \mathfrak{A}' = \mathbb{C}I$, i.e. if $\mathfrak{A}$ is as noncommutative as possible. It was one of the early surprises of the theory that there are other factors than $\mathbf{B}(\mathfrak{H})$. It is now known that there is an uncountable number of nonisomorphic factors on the separable, infinite-dimensional Hilbert space; and their classification plays an important role in group theory (especially for representations of semisimple Lie groups) and in the (tentative) mathematical formulation of quantum mechanics (relativistic statistical quantum mechanics and quantum field theory).

4.6.10. For the description of von Neumann algebras as dual spaces we need a fourth topology on $\mathbf{B}(\mathfrak{H})$. Recall from 3.4.12 and 3.4.13 that with $\mathbf{B}^1(\mathfrak{H})$ the Banach space of trace class operators, the bilinear form

$$\langle S, T \rangle = \operatorname{tr}(ST), \quad S \in \mathbf{B}(\mathfrak{H}),\ T \in \mathbf{B}^1(\mathfrak{H}),$$

implements an isometric isomorphism between $\mathbf{B}(\mathfrak{H})$ and the dual space $(\mathbf{B}^1(\mathfrak{H}))^*$. We define the *σ-weak topology* on $\mathbf{B}(\mathfrak{H})$ as the w^*-topology it inherits under the identification with $(\mathbf{B}^1(\mathfrak{H}))^*$.

4.6.11. Proposition. *For a functional φ on $\mathbf{B}(\mathfrak{H})$ the following conditions are equivalent:*

(i) *There are sequences (x_n) and (y_n) of vectors in $\mathfrak{H}$ with $\sum \|x_n\|^2 < \infty$ and $\sum \|y_n\|^2 < \infty$, such that*

$$\varphi(S) = \sum (Sx_n | y_n), \quad S \in \mathbf{B}(\mathfrak{H}).$$

(ii) *As in (i), but now with sequences (x_n) and (y_n) of pairwise orthogonal vectors.*

(iii) *There is a trace class operator T on $\mathfrak{H}$ such that $\varphi(S) = \operatorname{tr}(ST)$ for every S
in $\mathbf{B}(\mathfrak{H})$.*

(iv) *φ is σ-weakly continuous.*

PROOF. Since the σ-weak topology is induced from the family of functionals
described in (iii), it follows from 2.4.4 that (iii) $\Leftrightarrow$ (iv). Evidently, (ii) $\Rightarrow$ (i), so
we are left with the implications (i) $\Rightarrow$ (iii) and (iii) $\Rightarrow$ (ii).

(i) $\Rightarrow$ (iii). Let (e_n) be an orthonormal basis for the closed subspace $\mathfrak{H}_0$ of $\mathfrak{H}$
spanned by the vectors x_n and y_n, $n \in \mathbb{N}$. If $\dim(\mathfrak{H}_0) = d < \infty$, we can express
each vector x_n and y_n as a finite linear combination of the vectors $e_1, \ldots, e_d$
and obtain scalar constants α_{ij}, $1 \leq i, j \leq d$, such that

$$\varphi(S) = \sum \alpha_{ij}(Se_i | e_j) = \sum (STe_j | e_j) = \operatorname{tr}(ST),$$

where T is the operator of finite rank given by $Te_j = \sum \alpha_{ij} e_i$, $1 \leq j \leq d$, and
$T | \mathfrak{H}_0^\perp = 0$.

If $\dim \mathfrak{H}_0 = \infty$, we define T_1 and T_2 in $\mathbf{B}(\mathfrak{H})$ by $T_i | \mathfrak{H}_0^\perp = 0$ for $i = 1, 2$ and

$$T_1 e_n = x_n, \quad T_2 e_n = y_n, \quad n \in \mathbb{N}.$$

Then

$$\operatorname{tr}(T_1^* T_1) = \sum (T_1^* T_1 e_n | e_n) = \sum \|x_n\|^2 < \infty,$$

so that $T_1 \in \mathbf{B}^2(\mathfrak{H})$. Similarly, $T_2 \in \mathbf{B}^2(\mathfrak{H})$, and thus $T = T_1 T_2^* \in \mathbf{B}^1(\mathfrak{H})$. Finally,
by 3.4.11 we have for every S in $\mathbf{B}(\mathfrak{H})$ that

$$\operatorname{tr}(ST) = \operatorname{tr}(ST_1 T_2^*) = \operatorname{tr}(T_2^* ST_1)$$

$$= \sum (ST_1 e_n | T_2 e_n) = \sum (Sx_n | y_n) = \varphi(S).$$

(iii) $\Rightarrow$ (ii). Let $T = U|T|$ be the polar decomposition of T; cf. 3.2.17. Then
$|T|$ is a positive, compact operator (cf. 3.4.5), so by 3.3.8 there is a closed,
separable subspace $\mathfrak{H}_0$ of $\mathfrak{H}$ with orthonormal basis (e_n) and a sequence (λ_n)
in $\mathbb{R}_+$ such that $|T| \mathfrak{H}_0^\perp = 0$ and $|T| = \sum \lambda_n e_n \odot e_n$ (in the notation of 3.3.9).
Since $|T| \in \mathbf{B}^1(\mathfrak{H})$, we have $\sum \lambda_n < \infty$. Define $x_n = \lambda_n^{1/2} U e_n$ and $y_n = \lambda_n^{1/2} e_n$.
Then (x_n) consists of pairwise orthogonal vectors because U is an isometry of
$\mathfrak{H}_0$ (cf. 3.2.17). Finally, for every S in $\mathbf{B}(\mathfrak{H})$,

$$\varphi(S) = \operatorname{tr}(ST) = \sum (STe_n | e_n)$$

$$= \sum (S\lambda_n U e_n | e_n) = \sum (Sx_n | y_n). \qquad \square$$

4.6.12. Corollary. *If φ is a positive, σ-weakly continuous functional on $\mathbf{B}(\mathfrak{H})$,
there is an orthogonal sequence (x_n) in $\mathfrak{H}$ with $\sum \|x_n\|^2 = \|\varphi\|$ such that $\varphi(S) = \sum (Sx_n | x_n)$ for every S in $\mathbf{B}(\mathfrak{H})$.*

PROOF. By 4.6.11 we have $\varphi = \operatorname{tr}(\cdot \, T)$ for some T in $\mathbf{B}^1(\mathfrak{H})$. Since

$$(Tx | x) = \operatorname{tr}((x \odot x)T) = \varphi(x \odot x) \geq 0$$

for every x in $\mathfrak{H}$, it follows that $T \geq 0$. We can therefore choose an orthonormal basis for $\mathfrak{H}$ that diagonalizes T (cf. 3.3.8), from which the result is immediate. $\qquad\square$

4.6.13. It follows from 4.6.11 and 4.6.4 that the weak topology on $\mathbf{B}(\mathfrak{H})$ is weaker than the σ-weak topology. In fact, from the proof of the implication (i) $\Rightarrow$ (iii) in 4.6.11 we see (by 2.4.4) that $\mathbf{B}(\mathfrak{H})$ in the weak topology has $\mathbf{B}_f(\mathfrak{H})$ as its topological dual space (via the bilinear form given by the trace). Thus, if $\mathfrak{H}$ is infinite-dimensional, the weak topology is strictly weaker than the σ-weak topology [because $\mathbf{B}_f(\mathfrak{H}) \neq \mathbf{B}^1(\mathfrak{H})$]. Nevertheless, we have the following results.

4.6.14. Proposition. *The weak and the σ-weak topology coincide on every bounded subset of* $\mathbf{B}(\mathfrak{H})$. *Thus a functional φ on* $\mathbf{B}(\mathfrak{H})$ *is σ-weakly continuous if and only if;*

(v)   *The restriction $\varphi|\mathbf{B}$ is weakly continuous on the unit ball* $\mathbf{B}$ *of* $\mathbf{B}(\mathfrak{H})$.

PROOF. The identical map of $(\mathbf{B}, \sigma\text{-weak})$ onto $(\mathbf{B}, \text{weak})$ is continuous and injective, and since $\mathbf{B}$ is σ-weakly compact by 2.5.2, the map is a homeomorphism by 1.6.8. Replacing $\mathbf{B}$ by $n\mathbf{B}$, $n \in \mathbb{N}$, we see that the two topologies coincide on every bounded subset of $\mathbf{B}(\mathfrak{H})$.

The observation above makes it evident that condition (v) is necessary for φ to be σ-weakly continuous. The sufficiency follows neatly from the Krein–Smulian theorem, 2.5.11, q.v. $\qquad\square$

4.6.15. Proposition. *If $\mathfrak{A}$ is a self-adjoint, unital subalgebra of* $\mathbf{B}(\mathfrak{H})$ *and $S \in \mathfrak{A}''$, then for each sequence (x_n) of vectors in $\mathfrak{H}$ with $\sum \|x_n\|^2 < \infty$ and $\varepsilon > 0$ there is a T in $\mathfrak{A}$ with $\sum \|(S - T)x_n\|^2 < \varepsilon^2$. In particular, $\mathfrak{A}''$ is the σ-weak closure of $\mathfrak{A}$.*

PROOF. The amplification described in 4.6.3 can be defined also in the case $n = \infty$, as a $*$-isomorphism Φ of $\mathbf{B}(\mathfrak{H})$ into the "diagonal" of $\mathbf{B}(\mathfrak{H}^\infty)$—the latter identified with those infinite matrices with entries from $\mathbf{B}(\mathfrak{H})$ for which the corresponding operators on $\mathfrak{H}^\infty$ are bounded. Substituting this map in the proof of the implication (iii) $\Rightarrow$ (i) of 4.6.7, that argument becomes a proof of the first part of the proposition. For the second part, let φ be a σ-weakly continuous functional on $\mathbf{B}(\mathfrak{H})$ in the form given in 4.6.11(i). Then for each S in $\mathfrak{A}''$ and $\varepsilon > 0$ we obtain T in $\mathfrak{A}$ as above, and applying the Cauchy–Schwarz inequality we get

$$|\varphi(S - T)| = \left|\sum ((S - T)x_n | y_n)\right| \leq \sum \|(S - T)x_n\|\,\|y_n\|$$

$$\leq \left(\sum \|(S - T)x_n\|^2\right)^{1/2} \left(\sum \|y_n\|^2\right)^{1/2} < \varepsilon \left(\sum \|y_n\|^2\right)^{1/2}.$$

Thus S belongs to the σ-weak closure of $\mathfrak{A}$, as claimed. $\qquad\square$

4.6.16. Remarks. From the preceding result we see that the von Neumann algebras can also be characterized as the unital C^*-subalgebras of $\mathbf{B}(\mathfrak{H})$ that are σ-weakly closed. But, evidently, the content of 4.6.15 says more. To appreciate the extra information we must introduce the σ-*strong topology* on $\mathbf{B}(\mathfrak{H})$ as the locally convex vector space topology induced by seminorms of the form

$$T \to \left(\sum_{n=1}^{\infty} \| Tx_n \|^2 \right)^{1/2}, \qquad (x_n) \subset H, \qquad \sum_{n=1}^{\infty} \| x_n \|^2 < \infty.$$

The σ-strong topology is seen to be stronger than both the strong and the σ-weak topology (but of course weaker than the norm topology). Now 4.6.15 says that the σ-strong closure of a unital $*$-algebra $\mathfrak{A}$ is precisely $\mathfrak{A}''$.

It is not hard to show by direct computations that the strong and the σ-strong topology coincide on every bounded subset of $\mathbf{B}(\mathfrak{H})$. On the other hand, the σ-weak and the σ-strong topology are distinct (when $\mathfrak{H}$ is infinite-dimensional), because the involution fails to be σ-strongly continuous. Despite their difference, the two topologies have the same continuous functionals. This is seen by taking the proof of (iii) $\Rightarrow$ (i) in 4.6.4, and replacing $\mathfrak{H}^n$ by $\mathfrak{H}^{\infty}$ as described in the proof of 4.6.15.

4.6.17. Theorem. *For every von Neumann algebra* $\mathfrak{A}$ *in* $\mathbf{B}(\mathfrak{H})$ *there is a Banach space* $\mathfrak{A}_*$, *such that* $\mathfrak{A}$ *is isometrically isomorphic to the dual space* $(\mathfrak{A}_*)^*$. *The* w^*-*topology on* $\mathfrak{A}$ [*identified with* $(\mathfrak{A}_*)^*$] *is the* σ-*weak topology.*

PROOF. Identifying $\mathbf{B}(\mathfrak{H})$ and $(\mathbf{B}^1(\mathfrak{H}))^*$, we let $\mathfrak{A}^\perp$ denote the annihilator of $\mathfrak{A}$, i.e.

$$\mathfrak{A}^\perp = \{ T \in \mathbf{B}^1(\mathfrak{H}) \,|\, \mathrm{tr}(ST) = 0, \quad \forall S \in \mathfrak{A} \}.$$

Then $\mathfrak{A}^\perp$ is a norm closed subspace of $\mathbf{B}^1(\mathfrak{H})$ (equipped with the 1-norm, cf. 3.4.12), and we can form the Banach space $\mathfrak{A}_* = \mathbf{B}^1(\mathfrak{H})/\mathfrak{A}^\perp$; cf. 2.1.5. Since $\mathfrak{A}$ is a σ-weakly closed subspace of $\mathbf{B}(\mathfrak{H})$ (even weakly closed) it follows from 2.4.11 that $\mathfrak{A} = (\mathfrak{A}^\perp)^\perp$, and thus from 2.4.13 we see that the Banach space adjoint of the quotient map $Q: \mathbf{B}^1(\mathfrak{H}) \to \mathfrak{A}_*$ is the isometric embedding of $(\mathfrak{A}_*)^*$ onto $\mathfrak{A}$ in $\mathbf{B}(\mathfrak{H})$. □

4.6.18. Remarks. The space $\mathfrak{A}_*$ in 4.6.17 is called the *predual* of $\mathfrak{A}$, and was shown by S. Sakai to be the unique Banach space having $\mathfrak{A}$ as its dual. He at the same time gave the beautiful abstract (i.e. space-free) characterization of von Neumann algebras as being exactly those C^*-algebras that are dual spaces.

As long as one considers von Neumann algebras on a fixed Hilbert space, the weak and the strong topologies are most natural to work with. However, given two von Neumann algebras $\mathfrak{A}_1 \subset \mathbf{B}(\mathfrak{H}_1)$ and $\mathfrak{A}_2 \subset \mathbf{B}(\mathfrak{H}_2)$, we may have an isometric $*$-isomorphism $\Phi: \mathfrak{A}_1 \to \mathfrak{A}_2$, which is not weak–weak or strong–strong continuous. It suffices to take $\mathfrak{A}_1 = \mathbf{B}(\mathfrak{H})$ and Φ as the infinite analogue of the amplification map described in the proof of 4.6.14. If $\mathfrak{H}^\infty = \mathfrak{H} \oplus \mathfrak{H} \oplus \cdots$ and (x_n) is an orthogonal sequence in $\mathfrak{H}$ with $\sum \|x_n\|^2 < \infty$, then $x = x_1 \oplus x_2 \oplus \cdots$ is a vector in $\mathfrak{H}^\infty$, and thus gives a weakly continuous functional on $\mathbf{B}(\mathfrak{H}^\infty)$. But for each S in $\mathbf{B}(\mathfrak{H})$ we have

$$(\Phi(S)x|x) = \sum (Sx_n|x_n),$$

and this expression is not a weakly continuous functional. By contrast, every isomorphism between von Neumann algebras is a σ-weak homeomorphism, corresponding to the fact that the σ-weak topology is more "internal," less space-dependent than the weak and strong topologies.

EXERCISES

E 4.6.1. For each n, let V_n denote the rank one operator in $\mathbf{B}(\ell^2)$ given by $V_n e_1 = e_n$ and $V_n e_k = 0$ if $k \neq 1$, where (e_n) is the standard orthonormal basis for ℓ^2. Find V_n^* and show that $V_n^* \to 0$, strongly. Conclude that the adjoint operation is not strongly continuous and that the strong and the weak topologies on $\mathbf{B}(\ell^2)$ are different.

E 4.6.2. Show that the adjoint operation is strongly continuous when restricted to the set of normal elements in $\mathbf{B}(\mathfrak{H})$.

E 4.6.3. Show that the strong and the weak topologies coincide on the group $\mathbf{U}(\mathfrak{H})$ of unitary operators in $\mathbf{B}(\mathfrak{H})$.

E 4.6.4. Take V_n, $n \in \mathbb{N}$, as in E 4.6.1 and put

$$\mathfrak{S} = \{n^{1/2} V_n^* | n \in \mathbb{N}\}.$$

Show that 0 belongs to the strong closure of $\mathfrak{S}$, but that no sequence in $\mathfrak{S}$ is weakly or strongly convergent to 0. Conclude that neither the strong nor the weak topology on $\mathbf{B}(\ell^2)$ is metrizable.

Hint: Use E 3.1.9 and/or the principle of uniform boundedness (2.2.9).

E 4.6.5. If I is a compact subset of $\mathbb{R}$ and $f \in C(I)$, show that the map $T \to f(T)$ is strongly continuous from $\{T \in \mathbf{B}(\mathfrak{H})_{sa} | \mathrm{sp}(T) \subset I\}$ into $\{T \in \mathbf{B}(\mathfrak{H}) | T \text{ normal}\}$.

Hint: Use Weierstrass' approximation theorem and E 4.6.2.

E 4.6.6. Let $\mathscr{C}$ denote the set of continuous, real functions f on $\mathbb{R}$ for which the map $T \to f(T)$ is strongly continuous from $\mathbf{B}(\mathfrak{H})_{sa}$ into itself. Put $\mathscr{C}_b = \mathscr{C} \cap C_b(\mathbb{R})$ and $\mathscr{C}_0 = \mathscr{C} \cap C_0(\mathbb{R})$.

(i) Show that $\mathscr{C}$ is a uniformly closed vector space in $C(\mathbb{R})$.
(ii) Show that $\mathscr{C}_b \mathscr{C} \subset \mathscr{C}$ and that $\mathscr{C}_b$ is a unital Banach algebra in $C_b(\mathbb{R})$.
(iii) Show that $\mathscr{C} + i\mathscr{C}$ contains every "strongly continuous" complex function on $\mathbb{R}$.

(iv) Show that $f_1 \in \mathscr{C}_0$, where $f_1(t) = (1 + t^2)^{-1}t$

(v) Show that $f_\alpha \in \mathscr{C}_0$ for every $\alpha > 0$ and that $g_0 \in \mathscr{C}_0$, where $f_\alpha(t) = (1 + \alpha t^2)^{-1}t$ and $g_0(t) = (1 + t^2)^{-1}$.

(vi) Show that $\mathscr{C}_0 = C_0(\mathbb{R})$.

(vii) Show that if $f \in C(\mathbb{R})$ and $|f(t)| \leq a|t| + b$ for some a, b and all t, then $f \in \mathscr{C}$.

Hints: For (ii) use (*) in 4.6.1. For (iii) use E 4.6.2. For (iv) use the identity

$$f_1(S) - f_1(T) = (I + S^2)^{-1}(S - T)(I + T^2)^{-1}$$
$$+ (I + S^2)^{-1}S(T - S)(I + T^2)^{-1}T,$$

and the fact that when $S \to T$, strongly, the factors $(I + S^2)^{-1}$ and $(I + S^2)^{-1}S$ stay bounded, cf. (*) in 4.6.1. For (v) use that $g_0(t) = 1 - f_1(t)t$, together with (ii). For (vi) use the Stone–Weierstrass theorem (4.3.5) in conjunction with (ii) and (v). For (vii), note that $fg_0 \in C_0(\mathbb{R})$ whence $fg_0 \in \mathscr{C}_0$ by (vi). Therefore, $ff_1 (=fg_0 id)$ belongs to $\mathscr{C}_b$ by (ii) and, again by (ii), $ff_1 id \in \mathscr{C}$. Finally, $f = fg_0 + ff_1 id \in \mathscr{C}$.

E 4.6.7. (*Kaplansky's density theorem*.) If $\mathfrak{A}$ is a self-adjoint, unital subalgebra of $B(\mathfrak{H})$, prove the following statements:

(i) The unit ball of $\mathfrak{A}_{sa}$ is strongly dense in that of $\mathfrak{A}''_{sa}$.

(ii) The unit ball of $\mathfrak{A}_+$ is strongly dense in that of $\mathfrak{A}''_+$.

(iii) The unitary group in $\mathfrak{A}^=$ is strongly dense in that of $\mathfrak{A}''$.

(iv) The unit ball of $\mathfrak{A}$ is strongly dense in that of $\mathfrak{A}''$.

Hints: The four conditions are trivial for the norm topology, so we may assume that $\mathfrak{A}$ is norm closed, i.e. a C^*-algebra. For (i), use first 4.6.5 to show that $\mathfrak{A}_{sa}$ is strongly dense in $\mathfrak{A}''_{sa}$. Then if $T \in \mathfrak{A}''_{sa}$ and $\|T\| \leq 1$, find $(T_\lambda)_{\lambda \in \Lambda}$ in $\mathfrak{A}_{sa}$ with $T_\lambda \to T$, strongly, and replace T_λ with $f(T_\lambda)$, where $f(t) = (t \wedge 1) \vee (-1)$, using E 4.6.6. For (ii) do as in (i), but use the function $t \to (t \wedge 1) \vee 0$. For (iii), if U is unitary in $\mathfrak{A}''$, take $T = -i\log U$ as in E 4.5.5. Argue that $T \in \mathfrak{A}''_{sa}$ (or quote 4.5.4) and find $(T_\lambda)_{\lambda \in \Lambda}$ in $\mathfrak{A}_{sa}$ with $T_\lambda \to T$, strongly. Then $(\exp i\, T_\lambda)_{\lambda \in \Lambda}$ are unitaries in $\mathfrak{A}$ and converge strongly to $\exp i\, T(= U)$ by E 4.6.6. For (iv) take T in $\mathfrak{A}''$ with $\|T\| < 1$. Then use 3.2.23 to reduce the problem to case (iii).

E 4.6.8. A positive functional φ on a von Neumann algebra $\mathfrak{A} \subset B(\mathfrak{H})$ is *completely additive* if $\varphi(P) = \sum \varphi(P_j)$, for every family $\{P_j | j \in J\}$ of pairwise orthogonal projections in $\mathfrak{A}$ with $\sum P_j = P$ (strongly convergent sum, cf. 4.5.2). Show in this case that there is a family $\{Q_i | i \in I\}$ of pairwise orthogonal projections in $\mathfrak{A}$ with $\sum Q_i = I$, such that every functional $T \to \varphi(TQ_i)$ is weakly continuous on $\mathfrak{A}$.

Hints: Take $\{Q_i | i \in I\}$ to be a maximal family of pairwise orthogonal projections in $\mathfrak{A}$ such that each $\varphi(\cdot\, Q_i)$ is weakly continuous.

If $I - \sum Q_i = Q \neq 0$, take a unit vector x in $Q(\mathfrak{H})$, and, with $\psi(T) = 2(Tx|x)$, choose a maximal family $\{P_j | j \in J\}$ of pairwise orthogonal projections in $\mathfrak{A}$ such that $P_j \leq Q$ and $\psi(P_j) < \varphi(P_j)$ for every j. Let $P = Q - \sum P_j$ and use the complete additivity to show that $P \neq 0$. Deduce from the maximality of $\{P_i\}$ that $\varphi(P_0) \leq \psi(P_0)$ for every projection $P_0 \leq P$ in $\mathfrak{A}$, and use E4.5.3 to conclude that $\varphi(T) \leq \psi(T)$ for every $T \geq 0$ in $P\mathfrak{A}P$. Use the Cauchy–Schwarz inequality (E4.3.10) to show that

$$|\varphi(TP)|^2 \leq \|\varphi\|\varphi(PT^*TP) \leq \|\varphi\|\psi(PT^*TP)$$

$$= 2\|\varphi\| \|TPx\|^2,$$

for every T in $\mathfrak{A}$, and conclude that $\varphi(\cdot P)$ is strongly, hence weakly, continuous on $\mathfrak{A}$, in contradiction with the maximality of $\{Q_j\}$.

E4.6.9. Show that every completely additive, positive functional φ on a von Neumann algebra $\mathfrak{A} \subset \mathbf{B}(\mathfrak{H})$ is σ-weakly continuous.

Hint: Use E4.6.8 and the Cauchy–Schwarz inequality (E4.3.10) to show that φ can be approximated in norm by weakly continuous functionals [of the form $\varphi(\cdot Q')$, where $Q' = Q_1 + \cdots + Q_n$], and recall from 4.6.17 that the predual $\mathfrak{A}_*$ of $\mathfrak{A}$ is a norm closed subspace of the dual $\mathfrak{A}^*$ of $\mathfrak{A}$.

E4.6.10. Show that every *-isomorphism $\Phi: \mathfrak{A}_1 \to \mathfrak{A}_2$ between von Neumann algebras $\mathfrak{A}_1 \subset \mathbf{B}(\mathfrak{H}_1)$ and $\mathfrak{A}_2 \subset \mathbf{B}(\mathfrak{H}_2)$, is a σ-weak homeomorphism.

Hint: Show that Φ is completely additive on projections; cf. E4.6.8, because $\sum \rho(P_j) \leq \rho(\sum P_j)$ both for $\rho = \Phi$ and for $\rho = \Phi^{-1}$. Then use E4.6.9 to show that Φ^* (2.3.9) takes $(\mathfrak{A}_2)_*$ onto $(\mathfrak{A}_1)_*$, so that 2.4.12 may be applied to $\Phi = (\Phi_*)^*$, where $\Phi_* = \Phi^*|(\mathfrak{A}_2)_*$.

4.7. Maximal Commutative Algebras

Synopsis. The condition $\mathfrak{A} = \mathfrak{A}'$. Cyclic and separating vectors. $\mathscr{L}^\infty(X)$ as multiplication operators. A measure-theoretic model for MAÇA's. Multiplicity-free operators. MAÇA's as a generalization of orthonormal bases. The spectral theorem revisited. Exercises.

4.7.1. From the rich (and still growing) theory of von Neumann algebras we shall only consider a single item, because it provides a new insight in the measure-theoretic form of the spectral theorem.

We say that a commutative, self-adjoint algebra $\mathfrak{A}$ of operators in $\mathbf{B}(\mathfrak{H})$ is *maximal commutative* if it is not contained properly in any larger commutative *-subalgebra of $\mathbf{B}(\mathfrak{H})$. Now it is quite obvious that if $T \in \mathfrak{A}'$ and $T = T^*$, then the algebra generated by $\mathfrak{A}$ and T will be commutative and self-adjoint. So, just as the condition $\mathfrak{A} \subset \mathfrak{A}'$ characterizes the commutative *-subalgebras of

$\mathbf{B}(\mathfrak{H})$, the condition

$$\mathfrak{A} = \mathfrak{A}' \qquad\qquad (*)$$

characterizes the maximal commutative algebras in the class of $*$-subalgebras of $\mathbf{B}(\mathfrak{H})$. In particular, each MAÇA is weakly closed and contains I, and is therefore a von Neumann algebra.

4.7.2. Given a self-adjoint algebra $\mathfrak{A}$ of operators on a Hilbert space $\mathfrak{H}$, we say that a vector x is *cyclic* for $\mathfrak{A}$ if the subspace $\mathfrak{A}x$ is dense in $\mathfrak{H}$. We say that a projection P in $\mathbf{B}(\mathfrak{H})$ is *cyclic* relative to $\mathfrak{A}$ if $\mathfrak{A}x$ is dense in $P(\mathfrak{H})$ for some vector x in $\mathfrak{H}$. Thus the algebra $\mathfrak{A}$ has a cyclic vector iff I is a cyclic projection relative to $\mathfrak{A}$. Note from the proof that (iii) $\Rightarrow$ (i) in 4.6.7 that every cyclic projection relative to $\mathfrak{A}$ belongs to $\mathfrak{A}'$.

We say that a vector x in $\mathfrak{H}$ is *separating* for the algebra $\mathfrak{A}$ if $Tx = 0$ implies $T = 0$ for every T in $\mathfrak{A}$.

4.7.3. Lemma. *A vector x in $\mathfrak{H}$ is cyclic for a self-adjoint subalgebra $\mathfrak{A}$ of $\mathbf{B}(\mathfrak{H})$ iff x is separating for $\mathfrak{A}'$.*

PROOF. If x is cyclic for $\mathfrak{A}$ and $T \in \mathfrak{A}'$, then $Tx = 0$ implies $T(\mathfrak{H}) = T(\mathfrak{A}x)^= \subset (\mathfrak{A}Tx)^= = \{0\}$, whence $T = 0$. Conversely, if x is separating for $\mathfrak{A}'$, let P denote the cyclic projection on $(\mathfrak{A}x)^=$. Then $P \in \mathfrak{A}'$, so $I - P \in \mathfrak{A}'$. But $(I - P)x = 0$, whence $P = I$. $\qquad\square$

4.7.4. Lemma. *For every self-adjoint subalgebra $\mathfrak{A}$ of $\mathbf{B}(\mathfrak{H})$ there is a family $\{P_j | j \in J\}$ of pairwise orthogonal projections in $\mathfrak{A}'$, cyclic relative to $\mathfrak{A}$, such that $\sum P_j = I$. If $\mathfrak{H}$ is separable, J is countable.*

PROOF. Choose by Zorn's lemma (1.1.3) a maximal family $\{e_j | j \in J\}$ of unit vectors in $\mathfrak{H}$, such that $(Se_i | Te_j) = 0$ for all S, T in $\mathfrak{A}$ if $i \neq j$. Let P_j denote the cyclic projection on $(\mathfrak{A}e_j)^=$, and note that $P_i P_j = 0$ if $i \neq j$. Put $P = \sum P_j$ (strong convergence), so that $P(\mathfrak{H})$ is the orthogonal sum (3.1.5) of the subspaces $P_j(\mathfrak{H})$. If $P \neq I$, there is by 3.1.8 a unit vector e_0 orthogonal to $\mathfrak{A}e_j$ for every j in J. But then $\mathfrak{A}e_0 \perp \mathfrak{A}e_j$, because $\mathfrak{A}$ is self-adjoint, which contradicts the maximality of the family $\{e_j | j \in J\}$. Consequently, $P = I$. $\qquad\square$

4.7.5. Lemma. *On a separable Hilbert space $\mathfrak{H}$, every commutative, self-adjoint subalgebra $\mathfrak{A}$ of $\mathbf{B}(\mathfrak{H})$ has a separating vector.*

PROOF. Choose by 4.7.4 a sequence (P_n) of pairwise orthogonal cyclic projections in $\mathfrak{A}'$ with sum I, and let (e_n) be the corresponding sequence of unit vectors such that $(\mathfrak{A}e_n)^= = P_n(\mathfrak{H})$ for every n. Take $e = \sum 2^{-n}e_n$. If $T \in \mathfrak{A}$ and $Te = 0$, then

$$TP_n(\mathfrak{H}) = (T\mathfrak{A}e_n)^= = (\mathfrak{A}TP_n e)^=$$
$$= (\mathfrak{A}P_n Te)^= = \{0\},$$

whence $TP_n = 0$ for every n. Since $\sum P_n = I$, this implies that $T = 0$, so that e is separating for $\mathfrak{A}$. □

4.7.6. Proposition. *Let $\int$ be a Radon integral on a locally compact, σ-compact Hausdorff space X, and define for each f in $\mathscr{L}^\infty(X)$ the multiplication operator M_f in $\mathbf{B}(L^2(X))$ by*

$$M_f g = fg, \quad g \in \mathscr{L}^2(X).$$

*Then the map $f \to M_f$ gives an isometric *-isomorphism of $L^\infty(X)$ onto a maximal commutative subalgebra of $\mathbf{B}(L^2(X))$, which contains $\{M_f | f \in C_c(X)\}$ as a strongly dense subalgebra.*

PROOF. It is clear that $f \to M_f$ is a well-defined norm decreasing *-homorphism of $L^\infty(X)$ into $\mathbf{B}(L^2(X))$. To see that it is an isometry, take for each $\varepsilon > 0$ a compact subset C of the set $\{x \in X \,|\, |f(x)| \geq \|f\|_\infty - \varepsilon\}$ with $\int[C] > 0$. Then with $g = \bar{f}|f|^{-1}[C]$ and $h = [C]$ we have

$$(M_f g | h) = \int |f|[C] \geq (\|f\|_\infty - \varepsilon)\|g\|_2 \|h\|_2,$$

whence $\|M_f\| \geq \|f\|_\infty - \varepsilon$.

Let $\mathfrak{A}_c = \{M_f | f \in C_c(X)\}$ and consider an operator T in $\mathfrak{A}_c'$. If g_1, h_1, g_2, h_2 are elements in $C_c(X)$ such that $\bar{h}_1 g_1 = \bar{h}_2 g_2$, choose by 1.7.5 a function e in $C_c(X)_{sa}$ that is 1 on the support of all the four functions. Then compute

$$(Tg_1 | h_1) = (Tg_1 | h_1 e) = (Tg_1 | M_{h_1} e) = (M_{h_1}^* Tg_1 | e)$$
$$= (TM_{h_1}^* g_1 | e) = (T\bar{h}_1 g_1 | e) = (T\bar{h}_2 g_2 | e) = \cdots = (Tg_2 | h_2).$$

This shows that the map $\bar{h}g \to (Tg|h)$ is a well-defined function on $C_c(X)$. Choosing for each g in $C_c(X)$ a function e in $C_c(X)_{sa}$ that is 1 on the support of g, we can therefore define

$$\varphi(g) = (Tg | e), \quad g \in C_c(X).$$

Then φ is a (linear) functional. Indeed, given g_1 and g_2 we may assume that $eg_1 = g_1$ and $eg_2 = g_2$, whence $\varphi(g_1) + \varphi(g_2) = (T(g_1 + g_2)|e) = \varphi(g_1 + g_2)$. Finally, from the decomposition $g = g|g|^{-1/2}|g|^{1/2}$ we see that

$$|\varphi(g)| = |(Tg|g|^{-1/2} \,|\, |g|^{1/2})|$$
$$\leq \|T\| \|g|g|^{-1/2}\|_2 \||g|^{1/2}\|_2$$
$$= \|T\| \||g|^{1/2}\|_2^2 = \|T\| \|g\|_1.$$

Thus, by 6.4.11 and 2.1.11 φ extends by continuity to a functional on $L^1(X)$, bounded by $\|T\|$. By 6.5.11 there is therefore a function f in $\mathscr{L}^\infty(X)$ such that $\varphi(g) = \int fg$ for every g in $\mathscr{L}^1(X)$. In particular, if g, h belong to $C_c(X)$, then

$$(M_f g | h) = \int fg\bar{h} = \varphi(g\bar{h}) = (Tg | h),$$

whence $T = M_f$, as $C_c(X)$ is dense in $L^2(X)$.

If $\mathfrak{A} = \{M_f | f \in \mathscr{L}^\infty(X)\}$, then the argument above showed that $\mathfrak{A}_c' \subset \mathfrak{A}$.

Since $\mathfrak{A}_c \subset \mathfrak{A}$, we also have $\mathfrak{A} \subset \mathfrak{A}' \subset \mathfrak{A}'_c$, which shows that $\mathfrak{A} = \mathfrak{A}'$. Thus $\mathfrak{A}$ is maximal commutative; cf. 4.7.1. Moreover, $\mathfrak{A}''_c = \mathfrak{A}$, which by 4.6.8 means that $\mathfrak{A}_c + \mathbb{C}I$ is strongly dense in $\mathfrak{A}$. However, $\mathfrak{A}_c$ contains I in its strong closure (use an approximate unit from $C_c(X)$), so we conclude that $\mathfrak{A}_c$ is strongly dense in $\mathfrak{A}$. $\qquad\qquad\square$

4.7.7. Theorem. *For a self-adjoint algebra $\mathfrak{A}$ of operators on a separable Hilbert space $\mathfrak{H}$, the following conditions are equivalent*:

(i) *$\mathfrak{A}$ is maximal commutative.*
(ii) *$\mathfrak{A}$ is a commutative von Neumann algebra with a cyclic vector.*
(iii) *There is a Radon integral $\int$ on a compact, second countable Hausdorff space X, and an isometry U of $L^2(X)$ on $\mathfrak{H}$, such that the map $f \to UM_f U^*$ is an isometric $*$-isomorphism of $L^\infty(X)$ onto $\mathfrak{A}$.*

PROOF. (i) $\Rightarrow$ (ii). By 4.7.5 there is a separating vector e for $\mathfrak{A}$. Since $\mathfrak{A} = \mathfrak{A}'$ (cf. 4.7.1), it follows from 4.7.3 that e is cyclic for $\mathfrak{A}$.

(ii) $\Rightarrow$ (iii). Since $\mathbf{B}^1(\mathfrak{H})$ is separable, it follows from 4.6.17 that $\mathfrak{A}_*$ is separable, whence $\mathfrak{A}$ is σ-weakly separable (because the unit ball is metrizable, cf. E 2.5.3). But then $\mathfrak{A}$ is also strongly separable by 4.6.5. We can therefore find a separable C^*-subalgebra $\mathfrak{A}_0$ (generated by some strongly dense sequence in $\mathfrak{A}$), which is strongly dense in $\mathfrak{A}$. Assuming, as we may, that $I \in \mathfrak{A}_0$, we see from 4.3.13 that $\mathfrak{A}_0$ is isometrically $*$-isomorphic to $C(X)$ for some compact Hausdorff space X. Since $C(X)$ is separable, X is second countable [being a closed subset of the unit ball of $C(X)^*$, which is w^*-compact and metrizable, hence second countable].

If $f \to T_f$ denotes the isomorphism of $C(X)$ onto $\mathfrak{A}_0$, we define a Radon integral $\int$ on X by $\int f = (T_f e | e)$, where e is the cyclic vector for $\mathfrak{A}$. The map $U_0 \colon C(X) \to \mathfrak{H}$ given by $U_0 f = T_f e$ satisfies the equation

$$\| U_0 f \|_2^2 = (T_{|f|^2} e | e) = \int |f|^2 = \| f \|_2^2.$$

It therefore extends by continuity (2.1.11) to an isometry $U \colon L^2(X) \to \mathfrak{H}$, which is surjective since e is cyclic and $\mathfrak{A}_0$ is strongly dense in $\mathfrak{A}$. For f and g in $C(X)$ we have

$$UM_f g = Ufg = T_{fg} e = T_f T_g e = T_f Ug.$$

Since $C(X)$ is dense in $L^2(X)$, it follows that $UM_f U^* = T_f$. The map $S \to USU^*$ from $\mathbf{B}(L^2(X))$ onto $\mathbf{B}(\mathfrak{H})$ is evidently a strong homeomorphism, and since $\{ M_f | f \in C(X) \}$ is strongly dense in $\{ M_f | f \in L^\infty(X) \}$ by 4.7.6, and $\mathfrak{A}_0$ is strongly dense in $\mathfrak{A}$ by choice, it follows that $f \to UM_f U^*$ takes $L^\infty(X)$ isometrically onto $\mathfrak{A}$.

(iii) $\Rightarrow$ (i). Since $S \to USU^*$ is an isomorphism of $\mathbf{B}(L^2(X))$ onto $\mathbf{B}(\mathfrak{H})$, it maps maximal commutative subalgebras (e.g. $L^\infty(X)$, cf. 4.7.6] to maximal commutative subalgebras. $\qquad\qquad\square$

4.7.8. Proposition. *If two maximal commutative von Neumann algebras $\mathfrak{A}_1$ and $\mathfrak{A}_2$ in $\mathbf{B}(\mathfrak{H})$ are $*$-isomorphic and $\mathfrak{H}$ is separable, then $U\mathfrak{A}_1 U^* = \mathfrak{A}_2$ for some unitary U in $\mathbf{B}(\mathfrak{H})$.*

PROOF. By 4.7.7 we have Radon integrals $\int_1$ and $\int_2$ on compact Hausdorff spaces X_1 and X_2 and isometries $U_j : L^2(X_i) \to \mathfrak{H}$, $j = 1, 2$, that implement the isomorphisms between $L^\infty(X_j)$ and $\mathfrak{A}_j$, $j = 1, 2$. Therefore, if $\Phi : \mathfrak{A}_1 \to \mathfrak{A}_2$ is a ∗-isomorphism, we obtain a ∗-isomorphism $\Psi : L^\infty(X_1) \to L^\infty(X_2)$ by

$$\Phi(U_1 M_f U_1^*) = U_2 M_{\Psi(f)} U_2^*, \quad f \in L^\infty(X_1).$$

The map $f \to \int_2 \Psi(f)$, $f \in L^\infty(X_1)$, is evidently a Radon integral on X_1, which (by definition, cf. 6.5.3) is equivalent with $\int_1$. By Radon–Nikodym's theorem (6.5.4) there is therefore a locally integrable (= integrable here) Borel function $m \geq 0$ (actually $m > 0$ almost everywhere) such that

$$\int_2 \Psi(f) = \int_1 fm, \quad f \in \mathscr{L}^\infty(X_1).$$

Since $m \in \mathscr{L}^1(X_1)$, $m^{1/2} f \in \mathscr{L}^2(X_1)$ for every f in $\mathscr{L}^\infty(X_1)$, so we can define an operator W_0 from $m^{1/2} \mathscr{L}^\infty(X_1) \subset \mathscr{L}^2(X_1)$ into $\mathscr{L}^\infty(X_2) \subset \mathscr{L}^2(X_2)$ by $W_0(m^{1/2} f) = \Psi(f)$. We have

$$\| W_0(m^{1/2} f)\|_2^2 = \|\Psi(f)\|_2^2 = \int_2 \Psi(|f|^2)$$

$$= \int_1 |f|^2 m = \|m^{1/2} f\|_2^2;$$

and since $m^{1/2} \mathscr{L}^\infty(X_1)$ is dense in $\mathscr{L}^2(X_1)$ and $\mathscr{L}^\infty(X_2)$ is dense in $\mathscr{L}^2(X_2)$ it follows that W_0 extends by continuity to an isometry W of $L^2(X_1)$ onto $L^2(X_2)$. It is elementary to check that $W M_f W^* = M_{\Psi(f)}$ for every f in $\mathscr{L}^\infty(X_1)$, and setting $U = U_2 W U_1^*$ we obtain an isometry of $\mathfrak{H}$ onto itself, i.e. a unitary operator, such that $U \mathfrak{A}_1 U^* = \mathfrak{A}_2$. Indeed, for each T in $\mathfrak{A}_1$ [of the form $U_1 M_f U_1^*$ for some f in $\mathscr{L}^\infty(X_1)$] we have

$$UTU^* = U_2 W U_1^* U_1 M_f U_1^* (U_2 W U_1^*)^* = U_2 W M_f W^* U_2^*$$

$$= U_2 M_{\Psi(f)} U_2^* = \Phi(U_1 M_f U_1^*) = \Phi(T). \qquad \square$$

4.7.9. A normal operator T on a separable Hilbert space $\mathfrak{H}$ is *multiplicity-free* if there exists a cyclic vector for the C^*-algebra $C^*(T)$, i.e. if the subspace $\{ f(T)x \mid f \in C(\mathrm{sp}(T))\}$ is dense in $\mathfrak{H}$ for some vector x. Since the smallest von Neumann algebra, $W^*(T)$, containing T and I, is the strong closure of $C^*(T)$, it follows from 4.7.7 that T is multiplicity-free iff $W^*(T)$ is maximal commutative in $\mathbf{B}(\mathfrak{H})$.

As a justification for the name, note that any (possible) eigenvalue λ for T must have multiplicity one. Indeed, if $Tx = \lambda x$ and $Ty = \lambda y$, with $\|x\| = \|y\| = 1$ and $x \perp y$, then the unitary operator U on $\mathfrak{H}$, for which $Ux = y$, $Uy = -x$ and $U = I$ on $(\mathbb{C}x + \mathbb{C}y)^\perp$, will belong to $W^*(T)'$ (becuase $UT = TU$), but not to $W^*(T)$ [because $(S(x - y)|x + y) = 0$ for every S in $W^*(T)$, whereas $(U(x - y)|x + y) = 2$].

Replacing $\mathfrak{A}_0$ with $C^*(T)$ in the proof of 4.7.7 we obtain the following *spatial version* of the *spectral theorem*.

4.7.10. Corollary. *If T is a normal, multiplicity-free operator on a separable Hilbert space $\mathfrak{H}$, there is a Radon integral on $\mathrm{sp}(T)$ and an isometry U of $L^2(\mathrm{sp}(T))$ onto $\mathfrak{H}$, such that if we define*

$$f(T) = UM_f U^*, \quad f \in \mathscr{L}^\infty(\mathrm{sp}(T)),$$

then the map $f \to f(T)$ induces an isometric $*$-isomorphism of $L^\infty(\mathrm{sp}(T))$ onto $W^*(T)$ that extends the isomorphism of $C(\mathrm{sp}(T))$ on $C^*(T)$.

4.7.11. Proposition. *If S and T are normal, multiplicity-free operators on a separable Hilbert space $\mathfrak{H}$, the following conditions are equivalent:*

(i) *S and T are unitarily equivalent.*
(ii) *$\mathrm{sp}(S) = \mathrm{sp}(T)$, and there are cyclic vectors x and y for $C^*(S)$ and $C^*(T)$, respectively, such that the two Radon integrals on $\mathrm{sp}(S)$ given by $f \to (f(S)x|x)$ and $f \to (f(T)y|y)$ for f in $C(\mathrm{sp}(S))$, are equivalent.*
(iii) *There is an isometric $*$-isomorphism Φ of $W^*(S)$ on $W^*(T)$ such that $\Phi(S) = T$.*

PROOF. (i) $\Rightarrow$ (ii) is evident, because if $USU^* = T$ and x is cyclic for $C^*(S)$, then $y = Ux$ will be cyclic for $C^*(T)$ and the two Radon integrals will coincide.

(ii) $\Rightarrow$ (iii). Since equivalent Radon integrals have identical L^∞-spaces by 6.5.3, the desired isomorphism Φ is obtained by composing the isomorphism of $W^*(S)$ onto $L^\infty(\mathrm{sp}(S))$ $[= L^\infty(\mathrm{sp}(T)]$ with that of $L^\infty(\mathrm{sp}(T))$ onto $W^*(T)$, cf. 4.7.10.

(iii) $\Rightarrow$ (i). By 4.7.8 there is a unitary U such that $\Phi = U \cdot U^*$. In particular, $USU^* = \Phi(S) = T$. $\qquad\square$

4.7.12. We say that a maximal commutative algebra $\mathfrak{A}$ in $\mathbf{B}(\mathfrak{H})$ is *atomic*, if it is isomorphic to $L^\infty(X)$ for some atomic integral; cf. 6.4.4. In the separable case this evidently means that $\mathfrak{A}$ is isomorphic to ℓ^∞. Choosing an orthonormal basis $\{e_n | n \in \mathbb{N}\}$ for $\mathfrak{H}$ we see (directly or by using 4.7.7) that the set $\mathfrak{A}$ of diagonal operators relative to (e_n) is an atomic, maximal commutative algebra in $\mathbf{B}(\mathfrak{H})$; and by 4.7.8 there are (up to unitary equivalence) no other. However, using 4.7.7 it is easy to construct maximal commutative algebras in $\mathbf{B}(\mathfrak{H})$ that are *continuous*, in the sense that every nonzero projection P in the algebra dominates some nonzero projection in the algebra not equal to P. These are isomorphic to L^∞-spaces corresponding to continuous integrals. From the remarks in 6.4.4 it is not hard to show that every maximal commutative algebra $\mathfrak{A}$ in $\mathbf{B}(\mathfrak{H})$ can be decomposed (uniquely) as $\mathfrak{A} = \mathfrak{A}_c \oplus \mathfrak{A}_a$, where $\mathfrak{A}_c$ has unit P and is a continuous, maximal commutative algebra in $\mathbf{B}(P(\mathfrak{H}))$, whereas $\mathfrak{A}_a$ has unit I-P and is an atomic, maximal commutative algebra in $\mathbf{B}((I - P)\mathfrak{H})$.

From the discussion above it is clear that we may regard a maximal commutative algebra in $\mathbf{B}(\mathfrak{H})$ as a "generalized basis" for $\mathfrak{H}$. Doing that, the spectral theorem regains its finite-dimensional flavor of "finding a basis that diagonalizes the operator." The following result illustrates this point of view.

4.7.13. Proposition. *If $\{T_j | j \in J\}$ is a commuting family of normal operators on a separable Hilbert space $\mathfrak{H}$, there is a Radon integral on a compact, second countable Hausdorff space X, an isometry U of $L^2(X)$ on $\mathfrak{H}$, and a family $\{f_j | j \in J\}$ in $\mathscr{L}^\infty(X)$, such that $UM_{f_j}U^* = T_j$ for every j in J.*

PROOF. Let $\mathfrak{A}_J$ denote the C^*-algebra in $\mathbf{B}(\mathfrak{H})$ generated by the operators T_j. Then $\mathfrak{A}_J$ is commutative and thus, by Zorn's lemma (1.1.3), contained in a maximal commutative $*$-subalgebra $\mathfrak{A}$ of $\mathbf{B}(\mathfrak{H})$. Now 4.7.7 applies to complete the proof.   □

4.7.14. Remarks. The result above can be beautified in several manners (without altering its substance, of course).

If J is countable, we may take the functions f_j, $j \in J$, in $C(X)$ [and not just $\mathscr{L}^\infty(X)$] by choosing the separable, strongly dense C^*-subalgebra $\mathfrak{A}_0$ from the proof of 4.7.7 such that $\mathfrak{A}_J \subset \mathfrak{A}_0$.

In another direction, we may replace the abstract space X in 4.7.13 by, say, the unit interval $[0, 1]$. This is a consequence of the theory of standard Borel spaces, where the main result states that any two second countable, complete metric spaces ("Polish spaces") are Borel isomorphic if they have the same cardinality as sets.

None of these embellishments are completely satisfactory when 4.7.13 is applied to a single operator T, because we loose track of the fundamental invariant sp(T). If, instead of enlarging $C^*(T)$ to a maximal commutative algebra, we cut up the Hilbert space into subspaces on which T is multiplicity-free, we arrive at the following version of the *spatial spectral theorem*.

4.7.15. Theorem. *If T is a normal operator on a separable Hilbert space $\mathfrak{H}$, there is a sequence (P_n) of pairwise orthogonal projections commuting with T such that $\sum P_n = I$, and a sequence $(\int_n)$ of normalized Radon integrals on sp(T) such that for each n we have an isometry U_n of $L_n^2(\mathrm{sp}(T))$ on $P_n(\mathfrak{H})$. Moreover, with $f(T) = \sum U_n M_f U_n^*$ we obtain an isometric $*$-isomorphism $f \to f(T)$ of $L^\infty(\mathrm{sp}(T))$ with respect to the Radon integral $\int = \sum 2^{-n} \int_n$ onto $W^*(T)$, that extends the isomorphism of $C(\mathrm{sp}(T))$ onto $C^*(T)$.*

PROOF. By 4.7.4 there is an orthogonal sequence (P_n) of cyclic projections relative to $W^*(T)$ [thus belonging to $W^*(T)'$] with strong sum I. Therefore, TP_n is multiplicity-free on $P_n(\mathfrak{H})$; so if x_n denotes a cyclic unit vector for $W^*(TP_n)$ on $P_n(\mathfrak{H})$, then 4.7.7 applies to produce the isometry U_n of $L_n^2(\mathrm{sp}(T))$ with respect to the normalized Radon integral given by $\int_n f = (f(T)x_n | x_n)$, $f \in C(\mathrm{sp}(T))$, onto $P_n(\mathfrak{H})$.

The operator $\sum U_n M_f U_n^*$ (strongly convergent sum) is well-defined in $\mathbf{B}(\mathfrak{H})$ for every bounded Borel function f on sp(T). If $f \in C(\mathrm{sp}(T))$, then $f(T)P_n = f(TP_n) = U_n M_f U_n^*$ for every n, whence $\sum U_n M_f U_n^* = \sum f(T)P_n = f(T)$. We can therefore define $f(T) = \sum U_n M_f U_n^*$ for every f in $\mathscr{B}_b(\mathrm{sp}(T))$, to obtain a norm decreasing $*$-homomorphism $f \to f(T)$ from $\mathscr{B}_b(\mathrm{sp}(T))$ into $\mathbf{B}(\mathfrak{H})$, extending the isomorphism of $C(\mathrm{sp}(T))$ on $C^*(T)$. (We are, in fact, reconstructing the spectral map from 4.5.4.) Clearly $f(T) = 0$ iff $M_f = 0$ on $L_n^2(\mathrm{sp}(T))$ for every n, i.e. iff f is a null function for the integral $\int = \sum 2^{-n} \int_n$. Thus $f \to f(T)$ defines an isomorphism of $L^\infty(\mathrm{sp}(T))$ with respect to $\int$ into $\mathbf{B}(\mathfrak{H})$. We interrupt the proof to establish a continuity property of the map.

4.7.16. Proposition. *The spectral map $f \to f(T)$ in 4.7.15 is a homeomorphism, when $L^\infty(\mathrm{sp}(T))$ is equipped with the w^*-topology as the dual space of $L^1(\mathrm{sp}(T))$ and $\mathbf{B}(\mathfrak{H})$ is given the σ-weak topology. In particular, the map is a w^*-weak homeomorphism of every bounded subset of $L^\infty(\mathrm{sp}(T))$ onto its image in $\mathbf{B}(\mathfrak{H})$.*

PROOF. If φ is a positive, σ-weakly continuous functional on $\mathbf{B}(\mathfrak{H})$, there is by 4.6.12 an orthogonal sequence (x_m) in $\mathfrak{H}$ such that $\varphi(S) = \sum (Sx_m | x_m)$ for every S in $\mathbf{B}(\mathfrak{H})$. With $g_{nm} = U_n^* P_n x_m$ in $L_n^2(\mathrm{sp}(T))$ (notations as in the proof of 4.7.15) we have, for each bounded Borel function f,

$$\varphi(f(T)) = \sum_m (f(T)x_m | x_m) = \sum_{n,m} (U_n M_f U_n^* x_m | x_m)$$

$$= \sum_{n,m} \int_n f |g_{nm}|^2.$$

Since $\int_n \le 2^n \int$ there is by Radon–Nikodym's theorem (6.5.4) a Borel function $m_n \ge 0$ on $\mathrm{sp}(T)$ such that $\int_n = \int \cdot m_n$. Put $g = \sum |g_{nm}|^2 m_n$, so that g is a positive, extended-valued Borel function. Then for $f \ge 0$ we get

$$\varphi(f(T)) = \sum \int_n f|g_{nm}|^2 = \sum \int f|g_{nm}|^2 m_n$$

$$= \int \sum f|g_{nm}|^2 m_n = \int fg.$$

In particular, with $f = 1$ we see that $g \in \mathscr{L}^1(\mathrm{sp}(T))$ with $\|g\|_1 = \varphi(I)$. Thus $\varphi(f(T)) = \int fg$ for every f in $\mathscr{L}^\infty(\mathrm{sp}(T))$.

Conversely, if $g \in \mathscr{L}^1(\mathrm{sp}(T))$ and $g \ge 0$, we may for each n consider the function $2^{-n}g$ as an element in $\mathscr{L}_n^1(\mathrm{sp}(T))$. Indeed,

$$\int_n 2^{-n}g = \int 2^{-n} g m_n \le \int g,$$

because $m_n \le 2^n$, as $\int_n \le 2^n \int$. Thus, $y_n = U_n((2^{-n}g)^{1/2}) \in P_n(\mathfrak{H})$, and for each $f \ge 0$ in $\mathscr{L}^\infty(\mathrm{sp}(T))$ we compute

$$\sum (f(T)y_n | y_n) = \sum (U_n M_f U_n^* y_n | y_n)$$

$$= \sum 2^{-n} \int_n fg = \int fg.$$

Since both $\mathbf{B}^1(\mathfrak{H})$ and $L^1(\mathrm{sp}(T))$ are spanned by their positive elements, it follows from the first part of the proof that the spectral map is w^*-σ-weak continuous from $L^\infty(\text{sp}(T))$ into $\mathbf{B}(\mathfrak{H})$, and from the second part that it is a homeomorphism of $L^\infty(\text{sp}(T))$ onto its image.

As the weak and the σ-weak topology coincide on bounded subsets of $\mathbf{B}(\mathfrak{H})$ by 4.6.14, the last statement in the proposition is obvious. $\qquad\square$

PROOF. (of 4.7.15, completion). We know from 6.5.12 that $C(\text{sp}(T))$ is w^*-dense in $L^\infty(\text{sp}(T))$ and from 4.6.14 that $C^*(T)$ is σ-weakly dense in $W^*(T)$. Since the spectral map $f \to f(T)$ is a w^*-σ-weak homeomorphism by 4.7.16, and takes $C(\text{sp}(T))$ onto $C^*(T)$, it follows that the image of $L^\infty(\text{sp}(T))$ is $W^*(T)$. $\qquad\square$

EXERCISES

E 4.7.1. Show that if $\mathfrak{A}$ is a self-adjoint, unital, commutative subalgebra of $\mathbf{B}(\mathfrak{H})$ having a cyclic vector, then each element in $\mathfrak{A}'$ is normal. Conclude that $\mathfrak{A}'$ is commutative and that $\mathfrak{A}''$ $(=\mathfrak{A}')$ is maximal commutative.

Hints: If x is a cyclic vector for $\mathfrak{A}$ and $S \in \mathfrak{A}'$, choose (T_n) in $\mathfrak{A}$ such that $T_n x \to Sx$. Note that $\|T_n^* x\| = \|T_n x\|$ (3.2.7), and show that

$$(T_n^* x \,|\, Tx) = (T^* x \,|\, T_n x) \to (T^* x \,|\, Sx) = (S^* x \,|\, Tx)$$

for every T in $\mathfrak{A}$, so that $T_n^* x \to S^* x$, weakly. Compute that, in fact, $T_n^* x \to S^* x$ in norm. Show for each T in $\mathfrak{A}$ that

$$\|STx\| = \|TSx\| = \lim\|TT_n x\| = \lim\|T_n Tx\|$$
$$= \lim\|T_n^* Tx\| = \lim\|TT_n^* x\|$$
$$= \|TS^* x\| = \|S^* Tx\|,$$

so that S and S^* are metrically identical. Finally, use the fact that if A and B are self-adjoint elements in $\mathfrak{A}'$, then $A + iB$ is normal iff $AB = BA$.

E 4.7.2. Show that every commutative von Neumann algebra $\mathfrak{A}$ on a separable Hilbert space $\mathfrak{H}$ is isometrically $*$-isomorphic to $L^\infty(X)$ for some Radon integral on a compact, second countable Hausdorff space X.

Hints: Choose by 4.7.5 a separating vector x for $\mathfrak{A}$, and let P denote the projection (in $\mathfrak{A}'$) on $(\mathfrak{A}x)^=$. Show that the map $T \to TP$ is injective (since $x = Px$) so that $\mathfrak{A}P$ is isometrically $*$-isomorphic to $\mathfrak{A}$ (cf. E 4.3.9). If $\mathfrak{A}^1$ denotes the unit ball of $\mathfrak{A}$, then $\mathfrak{A}^1 P$ is therefore the unit ball of $\mathfrak{A}P$. The map $T \to TP$ being σ-weakly continuous, it follows that the unit ball of $\mathfrak{A}P$ is σ-weakly compact, whence $\mathfrak{A}P$ is σ-weakly closed (cf. 2.5.10 and 4.6.10). Now use the fact that $\mathfrak{A}P$ is a commutative von Neumann algebra on $P(\mathfrak{H})$ (cf. 4.6.15) with a cyclic vector.

E 4.7.3. Take $\mathfrak{A} = \{M_f | f \in \mathscr{L}^\infty(X)\} \subset \mathbf{B}(L^2(X))$ as in 4.7.6. Show that the weak and the σ-weak topology coincide on $\mathfrak{A}$.

E 4.7.4. Let $\mathfrak{H}$ be a separable Hilbert space with orthonormal basis (e_n), and let $\int$ be a Radon integral on a locally compact, σ-compact Hausdorff space X. Denote by $\mathscr{L}^2(X, \mathfrak{H})$ the space of functions $f: X \to \mathfrak{H}$, such that each function $t \to (f(t)|e_n)$ belongs to $\mathscr{L}^2(X)$, and such that

$$\|f\|_2^2 = \sum_n \|(f(\cdot)|e_n)\|_2^2 < \infty.$$

Denote by $\mathscr{N}$ the subspace of functions f with $\|f\|_2 = 0$ and show that $L^2(X, \mathfrak{H}) = \mathscr{L}^2(X, \mathfrak{H})/\mathscr{N}$ is a Hilbert space with the inner product

$$(f|g) = \int (f(t)|g(t)) = \sum \int (f(t)|e_n)\overline{(g(t)|e_n)}.$$

Show that $L^2(X, \mathfrak{H}) = L^2(X) \otimes \mathfrak{H}$ as defined in E 3.2.18.

E 4.7.5. Given $\mathfrak{H}$, X, and $\int$ as in E 4.7.4, let $\mathscr{L}^\infty(X, \mathbf{B}(\mathfrak{H}))$ denote the space of bounded functions $T: X \to \mathbf{B}(\mathfrak{H})$ such that $t \to (T(t)x|y)$ belongs to $\mathscr{L}^\infty(X)$ for all x, y in $\mathfrak{H}$. Show that $\mathscr{L}^\infty(X, \mathbf{B}(\mathfrak{H}))$ is an algebra with involution under the obvious pointwise operations:

$$(S + T)(t) = S(t) + T(t), \qquad ST(t) = S(t)T(t), \qquad T^*(t) = T(t)^*.$$

Show that if $T \in \mathscr{L}^\infty(X, \mathbf{B}(\mathfrak{H}))$, then $t \to \|T(t)\|$ belongs to $\mathscr{L}^\infty(X)$, and define $\|T\|$ to be the ∞-norm of that function. Denote by $\mathscr{N}$ the set of elements T with $\|T\| = 0$, and set $L^\infty(X, \mathbf{B}(\mathfrak{H})) = \mathscr{L}^\infty(X, \mathbf{B}(\mathfrak{H}))/\mathscr{N}$. Show that $L^\infty(X, \mathbf{B}(\mathfrak{H}))$ is a C^*-algebra.

Take $L^2(X, \mathfrak{H})$ as in E 4.7.4 and define

$$Tf(t) = T(t)f(t), \quad T \in \mathscr{L}^\infty(X, \mathbf{B}(\mathfrak{H})), f \in \mathscr{L}^2(X, \mathfrak{H}).$$

Show that this gives an isometric $*$-isomorphism of $L^\infty(X, \mathbf{B}(\mathfrak{H}))$ as a subalgebra of $\mathbf{B}(L^2(X, \mathfrak{H}))$. Identify $L^\infty(X)$ with the algebra $\mathfrak{A}$ of (equivalence classes of) functions in $\mathscr{L}^\infty(X, \mathbf{B}(\mathfrak{H}))$ whose range is some point in $\mathbb{C}I$. Show that $\mathfrak{A}$ is a commutative von Neumann algebra [with uniform multiplicity $= \dim(\mathfrak{H})$] and that

$$\mathfrak{A}' = L^\infty(X, \mathbf{B}(\mathfrak{H})).$$

CHAPTER 5

Unbounded Operators

Many problems in analysis lead irrevocably to unbounded operators. It suffices to mention the differential process, for early encounters, and, as a branch of functional analysis, the theory of partial differential equations (the final showdown). This chapter does not, by a long shot, cover the theory of unbounded operators (and a good excuse would be that there is no theory, only myriads of examples). A small area of this vast territory—dealing with a single unbounded, self-adjoint (or, maybe, normal) operator in a Hilbert space—can, however, be cultivated by the spectral theory of bounded operators; and this we propose to do in some detail.

The first major problem with unbounded operators is that two such may be equal on a dense subspace, and yet be quite different. There is no "extension by continuity." The second problem is the breakdown of the algebraic rules for sums and products of operators. Two perfectly nice (e.g. self-adjoint) unbounded operators may fail—spectacularly—to have a sum, because the intersection of their domains of definition may be zero. In the first section we develop the notions necessary to handle these two problems. In the second, we consider symmetric operators, and the obstructions to extend one such to a self-adjoint operator. Finally, concentrating on a self-adjoint operator S we use the Cayley transform to construct, via the bounded case, a spectral function calculus $f \rightarrow f(S)$ for arbitrary Borel functions and to describe the spectral family $\{f(S)\}$ as the algebra of normal operators affiliated with a certain commutative von Neumann algebra $W^*(S)$. As applications of the theory we prove Stone's theorem and extend the polar decomposition to arbitrary (closed and densely defined) operators.

5.1. Domains, Extensions, and Graphs

Synopsis. Densely defined operators. The adjoint operator. Symmetric and self-adjoint operators. The operator T^*T. Semibounded operators. The Friedrichs extension. Examples.

5.1.1. By an operator *in* a Hilbert space $\mathфرak{H}$ we mean an operator $T: \mathfrak{D}(T) \to \mathfrak{H}$, where $\mathfrak{D}(T)$—the *domain* of T—is a (linear) subspace of $\mathfrak{H}$. We shall be almost exclusively concerned with the case where T is *densely defined*, i.e. where $\mathfrak{D}(T)^= = \mathfrak{H}$. If S and T are operators in $\mathfrak{H}$ such that $\mathfrak{D}(S) \subset \mathfrak{D}(T)$ and $Sx = Tx$ for every x in $\mathfrak{D}(S)$, we say that T is an *extension* of S, and write $S \subset T$. For operators S and T in $\mathfrak{H}$ the statement $S = T$ is therefore equivalent with $S \subset T$ and $T \subset S$.

It will be convenient in the sequel to have a symbol for the range of an (unbounded) operator T in $\mathfrak{H}$. We set

$$\mathfrak{R}(T) = T\mathfrak{D}(T).$$

Computations with unbounded operators are considerably more involved than in the bounded case. If S and T are operators in $\mathfrak{H}$, the symbol $S + T$ will denote the operator with $\mathfrak{D}(S + T) = \mathfrak{D}(S) \cap \mathfrak{D}(T)$ and (of course) $(S + T)x = Sx + Tx$; whereas ST will denote the operator with $\mathfrak{D}(ST) = \mathfrak{D}(T) \cap T^{-1}\mathfrak{D}(S)$ and $STx = S(Tx)$. If these rules are observed, the associative laws will hold for sums and products (but not the distributive law).

The notion of inverse actually becomes simpler in the unbounded case: For every injective operator T in $\mathfrak{H}$ we define T^{-1} to be the operator with $\mathfrak{D}(T^{-1}) = \mathfrak{R}(T)$ and $\mathfrak{R}(T^{-1}) = \mathfrak{D}(T)$, such that $T^{-1}(Tx) = x$. Thus $T^{-1}T = I_{\mathfrak{D}(T)}$ and $TT^{-1} = I_{\mathfrak{R}(T)}$, where $I_{\mathfrak{D}}$ denotes the identical operator defined on the subspace $\mathfrak{D}$ of $\mathfrak{H}$. It follows that if both S and T are injective, then $(ST)^{-1} = T^{-1}S^{-1}$. Note that the symbol T^{-1}, applied in the set-theoretic sense, agrees with its new meaning as an operator (when T is injective).

5.1.2. For a densely defined operator T in $\mathfrak{H}$ we form the *adjoint operator* T^* in $\mathfrak{H}$ by letting $\mathfrak{D}(T^*)$ be the subspace of vectors x in $\mathfrak{H}$ for which the functional $y \to (Ty|x)$ on $\mathfrak{D}(T)$ is bounded (= continuous). Since $\mathfrak{D}(T)$ is dense in $\mathfrak{H}$, the functional extends by continuity (2.1.11) to $\mathfrak{H}$, and thus by 3.1.9 there is a unique vector T^*x in $\mathfrak{H}$ such that

$$(y|T^*x) = (Ty|x), \quad y \in \mathfrak{D}(T), x \in \mathfrak{D}(T^*). \tag{$*$}$$

It is clear that T^* is an operator in $\mathfrak{H}$, but it need not (in general) be densely defined.

Note that if $S \subset T$, then $T^* \subset S^*$. Note further that from the definitions of sum and product it follows that

$$S^* + T^* \subset (S + T)^* \quad \text{and} \quad T^*S^* \subset (ST)^*,$$

provided that $S, T, S + T$, and ST are all densely defined. Finally, we see from

(∗) that for every densely defined operator T in $\mathfrak{H}$ we have

$$\ker T^* = \mathfrak{R}(T)^{\perp}, \qquad\qquad (∗∗)$$

exactly as for bounded operators (3.2.5). Applied to the operator $T - \lambda I$, where $\lambda \in \mathbb{C}$, this gives information about the multiplicity of $\bar{\lambda}$ as an eigenvalue for T^*, viz.

$$\{x \in \mathfrak{D}(T^*) \mid T^*x = \bar{\lambda}x\} = \mathfrak{R}(T - \lambda I)^{\perp}. \qquad\qquad (∗∗∗)$$

5.1.3. We say that a densely defined operator S in $\mathfrak{H}$ is *symmetric* if

$$(Sx \mid y) = (x \mid Sy), \quad x, y \in \mathfrak{D}(S).$$

It follows from 5.1.2 that this happens iff $S \subset S^*$. Since we are (tacitly) dealing with complex Hilbert spaces, the equation (∗) in 3.1.1 applied to the form $(S \cdot \mid \cdot)$ on $\mathfrak{D}(S)$ shows that S is symmetric iff $(Sx \mid x) \in \mathbb{R}$ for every x in $\mathfrak{D}(S)$.

Given a symmetric operator S and $\lambda = \alpha + i\beta$ in $\mathbb{C}$ we have for each x in $\mathfrak{D}(S)$ that

$$\|(S - \lambda I)x\|^2 = \|(S - \alpha I)x\|^2 + \beta^2\|x\|^2,$$

because $S - \alpha I$ is symmetric. It follows that $S - \lambda I$ is injective whenever $\operatorname{Im} \lambda \,(= \beta)$ is nonzero, and that $(S - \lambda I)^{-1}$ [defined on $\mathfrak{R}(S - \lambda I)$] is bounded by $|\beta|^{-1}$.

A symmetric operator S, such that $S \subset T$ implies $S = T$ for every symmetric operator T in $\mathfrak{H}$, is called *maximal symmetric*. Furthermore, we say that S is *self-adjoint* if $S = S^*$. Note that every self-adjoint operator S is maximal symmetric (but not vice versa). Indeed, $S \subset T$ gives $T^* \subset S^*$; and if $S = S^*$ and $T \subset T^*$ this means that $T \subset S$, i.e. $T = S$.

5.1.4. The most important notion in the theory of unbounded operators is that of a *closed operator*, meaning an operator T in $\mathfrak{H}$ for which the *graph*

$$\mathfrak{G}(T) = \{(x, Tx) \mid x \in \mathfrak{D}(T)\}$$

is a closed subspace of $\mathfrak{H} \oplus \mathfrak{H}$. A closed operator, while not necessarily continuous, at least has some decent limit behavior: If (x_n) is a sequence in $\mathfrak{D}(T)$ converging to some x *and* if (Tx_n) converges to some y, then $x \in \mathfrak{D}(T)$ and $Tx = y$.

It follows from the closed graph theorem (2.2.7) that an everywhere defined, closed operator is bounded. Furthermore, we see from 2.3.11 that an operator T with $\mathfrak{D}(T) = \mathfrak{D}(T^*) = \mathfrak{H}$ is bounded (because it has closed graph). In particular, an everywhere defined symmetric operator is bounded (and self-adjoint).

An operator T in $\mathfrak{H}$ is *closable* (or *preclosed*) if the (norm) closure of $\mathfrak{G}(T)$ in $\mathfrak{H} \oplus \mathfrak{H}$ is the graph of an operator $\bar{T}$. In that case $\bar{T}$ is a closed operator and is the minimal closed operator that extends T. It is easy to verify that T is closable iff for each sequence (x_n) in $\mathfrak{D}(T)$ converging to zero, the only accumulation point of (Tx_n) is zero.

In the converse direction, if T is a closed operator in $\mathfrak{H}$ and $\mathfrak{D}_0$ is a subspace of $\mathfrak{D}(T)$ such that T is the closure of $T_0 = T|\mathfrak{D}_0$, we say that $\mathfrak{D}_0$ is a *core* for T. Evidently every dense subspace of $\mathfrak{G}(T)$ defines (by projection on the first coordinate) a core for T. The importance of this concept is obvious when more than one unbounded operator is involved. One may then hope to find a common core for the operators, and avoid the cumbersome bookkeeping of domains of linear combinations of operators prescribed in 5.1.1.

5.1.5. Theorem. *If T is a densely defined operator in $\mathfrak{H}$, then T^* is a closed operator and we have an orthogonal decomposition*

$$\mathfrak{H} \oplus \mathfrak{H} = \mathfrak{G}(T)^= \oplus U\mathfrak{G}(T^*), \qquad (*)$$

where U is the unitary operator on $\mathfrak{H} \oplus \mathfrak{H}$ given by $U(z,y) = (-y,z)$. Furthermore, T is closable iff T^* is densely defined, and in that case $\bar{T} = T^{**}$, i.e. $\mathfrak{G}(T)^= = \mathfrak{G}(T^{**})$.

PROOF. Consider a vector (z,y) in $\mathfrak{G}(T)^\perp$. Thus

$$0 = ((z,y)|(x,Tx)) = (z|x) + (y|Tx), \quad x \in \mathfrak{D}(T).$$

But this is equivalent to the demand that $y \in \mathfrak{D}(T^*)$ with $T^*y = -z$. Consequently, $\mathfrak{G}(T)^\perp = U\mathfrak{G}(T^*)$. Since U is unitary, it follows that T^* is closed and that we have the decomposition $(*)$, because $\mathfrak{G}(T)^= = \mathfrak{G}(T)^{\perp\perp}$ by 3.1.8.

Now take x in $\mathfrak{D}(T^*)^\perp$. Since $U(\mathfrak{G}^\perp) = (U\mathfrak{G})^\perp$ for any unitary U and every subspace $\mathfrak{G}$ we get

$$(0,x) = U(x,0) \in U(\mathfrak{G}(T^*)^\perp)$$
$$= (U\mathfrak{G}(T^*))^\perp = \mathfrak{G}(T)^=.$$

If T is closable, this means that $x = \bar{T}(0) = 0$, so that T^* is densely defined.

If T^* is densely defined, we may consider the closed operator T^{**}. Applying $(*)$, and noting that $U^* = -U$, we find the relations

$$\mathfrak{G}(T^{**}) = U^*(U\mathfrak{G}(T^{**})) = U^*(\mathfrak{G}(T^*)^\perp)$$
$$= (U^*\mathfrak{G}(T^*))^\perp = (U\mathfrak{G}(T^*))^\perp = \mathfrak{G}(T)^=,$$

from which we deduce that T is closable with $\bar{T} = T^{**}$. $\square$

5.1.6. Every symmetric operator S is closable, because S is densely defined and $S \subset S^*$, whence $S \subset \bar{S} \subset S^*$. We say that S is *essentially self-adjoint* if $\bar{S}$ is self-adjoint, i.e. if $\bar{S} = S^*$.

5.1.7. Proposition. *If T is a densely defined, closed operator in $\mathfrak{H}$, and T is injective with dense range, then the same properties hold for T^* and for T^{-1},* *and*

$$(T^*)^{-1} = (T^{-1})^*.$$

PROOF. Let W be the unitary operator on $\mathfrak{H} \oplus \mathfrak{H}$ given by $W(x, y) = (y, x)$, and note that

$$W(\mathfrak{G}(T)) = \mathfrak{G}(T^{-1}).$$

Consequently, T^{-1} is closed (and evidently it is densely defined with dense range). From 5.1.5 we know that T^* is closed and densely defined, and from (**) in 5.1.2 it follows that $\ker T^* = \{0\}$, i.e. T^* is injective. At the same time we get

$$\mathfrak{R}(T^*)^\perp = \ker T^{**} = \ker T = \{0\},$$

so that T^* has dense range.

With U as in 5.1.5 we have

$$UW\mathfrak{G}((T^*)^{-1}) = U\mathfrak{G}(T^*) = \mathfrak{G}(T)^\perp$$
$$= (W\mathfrak{G}(T^{-1}))^\perp = W(\mathfrak{G}(T^{-1})^\perp) = WU\mathfrak{G}((T^{-1})^*).$$

Since $UW = -WU$, this shows that $(T^*)^{-1} = (T^{-1})^*$.   □

5.1.8. Corollary. *If T is a self-adjoint operator in $\mathbf{B}(\mathfrak{H})$ and T is injective (equivalently, T has dense range), then T^{-1} is a self-adjoint operator in $\mathfrak{H}$.*

5.1.9. Theorem. *For a densely defined, closed operator T in $\mathfrak{H}$ we have the following statements:*

(i) *T^*T is self-adjoint and $\mathfrak{D}(T^*T)$ is a core for T.*
(ii) *$T^*T + I$ is bijective from $\mathfrak{D}(T^*T)$ onto $\mathfrak{H}$, so that $(T^*T + I)^{-1} \in \mathbf{B}(\mathfrak{H})$ with $0 \leq (T^*T + I)^{-1} \leq I$.*
(iii) *The closure of $(TT^* + I)^{-1}T$ is $T(T^*T + I)^{-1}$, which belongs to $\mathbf{B}(\mathfrak{H})$ with $\|T(T^*T + I)^{-1}\| \leq 1$.*
(iv) *The net $((\varepsilon^2 T^*T + I)^{-1})_{\varepsilon > 0}$ converges strongly to I as $\varepsilon \to 0$.*

PROOF. By 5.1.5 the vector $(x, 0)$ in $\mathfrak{H} \oplus \mathfrak{H}$ has an orthogonal decomposition along the closed subspaces $\mathfrak{G}(T)$ and $U\mathfrak{G}(T^*)$. This means that there are unique vectors Sx in $\mathfrak{D}(T)$ and Rx in $\mathfrak{D}(T^*)$, such that

$$(x, 0) = (Sx, TSx) \oplus (T^*Rx, -Rx) = (Sx + T^*Rx, TSx - Rx).$$

Note that

$$\|x\|^2 = (\|Sx\|^2 + \|TSx\|^2) + (\|T^*Rx\|^2 + \|Rx\|^2),$$

so that $\|Sx\| \leq \|x\|$ and $\|Rx\| \leq \|x\|$. The first equation shows that $TSx = Rx \in \mathfrak{D}(T^*)$ and that $x = Sx + T^*TSx$. Since clearly the assignments $x \to Sx$

and $x \to Rx$ are linear, we have constructed operators S and R in $\mathbf{B}(\mathfrak{H})$ (with norms ≤ 1), such that $(I + T^*T)S = I$ and $R = TS$. For each x in $\mathfrak{H}$ we have

$$(S^*x|x) = (S^*(T^*T + I)Sx|x) = \|Rx\|^2 + \|Sx\|^2 \geq 0,$$

whence $S = S^*$ and $S \geq 0$. Since S is injective, it follows from 5.1.8 that S^{-1} is self-adjoint. In particular, S^{-1} is maximal symmetric (5.1.3), and since $S^{-1} \subset T^*T + I$, which is symmetric, it follows that $S^{-1} = T^*T + I$. Consequently, T^*T is self-adjoint, and $(T^*T + I)^{-1} = (S^{-1})^{-1} = S$. Moreover, $R = T(T^*T + I)^{-1}$.

Using the arguments above on T^* we obtain the operators $TT^* + I$ and $(TT^* + I)^{-1}$, and since $T(T^*T + I) = (TT^* + I)T$ with domain $\mathfrak{D}(TT^*T)$ we compute

$$(TT^* + I)^{-1}T = (TT^* + I)^{-1}T(T^*T + I)(T^*T + I)^{-1}$$

$$= (TT^* + I)^{-1}(TT^* + I)T(T^*T + I)^{-1}$$

$$\subset T(T^*T + I)^{-1} = R.$$

For $\varepsilon > 0$ let $S_\varepsilon = (\varepsilon^2 T^*T + I)^{-1}$, and note that $\mathfrak{R}(S_\varepsilon) = \mathfrak{D}(T^*T)$. To show that $S_\varepsilon \to I$, strongly, note from the preceding that

$$I - S_\varepsilon = \varepsilon^2 T^*T(\varepsilon^2 T^*T + I)^{-1}$$

$$\supset \varepsilon^2 T^*(\varepsilon^2 TT^* + I)^{-1}T \supset \varepsilon^2(\varepsilon^2 T^*T + I)^{-1}T^*T.$$

This implies for every x in $\mathfrak{D}(T^*T)$ that

$$\|x - S_\varepsilon x\| = \varepsilon^2 \|(\varepsilon^2 T^*T + I)^{-1}T^*Tx\| \leq \varepsilon^2 \|T^*Tx\| \to 0.$$

Since the net $(S_\varepsilon)_{\varepsilon > 0}$ is bounded [actually it is monotone increasing in the unit ball of $\mathbf{B}(\mathfrak{H})_+$] and converges to I on a dense subset of $\mathfrak{H}$, it converges to I everywhere, as claimed.

To prove, finally, that $\mathfrak{D}(T^*T)$ is a core for T, take x in $\mathfrak{D}(T)$. Then $S_\varepsilon x \in \mathfrak{D}(T^*T)$ and $S_\varepsilon x \to x$. Moreover, with $\tilde{S}_\varepsilon = (\varepsilon^2 TT^* + I)^{-1}$, we have

$$TS_\varepsilon x = \tilde{S}_\varepsilon Tx \to Tx,$$

because $\tilde{S}_\varepsilon \to I$, strongly. Thus $\mathfrak{G}(T|\mathfrak{D}(T^*T))$ is dense in $\mathfrak{G}(T)$, and the proof is complete. $\square$

5.1.10. Proposition. *For a densely defined, closed operator T in $\mathfrak{H}$ the following conditions are equivalent:*

(i) $\mathfrak{D}(T) = \mathfrak{D}(T^*)$, *and* $\|Tx\| = \|T^*x\|$ *for every x in $\mathfrak{D}(T)$.*

(ii) $T^*T = TT^*$.

(iii) *There are self-adjoint operators A and B in $\mathfrak{H}$ such that $T = A + iB$, $T^* = A - iB$, and $\|Tx\|^2 = \|Ax\|^2 + \|Bx\|^2$ for every x in $\mathfrak{D}(T)$. In this case, A and B are the closures of the operators $\frac{1}{2}(T + T^*)$ and $\frac{1}{2}i(T^* - T)$, respectively.*

PROOF. (i) $\Rightarrow$ (ii). *If $x \in \mathfrak{D}(T^*T)$ and $y \in \mathfrak{D}(T)$, then by 3.1.2,*

$$4(T^*Tx|y) = 4(Tx|Ty) = \sum_{k=0}^{3} i^k \| T(x + i^k y)\|^2$$

$$= \sum_{k=0}^{3} i^k \| T^*(x + i^k y)\|^2 = 4(T^*x|T^*y).$$

This shows that $T^*x \in \mathfrak{D}(T^{**})$ $[= \mathfrak{D}(T)]$ and that $TT^*x = T^*Tx$. Thus $T^*T \subset TT^*$, and by symmetry $T^*T = TT^*$.

(ii) $\Rightarrow$ (i). Put $S_\varepsilon = (\varepsilon^2 T^*T + I)^{-1} = (\varepsilon^2 TT^* + I)^{-1}$ as in the proof of 5.1.9. The assumption (ii) implies that $\| Ty \| = \| T^*y \|$ for every y in $\mathfrak{D}(T^*T)$. Now if $x \in \mathfrak{D}(T)$, then $S_\varepsilon x \to x$; and as in the proof of 5.1.9,

$$TS_\varepsilon x = T(\varepsilon^2 T^*T + I)^{-1}x = (\varepsilon^2 TT^* + I)^{-1}Tx = S_\varepsilon Tx \to Tx.$$

It follows that $(T^*S_\varepsilon x)_{\varepsilon > 0}$ is a Cauchy net, hence convergent to a vector y with $\| y \| = \| Tx \|$. But T^* is a closed operator, whence $x \in \mathfrak{D}(T^*)$ with $T^*x = y$. Consequently, $\mathfrak{D}(T) \subset \mathfrak{D}(T^*)$ and $\| T^*x \| = \| Tx \|$ for every x in $\mathfrak{D}(T)$. By symmetry we see that $\mathfrak{D}(T) = \mathfrak{D}(T^*)$.

(i) $\Rightarrow$ (iii). Take S_ε as above and put $A_0 = \frac{1}{2}(T + T^*)$. Then A_0 is symmetric, i.e. $A_0 \subset A_0^*$. Now note that if $x \in \mathfrak{D}(A_0^*)$, then for each y in $\mathfrak{D}(T)$ we have

$$(S_\varepsilon A_0^* x | y) = (x|A_0 S_\varepsilon y) = \frac{1}{2}(x|TS_\varepsilon y) + \frac{1}{2}(x|T^*S_\varepsilon y)$$

$$= \frac{1}{2}(x|S_\varepsilon(T + T^*)y) = (S_\varepsilon x|A_0 y) = (A_0 S_\varepsilon x | y).$$

Thus, $S_\varepsilon A_0^* x = A_0 S_\varepsilon x$. As $S_\varepsilon x \to x$ and $S_\varepsilon A_0^* x \to A_0^* x$, it follows that A_0^* is the closure of A_0, i.e. A_0 is essentially self-adjoint, cf. 5.1.6. We may therefore take $A = A_0^*$. Similarly, B is taken as the closure ($=$ adjoint) of the essentially self-adjoint operator $\frac{1}{2}i(T^* - T)$, so that $B = B^*$. We have $\mathfrak{D}(T) \subset \mathfrak{D}(A)$ and $\mathfrak{D}(T) \subset \mathfrak{D}(B)$ so that $A + iB | \mathfrak{D}(T) = T$. However, if $x \in \mathfrak{D}(A) \cap \mathfrak{D}(B)$, then as above

$$(A + iB)x = \lim S_\varepsilon(A + iB)x$$

$$= \lim(A + iB)S_\varepsilon x = \lim TS_\varepsilon x.$$

Since T is closed, $x \in \mathfrak{D}(T)$ with $Tx = (A + iB)x$. Similarly, we show that $A - iB = T^*$.

Since $\frac{1}{2}(T + T^*) \subset A$, it follows that $\mathfrak{D}(T^*T) \subset \mathfrak{D}(A^2)$. Similarly, $\mathfrak{D}(T^*T) \subset \mathfrak{D}(B^2)$, and for every x in $\mathfrak{D}(T^*T)$ we have by straightforward (and legitimate) calculations that

$$(A^2 + B^2)x = \frac{1}{4}(T + T^*)^2 x - \frac{1}{4}(T^* - T)^2 x = T^*Tx.$$

Thus $T^*T \subset A^2 + B^2$. Incidentally, this implies that

$$T^*T = A^2 + B^2, \tag{*}$$

because $A^2 + B^2$ is symmetric and T^*T, being self-adjoint, is maximal symmetric. With S_ε as above we have for each x in $\mathfrak{D}(T)$ that

$$\|Tx\|^2 = \lim \|S_\varepsilon Tx\|^2 = \lim \|TS_\varepsilon x\|^2$$
$$= \lim(T^*TS_\varepsilon x|S_\varepsilon x) = \lim((A^2 + B^2)S_\varepsilon x|S_\varepsilon x)$$
$$= \lim(\|AS_\varepsilon x\|^2 + \|BS_\varepsilon x\|^2 = \lim(\|S_\varepsilon Ax\|^2 + \|S_\varepsilon Bx\|^2)$$
$$= \|Ax\|^2 + \|Bx\|^2.$$

(iii) $\Rightarrow$ (i). Since $T = A + iB$ and $T^* = A - iB$, we have

$$\mathfrak{D}(T) = \mathfrak{D}(A) \cap \mathfrak{D}(B) = \mathfrak{D}(T^*).$$

Moreover, for each x in $\mathfrak{D}(T)$ we set $y = Ax$ and $z = Bx$ and compute

$$\|Tx\|^2 = \|y + iz\|^2 = \|y\|^2 + \|z\|^2 + 2\,\mathrm{Im}(y|z).$$

Thus, by assumption $\mathrm{Im}(y|z) = 0$, whence

$$\|Tx\|^2 - \|T^*x\|^2 = 4\,\mathrm{Im}(y|z) = 0. \qquad \square$$

5.1.11. An operator satisfying the conditions in 5.1.10 is called *normal*. Since self-adjoint operators in the unbounded case are a good deal easier to work with than normal ones, it would have been preferable to define an operator to be normal if it had the form $T = A + iB$, with A and B a commuting pair of self-adjoint operators; cf. 5.1.10(iii). This is also the case, but the concept of commuting (unbounded) operators is *very* delicate. As a warning we mention Nelson's example of two self-adjoint operators A and B for which there is a dense subspace $\mathfrak{D}$ in $\mathfrak{H}$ that is an invariant core for both A and B, and such that $ABx = BAx$ for every x in $\mathfrak{D}$. Nevertheless, some of the bounded spectral functions of A and B, e.g. $\exp iA$ and $\exp iB$, do not commute.

5.1.12. A densely defined, symmetric operator S in $\mathfrak{H}$ is *semibounded* (or, more precisely, *bounded from below*) if we have

$$(Sx|x) \geq \alpha\|x\|^2, \quad x \in \mathfrak{D}(S),$$

for some α in $\mathbb{R}$. We express this in symbols as $S \geq \alpha I$. In particular, we say that S is *positive* if $S \geq 0$. Thus $T^*T \geq 0$ for every densely defined, closed operator T in $\mathfrak{H}$ by 5.1.9. Note that a densely defined, symmetric operator S is bounded [i.e. $\bar{S} \in \mathbf{B}(\mathfrak{H})$] iff both S and $-S$ are semibounded.

In search for self-adjoint extensions of a semibounded operator S, let us first examine the closure $\bar{S}$, i.e. let us look for essential self-adjointness; cf. 5.1.6. If $S \geq I$, then for each x in $\mathfrak{D}(S)$ we have

$$\|x\|^2 \leq (Sx|x) \leq \|Sx\|\,\|x\|.$$

Therefore, if $\mathfrak{R}(S)^= = \mathfrak{H}$, we see that S^{-1} is a densely defined, symmetric operator with $\|S^{-1}x\| \leq \|x\|$ for every x in $\mathfrak{D}(S^{-1})$ $[= \mathfrak{R}(S)]$. Consequently, the closure T of S^{-1} belongs to $\mathbf{B}(\mathfrak{H})_{sa}$ and satisfies $0 \leq T \leq I$. Since $\mathfrak{R}(T)^= = \mathfrak{H}$, it follows from 5.1.8 that T^{-1} is self-adjoint in $\mathfrak{H}$. Moreover, $T^{-1} \geq I$. Since T is the closure of S^{-1}, T^{-1} is the closure of S, because $W\mathfrak{G}(R) = \mathfrak{G}(R^{-1})$ for every injective operator R in $\mathfrak{H}$, where $W(x, y) = (y, x)$. Thus in this case $\bar{S}$ is a self-adjoint extension of S.

Now assume that $S \geq \alpha I$ for some α in $\mathbb{R}$ and that $\mathfrak{R}(S - \beta I)^= = \mathfrak{H}$ for some $\beta < \alpha$. [By (***) in 5.1.2 this means that S^* does not have every $\lambda < \alpha$ as an eigenvalue.] Replacing S by $(\alpha - \beta)^{-1}(S - \beta I)$ we are back in the assumptions above. Thus, also in this case we see that $\bar{S}$ is a self-adjoint extension of S.

A somewhat more sophisticated approach shows that every semibounded operator has a self-adjoint extension—the *Friedrichs extension*. It may, in general, have several; but if $\bar{S}$ is self-adjoint, as above, then clearly it is the only one.

5.1.13. Theorem. *Each semibounded operator S in $\mathfrak{H}$ has a self-adjoint extension with the same lower bound as S.*

PROOF. Assume first that $S \geq I$. Then the sesquilinear form

$$(x|y)_S = (Sx|y), \quad x, y \in \mathfrak{D}(S),$$

is an inner product, and the corresponding norm satisfies $\|x\|_S \geq \|x\|$ for every x in $\mathfrak{D}(S)$. The identical map of the pre-Hilbert space $(\mathfrak{D}(S), \|\cdot\|_S)$ into $\mathfrak{H}$ is therefore norm decreasing, and extends by continuity (2.1.11) to an operator $J: \mathfrak{H}_S \to \mathfrak{H}$ on the completion $\mathfrak{H}_S$, with $\|J\| \leq 1$. We claim that

$$(x|y)_S = (Sx|Jy), \quad x \in \mathfrak{D}(S), y \in \mathfrak{H}_S. \qquad (*)$$

Indeed, if (y_n) is a sequence in $\mathfrak{D}(S)$ converging to y in $\mathfrak{H}_S$, then $Jy_n \to Jy$, whence

$$(Sx|Jy) = \lim(Sx|Jy_n) = \lim(x|y_n)_S = (x|y)_S.$$

Now consider a vector y in $\mathfrak{H}_S$ such that $Jy \in \mathfrak{D}(S^*)$. Choosing a sequence (y_n) in $\mathfrak{D}(S)$ converging to y in $\mathfrak{H}_S$ and using (*) we get

$$\begin{aligned}(S^*Jy|Jy) &= \lim(S^*Jy|Jy_n) = \lim(Jy|Sy_n) \\ &= \lim(y|y_n)_S = \|y\|_S^2 \geq \|Jy\|^2.\end{aligned} \qquad (**)$$

Applying (**) to a vector in $\ker J$ we see that J is injective, so that $\mathfrak{H}_S$ may be regarded as a subspace of $\mathfrak{H}$. Furthermore, (**) shows that the operator

$$T = S^*|J(\mathfrak{H}_S) \cap \mathfrak{D}(S^*) \qquad (***)$$

is symmetric and semibounded in $\mathfrak{H}$ with $T \geq I$. In fact we have $S \subset T \subset S^*$. Let $J^*: \mathfrak{H} \to \mathfrak{H}_S$ be the adjoint of J; cf. E 3.2.15. Then JJ^* is a positive operator in $\mathbf{B}(\mathfrak{H})$ and for each x in $\mathfrak{H}$ and y in $\mathfrak{D}(S)$ we have [again by (*)] that

$$(Sy|JJ^*x) = (y|J^*x)_S = (Jy|x) = (y|x).$$

This shows that $JJ^*x \in \mathfrak{D}(S^*)$ and that $S^*JJ^*x = x$; or, by (***), $JJ^*x \in \mathfrak{D}(T)$ and $TJJ^* = I$. Thus $T \supset (JJ^*)^{-1}$. Since JJ^* is self-adjoint, $(JJ^*)^{-1}$ is self-adjoint by 5.1.8. In particular, $(JJ^*)^{-1}$ is maximal symmetric, and since T is symmetric, we conclude that $T = (JJ^*)^{-1}$, whence $T = T^*$.

The general case follows easily from the above. For if S is semibounded and α is the greatest lower bound for S, then $S + (1 - \alpha)I$ is densely defined and symmetric, and $S + (1 - \alpha)I \geq I$. There is therefore a self-adjoint operator $T_0 \geq I$ that extends $S + (1 - \alpha)I$, whence $T = T_0 + (\alpha - 1)I$ is a self-adjoint extension of S with $T \geq \alpha I$. $\square$

5.1.14. Example. The most accessible example of an unbounded operator obtains by taking a closed subset X of $\mathbb{C}$ (or of $\mathbb{R}$) with positive Lebesgue measure, and put $\mathfrak{H} = L^2(X)$. We can then define the operator M_{id} in $\mathfrak{H}$ by

$$(M_{id}f)(\lambda) = \lambda f(\lambda), \quad f \in \mathscr{L}^2(X), \lambda \in X,$$

with the domain

$$\mathfrak{D}(M_{id}) = \left\{ f \in \mathscr{L}^2(X) \,\middle|\, \int |\lambda f(\lambda)|^2 \, d\lambda < \infty \right\}.$$

If X is unbounded, the operator M_{id} is unbounded, but in all other respects it behaves much like before (see 4.4.3). Thus we easily show that $\mathfrak{D}(M_{id}^*) = \mathfrak{D}(M_{id})$ with

$$(M_{id}^*f)(\lambda) = \bar{\lambda} f(\lambda), \quad f \in \mathfrak{D}(M_{id}).$$

It follows that M_{id} is self-adjoint when $X \subset \mathbb{R}$, and, in general, M_{id} is a closed, densely defined operator satisfying $M_{id}M_{id}^* = M_{id}^*M_{id}$, so that M_{id} is normal by 5.1.10 (cf. 5.1.11). The operator M_{id} has no eigenvalues (because the Lebesgue measure is continuous), but if we choose X such that $(X^0)^- = X$ and define the spectrum as for bounded operators, see 5.2.10, then $\mathrm{sp}(M_{id}) = X$.

5.1.15. Example. Let $\int$ be a Radon integral on a locally compact Hausdorff space X and define for each Borel (or just measurable) function f on X the operator M_f in $L^2(X)$ by

$$M_f g = fg, \quad g \in \mathfrak{D}(M_f) = \{g \in \mathscr{L}^2(X) | fg \in \mathscr{L}^2(X)\}.$$

The example is obviously a simultaneous generalization of 4.7.6 and 5.1.14. It is easy to prove that M_f is bounded iff $f \in \mathscr{L}^\infty(X)$ and that $M_f = 0$ iff f is a null function. Moreover, $\mathfrak{D}(M_f^*) = \mathfrak{D}(M_f)$ and $M_f^* = M_{\bar{f}}$, so that M_f is a densely defined, closed operator in $L^2(X)$ with (the image of) $C_c(X)$ as a core; cf. 5.1.4. The operator M_f is normal (5.1.11), and if $f = \bar{f}$ almost everywhere, M_f is self-adjoint. The possible eigenvalues for M_f are those λ in $\mathbb{C}$ for which the set $f^{-1}\{\lambda\}$ is not a null set in X; cf. 4.5.10. The spectrum of M_f (defined in 5.2.10) is the set of complex λ's for which the (extended-valued) function $(f - \lambda)^{-1}$ is not in $\mathscr{L}^\infty(X)$.

5.1.16. Example. To illustrate the problems concerning extensions of unbounded operators we shall present three densely defined, closed operators D_1, D_2, and D_3, where $D_3 \subset D_2 \subset D_1$, $D_1^* = D_3$, $D_2^* = D_2$, and $D_3^* = D_1$. Thus, D_3 will be symmetric, D_2 will be self-adjoint, and D_1 will be an extension of D_2 that is not symmetric.

Let $\mathfrak{H} = L^2([0,1])$ with respect to Lebesgue measure. We define the integral operator (Volterra operator, see E 3.4.5) T on $\mathfrak{H}$ by

$$Tf(x) = i \int_0^x f(y)\,dy = \int_0^1 k(x,y)f(y)\,dy,$$

where k is the characteristic function for the triangle $\{(x,y) \in \mathbb{R}^2 | 0 \le y \le x \le 1\}$ multiplied by i. Since $k - k^*$ is i times the characteristic function for the unit square (almost everywhere), it follows from 3.4.16 that

$$Tf - T^*f = i(f|1)1, \quad f \in \mathfrak{H}, \tag{$*$}$$

where 1 denotes the constant function $1(x) = 1$ on $[0,1]$. It follows from the definition of T that for each f in $\mathfrak{H}$ the function Tf is continuous and vanishes at 0. Define the linear subspace

$$\mathfrak{D}_1 = \mathfrak{R}(T) + \mathbb{C}1 \subset C([0,1]),$$

and note that $C^1([0,1]) \subset \mathfrak{D}_1$, because

$$T\left(-i\frac{d}{dx}u\right) + u(0)1 = u \tag{$**$}$$

for every u in $C^1([0,1])$. In particular, $\mathfrak{D}_1$ is dense in $\mathfrak{H}$. Since C^1-functions vanishing at the endpoints also form a dense set in $\mathfrak{H}$, the smaller subspaces

$$\mathfrak{D}_2 = \{u \in \mathfrak{D}_1 | u(0) = u(1)\} = \{u = Tf + \alpha 1 | 0 = i(f|1)\};$$

$$\mathfrak{D}_3 = \{u \in \mathfrak{D}_1 | u(0) = u(1) = 0\} = \{u = Tf | (f|1) = 0\};$$

are still dense in $\mathfrak{H}$.

If $Tf = 0$, then with h as the characteristic function for the interval $[x,y]$ we have

$$(f|h) = \int_x^y f(s)\,ds = -i(Tf(y) - Tf(x)) = 0.$$

Thus f is orthogonal to every step function, and since these are dense in $\mathfrak{H}$, we conclude that $f = 0$, so that T is injective. Therefore, we can define the operators D_k, $k = 1, 2, 3$, uniquely by

$$D_k u = f \quad \text{if} \quad u = Tf + \alpha 1, \quad u \in \mathfrak{D}_k.$$

Note that if u is a C^1-function in $\mathfrak{D}_k$ we have $D_k u = -i(d/dx)u$ by $(**)$, so that the D_k's are differential operators in a generalized sense.

For any two vectors $u = Tf + \alpha 1$ and $v = Tg + \beta 1$ in $\mathfrak{D}_1$ we use $(*)$ to compute

$$(D_1 u|v) - (u|D_1 v) = (f|Tg + \beta 1) - (Tf + \alpha 1|g)$$

$$= ((T^* - T)f|g) + (f|\beta 1) - (\alpha 1|g)$$

$$= -i(f|1)(1|g) + (f|\beta 1) - (\alpha 1|g)$$

$$= -i((f - i\alpha 1|1)(1|g - i\beta 1) - \alpha\bar{\beta}).$$

Since $u(1) = i(f - i\alpha|1)$ and $u(0) = \alpha$, it follows that

$$(D_1 u|v) - (u|D_1 v) = -i(u\bar{v}(1) - u\bar{v}(0)).$$

From this identity we read off the inclusions

$$D_1 \subset D_3^*, \quad D_2 \subset D_2^*, \quad D_3 \subset D_1^*.$$

Now fix k and take u in $\mathfrak{D}(D_k^*)$. Then for every g in $\{1\}^\perp$ we have, again by $(*)$,

$$(TD_k^* u|g) = (T^* D_k^* u + i(D_k^* u|1)1|g)$$

$$= (D_k^* u|Tg) = (u|D_k Tg) = (u|g),$$

using the fact that $Tg \in \mathfrak{D}_3$. Since $\{1\}^{\perp\perp} = \mathbb{C}1$ we conclude that $TD_k^* u - u \in \mathbb{C}1$, whence

$$u = TD_k^* u + u(0)1.$$

If $k = 3$, we see that $\mathfrak{D}(D_3^*) \subset \mathfrak{D}_1$ with

$$D_1 u = D_1(TD_3^* u + u(0)1) = D_3^* u.$$

Thus $D_3^* \subset D_1$, i.e. $D_3^* = D_1$. If $k = 2$, we have

$$u(1) - u(0) = TD_2^* u(1) = i(D_2^* u|1) = i(u|D_2 1) = 0.$$

Thus $u \in \mathfrak{D}_2$ and $D_2 u = D_2^* u$. Hence $D_2^* = D_2$. If $k = 1$, we have

$$(u|1) - u(0) = (TD_1^* u|1) = (D_1^* u|T^*1)$$

$$= (D_1^* u|T1 - i1) = (u|D_1(T1 - i1)) = (u|1).$$

Therefore $u(0) = 0 = u(1)$ (as $D_1^* \subset D_2^*$), so that $u \in \mathfrak{D}_3$ with $D_3 u = D_1^* u$. Consequently, $D_1^* = D_3$, and the proof is complete.

5.1.17. Example. The *Laplace operator* Δ is defined on $C_c^\infty(X)$, where X is an open subset of $\mathbb{R}^n$, by

$$\Delta f = \sum_{k=1}^{n} \frac{\partial^2 f}{\partial x_k^2}, \quad f \in C_c^\infty(X).$$

Regarding $C_c^\infty(X)$ as a dense subspace in $L^2(X)$ (with respect to Lebesgue measure on X), we see that Δ is a densely defined operator in $L^2(X)$. If $n = 1$, we use partial integration to compute

$$\int_X \frac{d^2 f}{dx^2} \bar{f} = \left[\frac{df}{dx} \bar{f}\right]_{\delta X} - \int_X \frac{df}{dx} \frac{d\bar{f}}{dx} = -\int_X \frac{df}{dx} \frac{d\bar{f}}{dx} \leq 0,$$

as f vanishes near the boundary of X. Since for $n > 1$ we employ a product measure, it follows that in general $(\Delta f|f) \leq 0$, $f \in C_c^\infty(X)$; so that $-\Delta$ is a symmetric and positive operator. The Friedrichs extension, 5.1.13, ensures the existence of a self-adjoint, negative extension of the Laplace operator. The domain of the extension is an example of a Sobolev space.

When $X = \mathbb{R}^n$ (so that X has no boundary, and no boundary conditions can be forced on Δ), we are in the essentially self-adjoint situation discussed in 5.1.12. Indeed, $\mathfrak{R}(I - \Delta)$ is dense in $L^2(\mathbb{R}^n)$, so that the closure of Δ is *the* self-adjoint extension. The Laplace operator Δ becomes much easier to understand if we extend its domain to the Schwartz space $\mathscr{S}(\mathbb{R}^n)$ of C^∞-functions f on $\mathbb{R}^n$ such that $pf^{(m)} \in \mathscr{L}^1(\mathbb{R}^n)$ for every polynomial p (in n variables) and every multiindex m (cf. E 3.1.14). Indeed, the Fourier transform defines a unitary operator F of $L^2(\mathbb{R}^n)$ that takes $\mathscr{S}(\mathbb{R}^n)$ onto itself (cf. Plancherel's theorem, E 3.1.16), and which transforms the Laplace operator into the multiplication operator $M_{|id|^2}$, i.e.

$$(F \Delta F^* f)(x) = -\sum x_k^2 f(x), \quad f \in \mathscr{S}(\mathbb{R}^n).$$

From this it is easy to describe the domain of $\bar{\Delta}$.

5.2. The Cayley Transform

Synopsis. The Cayley transform of a symmetric operator. The inverse transformation. Defect indices. Affiliated operators. Spectrum of unbounded operators.

5.2.1. From elementary complex function theory we know the Möbius transformations $z \to (\alpha z + \beta)(\gamma z + \delta)^{-1}, \alpha\delta - \beta\gamma \neq 0$, as a transformation group of homeomorphisms of the compact space $\mathbb{C} \cup \{\infty\}$. One of them, the Cayley transformation

$$\kappa(z) = (z - i)(z + i)^{-1}$$

distinguishes itself by taking $\mathbb{R} \cup \{\infty\}$ homeomorphically onto the circle $\mathbb{T}$. The inverse transformation is given by $\kappa^{-1}(z) = i(1 + z)(1 - z)^{-1}$. Just as we before (in 4.3.12 and 4.4.12) used the map $z \to \exp(iz)$ to transform (bounded) self-adjoint operators into unitaries, we shall now use the Cayley transformation. The purpose is to extend the spectral theorems (4.4.1, 4.5.4, 4.7.10, and 4.7.15) to hold also for unbounded, self-adjoint operators.

5.2.2. Let S be a densely defined, symmetric operator in the (complex) Hilbert space $\mathfrak{H}$. By 5.1.3 we can then form the bounded operator $(S + iI)^{-1}$ on the subspace $(S + iI)\mathfrak{D}(S)$ with range $\mathfrak{D}(S)$. We define the *Cayley transformed* of S as the operator

$$\kappa(S) = (S - iI)(S + iI)^{-1},$$

and note that

$$\mathfrak{D}(\kappa(S)) = (S + iI)\mathfrak{D}(S), \qquad \mathfrak{R}(\kappa(S)) = (S - iI)\mathfrak{D}(S).$$

5.2.3. Lemma. *The Cayley transformed $\kappa(S)$ is an isometry of $(S + iI)\mathfrak{D}(S)$ onto $(S - iI)\mathfrak{D}(S)$. Moreover, $I - \kappa(S)$ is injective with $\mathfrak{R}(I - \kappa(S)) = \mathfrak{D}(S)$, and*

$$i(I + \kappa(S))(I - \kappa(S))^{-1} = S.$$

PROOF. If $y \in \mathfrak{D}(S)$, then by 5.1.3

$$\|(S + iI)y\|^2 = \|Sy\|^2 + \|y\|^2 = \|(S - iI)y\|^2.$$

Taking $x = (S + iI)y$ we therefore have $\|(\kappa(S))x\| = \|x\|$, which shows that $\kappa(S)$ is an isometry. Furthermore, we see that $\kappa(S)x = x$ is equivalent with $(S - iI)y = (S + iI)y$, whence $y = 0$ and $x = 0$, so that $I - \kappa(S)$ is injective. Finally, if we put $I_T = I | \mathfrak{D}(T)$ for each T in $\mathfrak{H}$, we have

$$I - \kappa(S) = I_{\kappa(S)} - \kappa(S)$$

$$= (S + iI)(S + iI)^{-1} - (S - iI)(S + iI)^{-1} = 2i(S + iI)^{-1}.$$

In particular,

$$\mathfrak{R}(I - \kappa(S)) = (S + iI)^{-1}(S + iI)\mathfrak{D}(S) = \mathfrak{D}(S).$$

Very similar computations give $I + \kappa(S) = 2S(S + iI)^{-1}$, and by composition this leads to

$$i(I + \kappa(S))(I - \kappa(S))^{-1} = 2iS(S + iI)^{-1}(2i)^{-1}(S + iI) = SI_S = S. \qquad \square$$

5.2.4. Theorem. *The Cayley transform* $S \to \kappa(S)$ *defines an order-preserving isomorphism (with respect to inclusion of operators in $\mathfrak{H}$) between the class of densely defined, symmetric operators in $\mathfrak{H}$, and the class of isometries U in $\mathfrak{H}$ such that $I - U$ has dense range. Under this isomorphism all or none of the following four objects (operators and vector spaces) are closed: S, $\mathfrak{R}(S + iI)$, $\mathfrak{R}(S - iI)$, and $\kappa(S)$.*

PROOF. It is clear that the transform is order preserving [i.e. $S_1 \subset S_2$ implies $\kappa(S_1) \subset \kappa(S_2)$]. Moreover, it is injective, because the inverse is given in 5.2.3.

To prove surjectivity, let U be an isometric operator in $\mathfrak{H}$ with $\mathfrak{R}(I - U)^= = \mathfrak{H}$. From the polarization identity [(**) in 3.1.2] it follows that $(Ux | Uy) = (x | y)$ for all x and y in $\mathfrak{D}(U)$. Therefore, if $Ux = x$ for some x in $\mathfrak{D}(U)$, then $(x | (I - U)y) = 0$ for all y in $\mathfrak{D}(U)$, whence by assumption $x = 0$. Thus, $I - U$ is injective, so that we can define an operator S on $\mathfrak{R}(I - U)$ by

$$S((I - U)y) = i(I + U)y, \quad y \in \mathfrak{D}(U).$$

By computation we get for each y in $\mathfrak{D}(U)$ that

$$(S((I - U)y)|(I - U)y) = (i(I + U)y | (I - U)y)$$

$$= i(\|y\|^2 - \|Uy\|^2 + (Uy | y) - (y | Uy))$$

$$= 2\operatorname{Im}(y | Uy) \in \mathbb{R}.$$

Thus, S is a densely defined, symmetric operator in $\mathfrak{H}$. Since $S = i(I + U)(I - U)^{-1}$, it follows from 5.2.3 that $U = \kappa(S)$, so that κ is surjective.

If S is closed, then $S + iI$ is closed. Since $\|(S + iI)^{-1}\| \leq 1$ (5.1.3), this implies that the subspace $\mathfrak{R}(S + iI)$ is closed. For if $(S + iI)y_n \to x$, then $y_n \to y$, whence $(y, x) \in \mathfrak{G}(S + iI)$, i.e. $(S + iI)y = x$. From 5.2.3 we see that the subspaces $\mathfrak{R}(S + iI)$ and $\mathfrak{R}(S - iI)$ are isometrically isomorphic; in particular, they are simultaneously closed. Again from 5.2.3 we see that when this happens, $\kappa(S)$ is closed (and a partial isometry). Assuming, finally, that $\kappa(S)$ is closed, we consider a sequence (x_n) in $\mathfrak{D}(S)$ such that $x_n \to x$ and $Sx_n \to y$. Then $(S \pm iI)x_n \to y \pm ix$, so that $(y + ix, y - ix) \in \mathfrak{G}(\kappa(S))$. There is therefore a z in $\mathfrak{D}(S)$ such that

$$y + ix = (S + iI)z, \qquad y - ix = (S - iI)z.$$

By elimination this gives $x = z$ and $y = Sz$, i.e. $y = Sx$. Consequently, S is a closed operator, and we have shown that the four objects under consideration are simultaneously closed. $\qquad \Box$

5.2.5. Proposition. *For a densely defined, symmetric operator S in $\mathfrak{H}$ the following conditions are equivalent:*

(i) *S is self-adjoint.*
(ii) *$S \pm iI$ are both surjective operators.*
(iii) *$\kappa(S)$ is unitary.*

PROOF. (i) $\Rightarrow$ (ii). Since $S = S^*$, it is closed (5.1.5), so that both $\mathfrak{R}(S \pm iI)$ are closed subspaces of $\mathfrak{H}$ by 5.2.4. Moreover, since $\pm i$ are not eigenvalues for S, it follows from (***) in 5.1.2 that these subspaces are also dense in $\mathfrak{H}$, and therefore both equal $\mathfrak{H}$.

(ii) $\Rightarrow$ (i). If $x \in \mathfrak{D}(S^*)$, there is a y in $\mathfrak{D}(S)$ such that $(S + iI)y = (S^* + iI)x$. Since $S \subset S^*$, this implies that

$$x - y \in \ker(S^* + iI) = \mathfrak{R}(S - iI)^\perp = \{0\}$$

[cf. (**) in 5.1.2]. Thus $\mathfrak{D}(S^*) = \mathfrak{D}(S)$, i.e. $S = S^*$.

(ii) $\Leftrightarrow$ (iii) are evident from 5.2.3, since a unitary operator by definition is an everywhere defined, surjective isometry. $\qquad \Box$

5.2.6. For a densely defined, symmetric operator S in a Hilbert space $\mathfrak{H}$ the *defect indices* Δ_+ and Δ_- are defined (in $\{0, 1, 2, \ldots, \infty\}$ if $\mathfrak{H}$ is separable, otherwise in the set of cardinal numbers, cf. E 1.1.6) as

$$\Delta_+ = \dim(\mathfrak{R}(S + iI)^\perp), \qquad \Delta_- = \dim(\mathfrak{R}(S - iI)^\perp).$$

The defect indices are thus seen to be the co-dimensions of the closures of the domain and the range of $\kappa(S)$. By (**) in 5.1.2 it follows that Δ_+ and Δ_- also can be viewed as the multiplicities of the eigenvalues $+i$ and $-i$ for S^*.

It follows from 5.2.4 and 5.2.5 that the problem of finding a self-adjoint extension of S is equivalent to finding a unitary U on $\mathfrak{H}$ such that $U|\mathfrak{R}(S + iI) = \kappa(S)$; and this (by 3.1.14) can be done exactly when $\Delta_+ = \Delta_-$. In the (interesting) case where S is closed, so that $\kappa(S)$ may be regarded as an honest partial isometry in $\mathbf{B}(\mathfrak{H})$, we see from 3.3.12 (or E 3.3.6) that S has a self-adjoint extension in $\mathfrak{H}$ iff index $\kappa(S) = 0$.

The extension problem does not always admit a solution. If we take U to be an isometry of $\mathfrak{H}$ onto a proper subspace, e.g. the unilateral shift (3.2.16), then by 5.2.4 the operator $S = \kappa^{-1}(U)$ is a closed, symmetric operator, which is maximal symmetric [because $S \subset T$ would imply $U = \kappa(S) \subset \kappa(T)$, whence $U = \kappa(T)$ and $S = T$]. However, S is not self-adjoint because U is not unitary (5.2.5), and the defect indices for S are 0 and $\dim(U(\mathfrak{H})^{\perp})$.

It is clear from the discussion above that a closed, symmetric operator S is maximal symmetric iff one of the defect indices is zero, and self-adjoint iff both defect indices are zero.

5.2.7. As mentioned before (5.1.11), commutativity among unbounded operators is a touchy business. Matters become a good deal easier if one of the operators involved is bounded. Note first that if S is an (unbounded) operator in $\mathfrak{H}$ and $T \in \mathbf{B}(\mathfrak{H})$, then $\mathfrak{D}(S + T) = \mathfrak{D}(S)$ and $\mathfrak{D}(TS) = \mathfrak{D}(S)$. Moreover, straightforward arguments show that

$$(S + T)^* = S^* + T^* \quad \text{and} \quad (TS)^* = S^*T^*; \tag{*}$$

cf. 5.1.2.

If S is an operator in $\mathfrak{H}$ and $T \in \mathbf{B}(\mathfrak{H})$, we say that S and T *commute* if $TS \subset ST$, i.e. if $T(\mathfrak{D}(S)) \subset \mathfrak{D}(S)$, and $TSx = STx$ for every x in $\mathfrak{D}(S)$. Using the graph of S, the commutativity is conveniently expressed by the relation

$$(T \oplus T)\mathfrak{G}(S) \subset \mathfrak{G}(S). \tag{**}$$

It is clear that the operators commuting with S form an algebra $\{S\}'$ in $\mathbf{B}(\mathfrak{H})$, and we see from (**) that $\{S\}'$ is strongly closed if S is a closed operator. Again from (**) we see from the decomposition (*) in 5.1.5 that

$$T \in \{S\}' \Rightarrow T^* \in \{S^*\}' \tag{***}$$

(because $T \oplus T$ certainly commutes with the unitary U in 5.1.5). Thus, for a densely defined, closed operator S, the set $\{S\}' \cap \{S^*\}'$ is a strongly closed, unital, *-subalgebra of $\mathbf{B}(\mathfrak{H})$, i.e. a von Neumann algebra (4.6.9).

For a densely defined, closed operator S we put

$$W^*(S) = (\{S\}' \cap \{S^*\}')',$$

and we see from the double commutant theorem (4.6.7) that when $S \in \mathbf{B}(\mathfrak{H})$, then $W^*(S)$ is the von Neumann algebra generated by S, in accordance with our previous use of the notation $W^*(S)$ in 4.5.3 and 4.7.9. We say that S is *affiliated* with a von Neumann algebra $\mathfrak{A}$ if $W^*(S) \subset \mathfrak{A}$. By 4.6.7 this is equivalent with the demand that $\mathfrak{A}' \subset \{S\}' \cap \{S^*\}'$. In the literature the affiliation of S with $\mathfrak{A}$ is often expressed by the symbol $S\eta\mathfrak{A}$.

5.2.8. Lemma. *If S is a self-adjoint operator in $\mathfrak{H}$ and $T \in \mathbf{B}(\mathfrak{H})$, then $T \in \{S\}'$ iff $T \in \{\kappa(S)\}'$. In particular, $W^*(S) = W^*(\kappa(S))$, so that S is affiliated with a von Neumann algebra $\mathfrak{A}$ iff $\kappa(S) \in \mathfrak{A}$.*

PROOF. If $T \in \{S\}'$, then $T(S \pm iI) \subset (S \pm iI)T$, whence $T(S \pm iI)^{-1} = (S \pm iI)^{-1}T$ (cf. 5.2.5), and thus $T\kappa(S) = \kappa(S)T$.

Conversely, if $T \in \{\kappa(S)\}'$ then, since $I - \kappa(S)$ is injective with range $\mathfrak{D}(S)$ by 5.2.3, we have

$$TS(I - \kappa(S)) = iT(I + \kappa(S)) = i(I + \kappa(S))T$$
$$= i(I + \kappa(S))(I - \kappa(S))^{-1}(I - \kappa(S))T = ST(I - \kappa(S)),$$

whence $TS \subset ST$. □

5.2.9. Lemma. *If a normal operator T in $\mathfrak{H}$ is affiliated with a von Neumann algebra $\mathfrak{A}$, then so are the self-adjoint operators A and B in its decomposition $T = A + iB$ described in 5.1.10.*

PROOF. By assumption $R \in \{T\}' \cap \{T^*\}'$ for every R in $\mathfrak{A}'$. Thus, with $A_0 = \frac{1}{2}(T + T^*)$ we have $RA_0 \subset A_0 R$. By (***) in 5.2.7 this implies that $R^*A_0^* \subset A_0^* R^*$. As $A = A_0^*$ (cf. the proof of 5.1.10), and $\mathfrak{A}'$ is self-adjoint, we conclude that $\mathfrak{A}' \subset \{A\}'$, so that A is affiliated with $\mathfrak{A}$. The argument for B is quite analogous. □

5.2.10. For an operator T in $\mathfrak{H}$ we define the *resolvent set* as those λ in $\mathbb{C}$ for which the operator $\lambda I - T$ is bijective from $\mathfrak{D}(T)$ onto $\mathfrak{H}$ and

$$\|(\lambda I - T)x\| \geq t\|x\|, \quad x \in \mathfrak{D}(T)$$

for some $t > 0$. This obviously means that there is an operator $S [=(\lambda I - T)^{-1}]$ in $\mathbf{B}(\mathfrak{H})$ such that

$$S(\lambda I - T) \subset (\lambda I - T)S = I.$$

The complement of the resolvent set is called the *spectrum* of T, and is denoted by $\mathrm{sp}(T)$. These definitions extend the notions of resolvent set and spectrum defined for elements in $\mathbf{B}(\mathfrak{H})$, or any other Banach algebra; cf. 4.1.10. As in the bounded case we have the *resolvent function* $R(\lambda) = (\lambda I - T)^{-1}$ defined on $\mathbb{C} \backslash \mathrm{sp}(T)$ with values in $\mathbf{B}(\mathfrak{H})$.

5.2.11. Proposition. *If T is an operator in $\mathfrak{H}$ and $\lambda \notin \mathrm{sp}(T)$, then for each ζ in $\mathbb{C}$ with $|\zeta| < \|R(\lambda)\|^{-1}$ we have $\lambda - \zeta \notin \mathrm{sp}(T)$ and*

$$R(\lambda - \zeta) = \sum_{n=0}^{\infty} R(\lambda)^{n+1}\zeta^n.$$

In particular, $\mathrm{sp}(T)$ is a closed subset of $\mathbb{C}$.

PROOF. By 4.1.7 we have

$$(I - \zeta R(\lambda))^{-1} = \sum_{n=0}^{\infty} R(\lambda)^n \zeta^n,$$

and the desired result follows by multiplication with $R(\lambda)$, since

$$(I - \zeta R(\lambda))^{-1} R(\lambda) = ((\lambda I - T)(I - \zeta R(\lambda)))^{-1} = ((\lambda - \zeta)I - T)^{-1}. \qquad \square$$

5.2.12. An unbounded operator may have empty spectrum. It suffices to find an injective operator T in $\mathbf{B}(\mathfrak{H})$ with dense range and spectral radius zero, e.g. the Volterra operator used in 5.1.16. Then T^{-1} is densely defined and for every λ in $\mathbb{C}$ we have

$$(\lambda I - T^{-1})^{-1} = -\sum_{n=0}^{\infty} \lambda^n T^{n+1} \in \mathbf{B}(\mathfrak{H}),$$

because $\|T^n\| \le |2\lambda|^{-n}$ for all large n by 4.1.13. In the specific example in 5.1.16, T^{-1} will be differentiation on the subspace of absolutely continuous functions on $[0, 1]$, vanishing at zero, whose derivatives belong to $L^2([0, 1])$.

At the other extreme, the spectrum of an unbounded operator may be all of $\mathbb{C}$. Just consider the multiplication operator M_{id} on $L^2(\mathbb{C})$; cf. 5.1.14.

5.2.13. Proposition. *If S is a self-adjoint operator in $\mathfrak{H}$, then $\operatorname{sp}(S)$ is a nonempty, closed subset of $\mathbb{R}$. Moreover, $\lambda \in \operatorname{sp}(S)$ iff $\kappa(\lambda) \in \operatorname{sp}(\kappa(S))$, and λ is an eigenvalue for S iff $\kappa(\lambda)$ is an eigenvalue for $\kappa(S)$.*

PROOF. If $\lambda \in \mathbb{C} \setminus \mathbb{R}$, then $(S - \lambda I)\mathfrak{D}(S) = \mathfrak{H}$ by 5.2.5, and $\|(S - \lambda I)^{-1}\| \le |\operatorname{Im} \lambda|^{-1}$ by 5.1.3, so that $\lambda \notin \operatorname{sp}(S)$. Thus $\operatorname{sp}(S) \subset \mathbb{R}$, just as $\operatorname{sp}(\kappa(S)) \subset \mathbb{T}$ by 4.3.12 and 5.2.5.

Take λ in $\mathbb{R}$ and use 5.2.3 to compute

$$\begin{aligned}
\lambda I - S &= (\lambda(I - \kappa(S)) - i(I + \kappa(S)))(I - \kappa(S))^{-1} \\
&= ((\lambda - i)I - (\lambda + i)\kappa(S))(I - \kappa(S))^{-1} \\
&= (\lambda + i)(\kappa(\lambda)I - \kappa(S))(I - \kappa(S))^{-1}.
\end{aligned}$$

It follows that a vector x in $\mathfrak{D}(S)$ is an eigenvector for S with respect to λ iff $x = (I - \kappa(S))y$, where y is an eigenvector for $\kappa(S)$ with respect to $\kappa(\lambda)$. Furthermore, we see that if $\kappa(\lambda) \notin \operatorname{sp}(\kappa(S))$, then $\lambda \notin \operatorname{sp}(S)$, since

$$(\lambda I - S)^{-1} = (I - \kappa(S))(\kappa(\lambda)I - \kappa(S))^{-1}(\lambda + i)^{-1}.$$

Finally, if $\kappa(\lambda) \in \operatorname{sp}(\kappa(S))$, there is by 4.4.4 a sequence (x_n) of unit vectors in $\mathfrak{H}$ such that $\|(\kappa(\lambda)I - \kappa(S))x_n\| \to 0$. With $y_n = (I - \kappa(S))x_n$ we know from 5.2.3 that $(y_n) \subset \mathfrak{D}(S)$, and

$$\lim \|y_n\| = \lim \|x_n - \kappa(S)x_n\| = |1 - \kappa(\lambda)| > 0.$$

However,

$$(\lambda I - S)y_n = (\lambda + i)(\kappa(\lambda)I - \kappa(S))x_n \to 0,$$

so that $(\lambda I - S)^{-1}$ is unbounded and $\lambda \in \operatorname{sp}(S)$. $\qquad \square$

5.3. Unlimited Spectral Theory

Synopsis. Normal operators affiliated with a MAÇA. The multiplicity-free case. The spectral theorem for an unbounded, self-adjoint operator. Stone's theorem. The polar decomposition.

5.3.1. Recall from 6.4.13 that if $\int$ is a Radon integral on a locally compact, σ-compact Hausdorff space X and if $\mathscr{L}(X)$, $\mathscr{B}(X)$, and $\mathscr{N}(X)$ denote the classes of measurable, Borel, and null functions, respectively, then $\mathscr{L}(X) = \mathscr{B}(X) + \mathscr{N}(X)$, so that we have a well-defined $*$-algebra

$$L(X) = \mathscr{L}(X)/\mathscr{N}(X) = \mathscr{B}(X)/\mathscr{N}(X).$$

To ease the notation let us further agree that a homogeneous map Φ from $\mathscr{L}(X)$ into the class of normal operators in a Hilbert space is an *essential homomorphism* if for any pair f, g of functions in $\mathscr{L}(X)$ the operators $\Phi(f + g)$ and $\Phi(fg)$ are the closures (cf. 5.1.4) of the operators $\Phi(f) + \Phi(g)$ and $\Phi(f)\Phi(g)$, respectively.

5.3.2. Proposition. *Let $\int$ be a Radon integral on a locally compact, σ-compact Hausdorff space X, and for each measurable function f define the multiplication operator M_f in $L^2(X)$ as in 5.1.15. Then the map $f \to M_f$ induces an essential $*$-isomorphism between the $*$-algebra $L(X)$ and the class of normal operators in $L^2(X)$ affiliated with the von Neumann algebra $\mathfrak{A} = \{M_f | f \in \mathscr{L}^\infty(X)\}$.*

PROOF. Each operator M_f, $f \in \mathscr{L}(X)$ is normal with $\mathfrak{D}(M_f) = \{h \in L^2(X) | fh \in L^2(X)\}$; cf. 5.1.15. To show that the map $f \to M_f$ is an essential homomorphism, take h in $\mathfrak{D}(M_{f+g})$ and let $h_n = (1 + n^{-1}(|f| + |g|))^{-1}h$. Then $h_n \in \mathfrak{D}(M_f) \cap \mathfrak{D}(M_g)$, and by Lebesgue's monotone convergence theorem (6.1.13) it follows that $h_n \to h$ and $(M_f + M_g)h_n \to M_{f+g}h$. Thus

$$\mathfrak{G}(M_f + M_g)^= = \mathfrak{G}(M_{f+g})$$

as claimed. The proof for the product is quite similar since (in this case) $\mathfrak{D}(M_f M_g) = \mathfrak{D}(M_{fg}) \cap \mathfrak{D}(M_g)$. Clearly, $M_f = 0$ iff $f \in \mathscr{N}(X)$, so that we have an essential algebra isomorphism $f \to M_f$ from $L(X)$ into a commutative $*$-algebra of normal operators in $L^2(X)$. Also, it is evident that the map is $*$-preserving, i.e. $M_f^* = M_{\bar{f}}$. Since $M_g M_f \subset M_f M_g$ for every f in $\mathscr{L}(X)$ and g in $\mathscr{L}^\infty(X)$, it follows that each multiplication operator M_f is affiliated with $\mathfrak{A}'$. Since $\mathfrak{A}' = \mathfrak{A}$ (4.7.6), we have the desired conclusion.

Now consider a self-adjoint operator S in $L^2(X)$ affiliated with $\mathfrak{A}$. Thus $\kappa(S) \in \mathfrak{A}$ by 5.2.8, whence $\kappa(S) = M_u$ for some Borel function u on X such that $|u(x)| = 1$ but $u(x) \neq 1$ for (almost) all x in X. Take $f = i(1 + u)(1 - u)^{-1} = \kappa^{-1}(u)$ and compute

$$M_f = i(1 + \kappa(S))(1 - \kappa(S))^{-1} = S.$$

Finally, if T is a normal operator in $L^2(X)$ affiliated with $\mathfrak{A}$, then by 5.2.9 (and 5.1.10) we have $T = A + iB$, where A and B are self-adjoint operators affiliated with $\mathfrak{A}$. From the preceding result $A = M_f$ and $B = M_g$ for some real-valued functions f and g in $\mathscr{L}(X)$, and since M_{f+ig} is the closure of $M_f + iM_g$ ($= A + iB = T$), which is already closed, we have $T = M_{f+ig}$ as desired. $\square$

5.3.3. Theorem. *If S is a self-adjoint operator in a separable Hilbert space $\mathfrak{H}$, and $\kappa(S)$ is multiplicity-free, there is a finite Radon integral on* sp(S) *and an isometry U of $L^2(\mathrm{sp}(S))$ onto $\mathfrak{H}$, such that if we define*

$$f(S) = UM_f U^*, \quad f \in \mathscr{L}(\mathrm{sp}(S)),$$

then the map $f \to f(S)$ induces an essential $*$-isomorphism between $L(\mathrm{sp}(S))$ and the $*$-algebra of normal operators in $\mathfrak{H}$ affiliated with $W^*(S)$. Moreover, id$(S) = S$ *and* $1(S) = I$, *and the two meanings of $\kappa(S)$ coincide.*

PROOF. Since $\kappa(S)$ is multiplicity-free, there is by 4.7.9 a unit vector x in $\mathfrak{H}$ that is cyclic for the algebra $C^*(\kappa(S))$. Now note that when f ranges over $C_0(\mathbb{R})$, then $f \circ \kappa^{-1}$ ranges over $C_0(\mathbb{T} \setminus \{1\})$. We can therefore define

$$\int f = (f \circ \kappa^{-1}(\kappa(S))x|x), \quad f \in C_0(\mathrm{sp}(S))$$

as a finite Radon integral on $\kappa^{-1}(\mathrm{sp}(\kappa(S))) = \mathrm{sp}(S)$; cf. 5.2.13. As in the proof of 4.7.7 we obtain the isometry U of $L^2(\mathrm{sp}(S))$ onto $\mathfrak{H}$ by starting with

$$Uf = (f \circ \kappa^{-1}(\kappa(S)))x, \quad f \in C_0(\mathrm{sp}(S)),$$

and extending by continuity. It follows from 5.3.2 that the map $f \to f(S)$ is an essential $*$-isomorphism of $L(\mathrm{sp}(S))$ onto the $*$-algebra of normal operators in $\mathfrak{H}$ affiliated with the maximal commutative von Neumann algebra

$$\mathfrak{A} = \{UM_f U^* | f \in \mathscr{L}^\infty(\mathrm{sp}(S))\}.$$

It only remains to show that id$(S) = S$ and that $\mathfrak{A} = W^*(S)$.

For each f in $C_0(\mathbb{R})$ we have

$$U(M_\kappa f) = U(\kappa \cdot f) = (\kappa \cdot f \circ \kappa^{-1}(\kappa(S)))x$$
$$= (\kappa \circ \kappa^{-1} \cdot f \circ \kappa^{-1}(\kappa(S)))x = \kappa(S)f \circ \kappa^{-1}\kappa(S)x = \kappa(S)Uf.$$

Thus, $UM_\kappa U^* = \kappa(S)$. Since $M_{id} = i(I + M_\kappa)(I - M_\kappa)^{-1}$, it follows that

$$UM_{id}U^* = i(I + \kappa(S))(I - \kappa(S))^{-1} = S.$$

When f ranges over the algebra $\mathscr{B}_b(\mathbb{R})$ of bounded Borel functions on $\mathbb{R}$, then $f \circ \kappa^{-1}$ ranges over the algebra $\mathscr{B}_b(\mathbb{T} \setminus \{1\})$. Since 1 is not an eigenvalue for $\kappa(S)$, this implies that

$$\mathfrak{A} = \{g(\kappa(S)) | g \in \mathscr{B}_b(\mathrm{sp}(\kappa(S)))\} = W^*(\kappa(S)) = W^*(S)$$

by 4.7.10 and 5.2.8. $\square$

5.3.4. It is perhaps a bit unsatisfactory that the main hypothesis in 5.3.3 concerns the Cayley transformed, and not the operator itself. To remedy this defect we make the following definition: A self-adjoint operator S in a (necessarily) separable Hilbert space $\mathfrak{H}$ is *multiplicity-free* if there is a vector x in $\bigcap \mathfrak{D}(S^n)$, $n \geq 0$, such that the subspace spanned by the vectors $\{S^n x \mid n \geq 0\}$ is dense in $\mathfrak{H}$.

5.3.5. Proposition. *A self-adjoint operator S is multiplicity-free iff the same is true for $\kappa(S)$.*

PROOF. If S is multiplicity-free with a cyclic vector x in $\bigcap \mathfrak{D}(S^n)$, we let P denote the projection on the subspace $(C^*(\kappa(S))x)^=$. Then $Px = x$ and $P \in C^*(\kappa(S))'$; cf. 4.7.2. By 5.2.8 this means that $P \in \{S\}'$, whence

$$(I - P)S^n x = S^n(I - P)x = 0$$

for every $n \geq 0$. It follows that $P = I$ so that x is cyclic for $C^*(\kappa(S))$. Consequently, $\kappa(S)$ is multiplicity-free (4.7.9).

Conversely, if $\kappa(S)$ is multiplicity-free, we may apply 5.3.3 to define $y = Ug$ in $\mathfrak{H}$, where $g(\lambda) = \exp(-\lambda^2)$, $\lambda \in \mathrm{sp}(S)$. Since $\lambda \to \lambda^n g(\lambda)$ is bounded for every n, it follows that $y \in \bigcap \mathfrak{D}(S^n)$; and since each element h in $C_c(\mathbb{R})$ can be approximated uniformly by a function pg, where p is a polynomial, it follows that the subspace of vectors $p(S)y$ is dense in $\mathfrak{H}$. $\qquad\square$

5.3.6. Remarks. One may use 5.3.3 to prove spectral theorems for operators with multiplicities just as in the bounded case (4.7.13). Thus, for a given family $\{S_j \mid j \in J\}$ of commuting, self-adjoint operators [i.e. $\kappa(S_i)\kappa(S_j) = \kappa(S_j)\kappa(S_i)$ for all i and j] we may embed the family $\{\kappa(S_j) \mid j \in J\}$ in a maximal commutative $*$-subalgebra of $\mathbf{B}(\mathfrak{H})$ to obtain a representation of the S_j's as a commuting family of multiplication operators $\{M_{f_j} \mid j \in J\}$.

In another direction we can, for a given self-adjoint operator S, decompose the Hilbert space in orthogonal subspaces on which S is multiplicity-free. The following version of the spectral theorem is then the unbounded analogue of 4.7.15, and is derived from 5.3.3 exactly as 4.7.15 is derived from 4.7.10. Recall that for a Radon integral on a locally compact Hausdorff space X we write $\mathscr{L}(X)$, $\mathscr{B}(X)$, and $\mathscr{N}(X)$ for the classes of measurable, Borel, and null functions, and that $\mathscr{L}(X) = \mathscr{B}(X) + \mathscr{N}(X)$ if X is σ-compact. Furthermore. $L(X) = \mathscr{L}(X)/\mathscr{N}(X)$, cf. 5.3.1.

First, however, a semitrivial lemma to clear away the underbrush.

5.3.7. Lemma. *If $(\mathfrak{H}_n)$ is a sequence of Hilbert spaces, and if for each n, T_n is a self-adjoint (respectively normal) operator in $\mathfrak{H}_n$, there is a unique self-adjoint (respectively normal) operator T in the orthogonal sum $\mathfrak{H} = \bigoplus \mathfrak{H}_n$ such that $T \mid \mathfrak{D}(T_n) = T_n$ for every n. Moreover, $\mathfrak{D}(T)$ consists of those vectors $x = (x_n)$ in $\mathfrak{H}$, such that $x_n \in \mathfrak{D}(T_n)$ for every n, and $\sum \| T_n x_n \|^2 < \infty$.*

PROOF. Let $\mathfrak{D}_0$ denote the linear span in $\mathfrak{H}$ of the orthogonal subspaces $\mathfrak{D}(T_n)$, and let T_0 be the operator in $\mathfrak{H}$ given by

$$T_0(\sum x_n) = \sum T_n x_n, \quad x = (x_n) \in \mathfrak{D}_0.$$

Furthermore, let T be the extension of T_0 indicated in the lemma, i.e.

$$T(\sum x_n) = \sum T_n x_n, \quad x = (x_n) \in \mathfrak{D}(T).$$

By construction $\mathfrak{D}(T_0)$ is dense in $\mathfrak{D}(T)$, so that $\mathfrak{D}_0$ is a core for T, which is easily seen to be closed.

In the self-adjoint case, both T_0 and T are symmetric operators. To prove that T is self-adjoint, take $x = (x_n)$ in $\mathfrak{D}(T^*)$. For each n we then know that the map

$$y_n \to (Ty_n|x) = (T_n y_n|x_n), \quad y_n \in \mathfrak{D}(T_n),$$

is bounded. Thus, $x_n \in \mathfrak{D}(T_n^*) = \mathfrak{D}(T_n)$. Moreover, for every finite sum $y = \sum y_n$ in $\mathfrak{D}_0$ we have

$$(Ty|x) = \sum (T_n y_n|x_n) = \sum (y_n|T_n x_n).$$

Since $\mathfrak{D}_0$ is dense in $\mathfrak{H}$ this implies that $\sum T_n x_n$ belongs to $\mathfrak{H}$ (and equals T^*x), i.e. $\sum \|T_n x_n\|^2 < \infty$. But then $x \in \mathfrak{D}(T)$ so that $T = T^*$, as desired.

In the normal case we have $T_n = A_n + iB_n$ for every n by 5.1.10, and from the first part of the proof we obtain self-adjoint operators A and B in $\mathfrak{H}$ extending the sequences (A_n) and (B_n). Note that $x = (x_n) \in \mathfrak{D}(A) \cap \mathfrak{D}(B)$ iff $\sum \|A_n x_n\|^2 < \infty$ and $\sum \|B_n x_n\|^2 < \infty$, i.e. iff

$$\sum \|A_n x_n\|^2 + \|B_n x_n\|^2 = \sum \|T_n x_n\|^2 < \infty;$$

cf. (iii) in 5.1.10. Thus, $\mathfrak{D}(T) = \mathfrak{D}(A) \cap \mathfrak{D}(B)$ and for each x in $\mathfrak{D}(T)$ we have $Tx = Ax + iBx$, $T^*x = Ax - iBx$, and $\|Tx\|^2 = \|Ax\|^2 + \|Bx\|^2$, whence T is normal by 5.1.10. $\square$

5.3.8. Theorem. *If S is a self-adjoint operator in a separable Hilbert space $\mathfrak{H}$, there is a finite Radon integral on* sp(S) *and an essential $*$-isomorphism $f \to f(S)$ from $L(\mathrm{sp}(S))$ onto the $*$-algebra of normal operators affiliated with $W^*(S)$. Moreover, id$(S) = S$ and $1(S) = I$, and the two meanings of $\kappa(S)$ coincide.*

PROOF. By 4.7.4 there is an orthogonal sequence (P_n) of cyclic projections relative to $W^*(S)$, with strong sum I. Thus, each operator $\kappa(S)P_n$ is multiplicity-free on $P_n(\mathfrak{H})$. Since P_n commutes with $\kappa(S)$, it commutes with S (5.2.8), and, regarding SP_n as a self-adjoint operator in $P_n(\mathfrak{H})$, we easily show that $\kappa(SP_n) = \kappa(S)P_n$. As in the proof of 4.7.15 we now (tacitly) identify sp(SP_n) with a closed subset of sp(S), and have by 5.3.3 a normalized Radon integral $\int_n$ on sp(S), an isometry U_n from $L^2(\mathrm{sp}(S))$ onto $P_n(\mathfrak{H})$, and an essential isomorphism $f \to U_n M_f U_n^*$ from $L_n(\mathrm{sp}(S))$ onto the class of normal operators in $P_n(\mathfrak{H})$ affiliated with $W^*(SP_n)$.

Define $\int = \sum 2^{-n} \int_n$ as a normalized Radon integral on sp(S), and for each

f in $\mathcal{L}(\mathrm{sp}(S))$ define $f(S)$ as the normal extension in $\mathfrak{H}$ of the sequence of operators $(U_n M_f U_n^*)$ as described in 5.3.7. Clearly, $f(S) = 0$ iff f is a null function with respect to every $\int_n$, i.e. iff f is a null function with respect to $\int$. Moreover, since for each n the map $f \to U_n M_f U_n^*$ is an essential *-homomorphism (5.3.1), it follows from 5.3.7 that the map $f \to f(S)$ is an essential *-homomorphism, and thus induces an essential *-isomorphism on $L(\mathrm{sp}(S))$. Also, $id(S) = S$, $1(S) = I$, and $\kappa(S) = \kappa(S)$, because these equations hold for every n.

Finally, if T is a self-adjoint operator in $\mathfrak{H}$ affiliated with $W^*(S)$, then $\kappa(T) \in W^*(\kappa(S))$ by 5.2.8. There is therefore by 4.7.15 a unitary function u on $\mathrm{sp}(\kappa(S))$ such that $\kappa(T) = u(\kappa(S))$. With $w = u \circ \kappa$ we have for each n that

$$\kappa(T)P_n = u(\kappa(S))P_n = u(\kappa(SP_n)) = w(SP_n) = U_n M_w U_n^*$$

by 5.3.3. Moreover, identifying $\kappa(T)P_n$ with $\kappa(TP_n)$ we get

$$TP_n = \kappa^{-1}(U_n M_w U_n^*) = U_n M_f U_n^*,$$

with $f = \kappa^{-1} \circ w$. Thus T and $f(S)$ agree on a common core, cf. 5.3.7, whence $T = f(S)$. If T is merely normal, then $T = A + iB$, where A and B are self-adjoint operators affiliated with $W^*(S)$ by 5.2.9. Thus $A = f(S)$, $B = g(S)$ for some real-valued functions f and g in $\mathcal{L}(\mathrm{sp}(S))$ from the preceding, and $(f + ig)(S)$, being the closure of $f(S) + ig(S)\,(= A + iB = T$ by 5.1.10) is equal to T. $\square$

5.3.9. The preceding result shows that we may compute formally inside the class of normal operators affiliated with $W^*(S)$, without worrying about domains, closedness, normality, or self-adjointness. If for some reason a common core for the operators is desirable, one may restrict attention to the class $\mathcal{L}^{\infty}_{loc}(\mathrm{sp}(S))$ of measurable functions f that are locally (essentially) bounded, which means that $[C]f \in \mathcal{L}^{\infty}(\mathrm{sp}(S))$ for every compact subset C of $\mathrm{sp}(S)$. Indeed, if $\mathfrak{H} = \bigoplus P_n(\mathfrak{H})$ and $P_n(\mathfrak{H}) = U_n L^2_n(\mathrm{sp}(S))$ as in the proof of 5.3.8, let $\mathfrak{D}$ be the linear span of the orthogonal subspaces $U_n(C_c(\mathrm{sp}(S)))$, $n \in \mathbb{N}$. Then $\mathfrak{D}$ is dense in $\mathfrak{H}$, and $\mathfrak{D}$ is a core for $f(S)$ for every f in $\mathcal{L}^{\infty}_{loc}(\mathrm{sp}(S))$.

We have not assigned any continuity properties to the spectral map $f \to f(S)$ in 5.3.8. However, the continuity concepts from integration theory carry over almost unchanged.

5.3.10. Proposition. *If S is a self-adjoint operator in a separable Hilbert space $\mathfrak{H}$, then each vector x in $\mathfrak{H}$ induces a finite Radon integral $\int_x$ on $\mathrm{sp}(S)$, absolutely continuous with respect to the integral mentioned in 5.3.8, and satisfying*

$$\int_x f = (f(S)x|x), \quad f \in \mathcal{B}(\mathrm{sp}(S))_+. \tag{*}$$

PROOF. When g ranges over the functions in $C_0(\mathbb{T} \setminus \{1\})$, the function $f = g \circ \kappa$ ranges over $C_0(\mathbb{R})$. Thus,

$$\int_x f = (g(\kappa(S))x|x), \quad f = g \circ \kappa,$$

defines a finite Radon integral $\int_x$ on $\mathrm{sp}(S)$ with $\int_x 1 = \|x\|^2$; cf. the proof of 4.5.4. That the relation $(*)$ is also valid for Borel functions, even unbounded ones, is evident from 5.3.3, if S is multiplicity-free. But in the general case (5.3.8) we have an orthogonal decomposition $x = \sum x_n$, where $x_n = P_n x$ (notation as in the proof of 5.3.8), and thus

$$\int_x f = (f(S)x|x) = \sum (f(S)x_n|x_n) = \sum \int_{x_n} f$$

if $f \in C_0(\mathrm{sp}(S))$. Thus $\int_x = \sum \int_{x_n}$ and the equation $(*)$, which holds for every x_n, is obtained for x by addition.

If $\mathcal{N}(\mathrm{sp}(S))$ denotes the class of null functions for the integral $\int$ mentioned in 5.3.8, then $f(S) = 0$ iff $f \in \mathcal{N}(\mathrm{sp}(S))$. In particular $\int_x f = 0$ for every f in $\mathcal{N}(\mathrm{sp}(S))$, so that $\int_x \ll \int$ by 6.5.2. $\square$

5.3.11. Corollary. *If (f_n), f and g are functions in $\mathcal{B}(\mathrm{sp}(S))$ such that $f_n \to f$, pointwise, and $|f_n| \le g$ for all n, then $f_n(S)x \to f(S)x$ for every x in $\mathfrak{D}(g(S))$.*

PROOF. Since $\int_x g^2 < \infty$, we see from Lebesgue's dominated convergence theorem (6.1.15) that

$$\|f_n(S)x - f(S)x\|^2 = \int_x |f_n - f|^2 \to 0.$$ $\square$

5.3.12. A *strongly continuous one-parameter unitary group* (not a SCOPUG, please!) is a strongly continuous function $U: \mathbb{R} \to \mathbf{U}(\mathfrak{H})$, such that $U_{s+t} = U_s U_t$ for all real s and t. Such groups turn up frequently in the applications to quantum mechanics, where they may represent time evolution of a physical system.

The following result explains the link between one-parameter unitary groups and (the spectral theory of) unbounded self-adjoint operators. It also justifies the use of the word "infinitesimal generator" for the group, as applied to the operator $dU/dt|_{t=0}$.

5.3.13. Proposition. *If S is a self-adjoint operator in the separable Hilbert space $\mathfrak{H}$, then the family of spectral functions $U_t = \exp itS$, $t \in \mathbb{R}$, is a strongly continuous one-parameter unitary group, and for each x in $\mathfrak{D}(S)$ we have*

$$Sx = \lim_{t \to 0} (it)^{-1}(U_t x - x). \tag{$*$}$$

Conversely, if $x \in \mathfrak{H}$ such that $(it)^{-1}(U_t x - x)$ has a limit in $\mathfrak{H}$, then $x \in \mathfrak{D}(S)$ and the limit is Sx.

PROOF. The functions $\lambda \to \exp it\lambda$, $t \in \mathbb{R}$, are unitary on $\mathrm{sp}(S)$ ($\subset \mathbb{R}$ by 5.2.13), so $(U_t)_{t \in \mathbb{R}}$ is a one-parameter family of unitary operators in $W^*(S)$ by 5.3.8.

The group property, $U_{s+t} = U_s U_t$, follows from the nature of the exponential function. To prove continuity, note that with $e_t(\lambda) = \exp it\lambda$ we have $e_s \to e_t$ pointwise and boundedly when $s \to t$, whence $U_s \to U_t$, strongly, by 5.3.11. Thus $(U_t)_{t \in \mathbb{R}}$ is a strongly continuous one-parameter unitary group.

If $x \in \mathfrak{D}(S)$, the function $\lambda \to \lambda^2$ is integrable with respect to the integral $\int_x$ defined in 5.3.10. Furthermore,

$$\|(it)^{-1}(U_t x - x) - Sx\|^2 = \int_x |f_t|^2,$$

where $f_t(\lambda) = (it)^{-1}(\exp it\lambda - 1 - it\lambda)$. Clearly, $f_t \to 0$, pointwise, as $t \to 0$ and, moreover, $|f_t(\lambda)| \leq a|\lambda| + b$ for suitable constants a and b and all real λ and all t with $|t| \leq 1$. It follows from Lebesgue's dominated convergence theorem that $\int_x |f_t|^2 \to 0$ as $t \to 0$, as desired.

In the converse direction, let $\mathfrak{D}$ be the linear subspace of $\mathfrak{H}$ consisting of vectors x for which $\lim(it)^{-1}(U_t x - x)$ exists, and let Tx denote the limit. Then T is an operator in $\mathfrak{H}$, and is an extension of S from what we proved above. As

$$\overline{(it)^{-1}(U_t x - x | x)} = (-it)^{-1}(x | U_t x - x) = (-it)^{-1}(U_{-t} x - x | x),$$

we see that T is symmetric, and since S is self-adjoint, this implies that $S = T$, so that $\mathfrak{D} = \mathfrak{D}(S)$. $\square$

5.3.14. Examples. Take $\mathfrak{H} = L^2(\mathbb{R})$ with respect to Lebesgue measure, and define

$$U_t f(s) = \exp(its) f(s), \quad f \in \mathscr{L}^2(\mathbb{R}).$$

Routine calculations show that this gives a strongly continuous one-parameter unitary group. Consider next the self-adjoint multiplication operator M_{id} in $L^2(\mathbb{R})$ defined in 5.1.14. Since $f(M_{id}) = M_f$ for every Borel function f on $\mathbb{R}$, it follows that $U_t = \exp(it M_{id})$, so that M_{id} is the infinitesimal generator for the unitary group as described in 5.3.13.

Now define

$$V_t f(s) = f(s - t), \quad f \in \mathscr{L}^2(\mathbb{R}).$$

Again it is routine to check that we have a strongly continuous one-parameter unitary group. To find the infinitesimal generator it is well to remember (E 3.1.16), that the Fourier transform induces a unitary F on $L^2(\mathbb{R})$ (of period four) such that

$$(Ff)(t) = \int f(s) \exp(-its) \, ds, \quad f \in \mathscr{L}^2(\mathbb{R}) \cap \mathscr{L}^1(\mathbb{R}).$$

It follows that in this case

$$V_t Ff(s) = Ff(s - t) = \int f(r) \exp(-ir(s - t)) \, dr$$

$$= \int (U_t f)(r) \exp(-irs) \, dr = FU_t f(s).$$

Thus $V_t = FU_t F^*$, so that the infinitesimal generator of $(V_t)_{t \in \mathbb{R}}$ is the self-adjoint operator $D = FM_{id}F^*$. With $\mathscr{S}$ as the space of Schwartz functions—see E 3.1.14—we have $F(\mathscr{S}) = \mathscr{S}$, and since $\mathscr{S}$ is a core for M_{id}, it is therefore also a core for D. Moreover, we have, by straightforward computations that

$$Df(t) = FM_{id}F^*f(t) = if'(t), \quad f \in \mathscr{S}.$$

Using (*) in 5.3.13 we see that actually $Df = if'$ for every f in $\mathfrak{D}(D)$.

5.3.15. Theorem. *If $(U_t)_{t \in \mathbb{R}}$ is a strongly continuous one-parameter unitary group, there is a self-adjoint operator S in $\mathfrak{H}$ such that $U_t = \exp itS$ for all t.*

PROOF. Recall from 4.2.8 that $L^1(\mathbb{R})$ is a Banach algebra (with convolution product) having a symmetric involution (cf. 4.3.8), so that the Gelfand transform (= Fourier transform) $f \to \hat{f}$ is a norm decreasing *-homomorphism of $L^1(\mathbb{R})$ onto a dense *-subalgebra of $C_0(\mathbb{R})$.

Take f in $L^1(\mathbb{R})$ and consider the expression

$$\int (U_s x | y) f(s) \, ds = (T_f x | y), \quad x, y \in \mathfrak{H}. \tag{*}$$

The left-hand side determines a sesquilinear form on $\mathfrak{H}$, bounded by $\|f\|_1$, and by 3.2.2 there is therefore a unique operator T_f in $\mathbf{B}(\mathfrak{H})$ satisfying the equation (*) above. Straightforward computations with (*) shows that the map $f \to T_f$ is a normdecreasing *-homomorphism of $L^1(\mathbb{R})$ into $\mathbf{B}(\mathfrak{H})$. Since $L^1(\mathbb{R})$ is commutative, its image in $\mathbf{B}(\mathfrak{H})$ is a commutative *-algebra; in particular, it consists of normal operators. Any homomorphism between Banach algebras diminishes the spectrum and therefore decreases the spectral radius r. Thus, for every f in $L^1(\mathbb{R})$ we have

$$\|T_f\| = r(T_f) \le r(f) = \|\hat{f}\|_\infty,$$

by 4.3.11 and 4.2.3. Since the image of $L^1(\mathbb{R})$ under Gelfand transformation is dense in $C_0(\mathbb{R})$, the map $\hat{f} \to T_f$ extends by continuity to a norm decreasing *-homomorphism $h \to T_h$ of $C_0(\mathbb{R})$ into $\mathbf{B}(\mathfrak{H})$.

For each vector x in $\mathfrak{H}$ we obtain a finite Radon integral $\int_x$ on $\mathbb{R}$ by

$$\int_x h = (T_h x | x), \quad h \in C_0(\mathbb{R}).$$

In particular,

$$\int_x \hat{f} = \int (U_s x | x) f(s) \, ds, \quad f \in L^1(\mathbb{R}).$$

Taking $f \ge 0$, $\|f\|_1 = 1$, and the support of f concentrated near 0, we see from the strong continuity of the unitary group that the right-hand side approaches $(Ix|x)$, so that $\int_x 1 = \|x\|^2$. Moreover, for t in $\mathbb{R}$ and f in $L^1(\mathbb{R})$ we have

$$(U_{-t}T_{\hat{f}}x|x) = \int (U_s x|x)f(s+t)\,ds = \int_x e_t\hat{f},$$

where $e_t(\lambda) = \exp it\lambda$. Since this holds for all f, we conclude as above that

$$(U_{-t}x|x) = \int_x e_t, \quad t \in \mathbb{R}, x \in \mathfrak{H}.$$

Defining $T_1 = I$ we extend the map $h \to T_h$ to a homomorphism from $C(\mathbb{R} \cup \{\infty\})$ into $\mathbf{B}(\mathfrak{H})$, and since $\kappa \in C(\mathbb{R} \cup \{\infty\})$, we obtain a unitary $W = T_\kappa$. If we had $W^*x = x$ for some x in $\mathfrak{H}$, then

$$0 = ((I - W)x|x) = \int_x 1 - \kappa.$$

Since $\operatorname{Re}(1 - \kappa)(\gamma) = \operatorname{Re}(1 - (\gamma - i)(\gamma + i)^{-1}) = \operatorname{Re}(2i(\gamma + i)^{-1}) = 2(\gamma^2 + 1)^{-1}$, which is strictly positive on $\mathbb{R}$, this implies that $\int_x = 0$, whence $x = 0$. Thus W^* does not have 1 as an eigenvalue, so that $S = -\kappa^{-1}(W)$ is a self-adjoint operator in $\mathfrak{H}$ by 5.2.4 and 5.2.5.

We define the unitary group $\exp itS$, $t \in \mathbb{R}$, as in 5.3.13 and have for each x in $\mathfrak{H}$ that

$$(\exp(itS)x|x) = (\exp(-it\kappa^{-1}(W))x|x)$$

$$= (\exp(-it\kappa^{-1}(T_\kappa))x|x) = \int_x e_{-t} = (U_t x|x).$$

By 3.2.25 this implies that $\exp itS = U_t$, as desired. $\qquad\square$

5.3.16. Remarks. The proof given above for Stone's theorem (1932) is not the shortest. One may prove directly that the symmetric operator defined by (*) in 5.3.13 is actually self-adjoint and is the infinitesimal generator for $(U_t)_{t \in \mathbb{R}}$. However, the proof at hand is capable of considerable generalization. Leaving aside the noncommutative case, the approach in 5.3.15 shows that if $t \to U_t$ is a strongly continuous unitary homomorphism from a locally compact abelian group G into $\mathbf{B}(\mathfrak{H})$—a *unitary representation*—it can first be integrated to an algebra homomorphism of $L^1(G)$ into $B(\mathfrak{H})$ and then—via Gelfand transformation—be extended to a homomorphism of the function algebra $C_0(\hat{G})$ into $\mathbf{B}(\mathfrak{H})$. Here, as in 4.2.6, $\hat{G}$ denotes the dual group of G, consisting of the continuous homomorphisms (characters) $\gamma: G \to \mathbb{T}$. Neglecting multiplicity, such a homomorphism is determined by a Radon integral on $\hat{G}$. The end result is that unitary representations of G, up to a certain equivalence (multiplicity), are parametrized by equivalence classes of Radon integrals on $\hat{G}$ or by the corresponding systems of null sets in $\hat{G}$.

For a more immediate gratification, the proof of 5.3.15 applies (almost verbatim) to show that if $(U_t)_{t \in \mathbb{R}^n}$ is a strongly continuous n-parameter unitary group, there are *commuting* self-adjoint operators $S_1, \ldots, S_n$ in $\mathfrak{H}$ such that

$$U_t = \exp it_1 S_1 \exp it_2 S_2 \cdots \exp it_n S_n, \quad t = (t_1, \ldots, t_n).$$

5.3.17. We shall finally extend the polar decomposition (3.2.17) to unbounded operators. If T is a densely defined, closed operator in $\mathfrak{H}$, we know from 5.1.9 that T^*T is self-adjoint and that $(T^*T + \lambda I)^{-1} \in \mathbf{B}(\mathfrak{H})$ for every $\lambda > 0$. It follows from 5.2.10 that $\mathrm{sp}(T^*T) \subset \mathbb{R}_+$, so that the function $\lambda \to \lambda^{1/2}$ is well defined and continuous on $\mathrm{sp}(T^*T)$. Using 5.3.8 we can therefore define the *absolute value* of T as the positive, self-adjoint operator

$$|T| = (T^*T)^{1/2}.$$

We note in passing that as in the bounded case (3.2.11 or 4.4.8) the square root is unique: If S is any positive, self-adjoint operator in $\mathfrak{H}$ such that $S^2 = T^*T$, then $S = |T|$. Indeed, since T^*T is affiliated with $W^*(S)$, so is $|T|$. Thus, by 5.3.8, S and $|T|$ are represented as positive multiplication operators with the same square, whence $S = |T|$.

5.3.18. Proposition. *For each densely defined, closed operator T in $\mathfrak{H}$ with absolute value $|T|$ we have $\mathfrak{D}(|T|) = \mathfrak{D}(T)$ and*

$$\||T|x\| = \|Tx\|, \quad x \in \mathfrak{D}(T).$$

Moreover, there is a unique partial isometry U with $\ker U = \ker T$ and $T = U|T|$. In particular, $U^*U|T| = |T|$, $U^*T = |T|$, and $UU^*T = T$.

PROOF. If $x \in \mathfrak{D}(T^*T)$, we clearly have

$$\|Tx\|^2 = (T^*Tx|x) = (|T|^2x|x) = \||T|x\|^2.$$

Since $\mathfrak{D}(T^*T)$ is a core for both T and $|T|$ by 5.1.9, it follows that $\mathfrak{D}(T) = \mathfrak{D}(|T|)$ and that $\|Tx\| = \||T|x\|$ for every x in $\mathfrak{D}(T)$ [cf. the implication (ii) $\Rightarrow$ (i) in the proof of 5.1.10].

We define an operator U_0 from $\mathfrak{R}(|T|)$ onto $\mathfrak{R}(T)$ by

$$U_0|T|x = Tx, \quad x \in \mathfrak{D}(T).$$

Then U_0 is a well-defined isometry and extends by continuity to an isometry U from $\mathfrak{R}(|T|)^=$ onto $\mathfrak{R}(T)^=$. We extend U to a partial isometry on $\mathfrak{H}$ by defining

$$\ker U = (\mathfrak{R}(|T|)^=)^\perp = \ker|T| = \ker T;$$

cf. 5.1.2. The equation $U|T| = T$ follows from the definition of U. The unicity of U and the last statements are proved exactly as in the bounded case (3.2.17). $\qquad\square$

5.3.19. Remarks. As in the bounded case we observe that if $T = U|T|$ is the polar decomposition of an operator T, then U is an isometry from $\mathfrak{R}(T^*)^=$ $[=(\ker T)^\perp]$ onto $\mathfrak{R}(T)^=$ $[=(\ker T^*)^\perp]$, so that U is unitary iff both T and T^* are injective. In particular, U is unitary if T is invertible.

If T is normal (5.1.11), then $\ker T = \ker T^*$, so that the partial isometry U in its polar decomposition is normal and commutes with $|T|$ (by the same arguments as in 3.2.20). It can therefore be enlarged to a unitary W commuting with $|T|$, such that $T = W|T|$. Define

$$T_0 = T(|T| + I)^{-1} = W|T|(|T| + I)^{-1}.$$

Then T_0 is normal with $\|T_0\| \leq 1$, and $I - |T_0|$ is injective. Conversely, if T_0 is such an operator, then $T = (I - |T_0|)^{-1}T_0$ is a normal operator in $\mathfrak{H}$. Using the bijective correspondence $T \leftrightarrow T_0$, the spectral theory for normal operators is easily established. Define ζ on the open unit disk by $\zeta(\lambda) = \lambda(1 - |\lambda|)^{-1}$. Then let

$$f(T) = f \circ \zeta(T_0) \qquad\qquad (*)$$

for any bounded Borel function f on $\zeta(\text{sp}(T_0))$ $[=\text{sp}(T)]$. The resulting spectral theory is the exact analogue of the bounded case, and satisfies 4.5.4 and 4.7.15. Proceeding with care, the formula $(*)$ can be extended to unbounded functions, giving the analogue of 5.3.8 for normal operators.

CHAPTER 6

Integration Theory

This chapter has two functions: Throughout the book it has served as an Appendix, to which the reader was referred for definitions, arguments, and results about measures and integrals. It will now serve as a functional analyst's dream of the ideal short course in measure theory. Thus, we shall develop the theory of Radon integrals (= Radon measures, cf. 6.3.4) on a locally compact Hausdorff space, assuming full knowledge of topology and topological vector spaces. This theory takes as point of departure an integral (a positive linear functional) on the minimal class of topologically relevant functions on X, namely, the class $C_c(X)$ of continuous functions with compact supports. The integral is extended by monotonicity to a larger class of (integrable) functions and the measure appears, post festum, as the value of the integral on characteristic functions.

In all honesty the author will admit that the reader should have had an ordinary course (however dull) in measure and integration theory in order to appreciate fully the high-tech approach here. He should also be aware that the theory is richer than the spartan exposition might lead to believe. A study of one or more of the classical areas of application, harmonic analysis, probability, potential theory, and ergodic theory, is advisable, in order to understand the significance of integration theory as a cornerstone in that dread Temple of Our Worth.

6.1. Radon Integrals

Synopsis. Upper and lower integral. Daniell's extension theorem. The vector lattice $\mathscr{L}^1(X)$. Lebesgue's theorems on monotone and dominated convergence. Stieltjes integrals.

6.1.1. Throughout this chapter X will denote a locally compact Hausdorff space. On this we consider the minimal class of topologically significant functions on X, namely, the algebra $C_c(X)$ of continuous real-valued functions on X with compact supports. From that we define the class $C_c(X)^m$ of functions $f: X \to \mathbb{R} \cup \{\infty\}$, for which there exist a monotone increasing net $(f_\lambda)_{\lambda \in \Lambda}$ [i.e., $\lambda \leq \mu$ implies $f_\lambda(x) \leq f_\mu(x)$ for every x in X] in $C_c(X)$, such that $f(x) = \sup f_\lambda(x)$ for every x in X. We use the symbols $f_\lambda \nearrow f$ to describe this situation. Analogously we define $C_c(X)_m$ as those functions $f: X \to \mathbb{R} \cup \{-\infty\}$, for which there is a monotone decreasing net $(f_\lambda)_{\lambda \in \Lambda}$ in $C_c(X)$ with $f_\lambda \searrow f$. It follows that $C_c(X)_m = -C_c(X)^m$.

Since $C_c(X)$ is stable under the lattice operations $\bigvee$ and $\bigwedge$, it is immediate that $C_c(X)^m$ is a function lattice that is even stable under formation of the supremum of an arbitrary family of elements. Moreover, $C_c(X)^m$ is an additive cone and contains the product of any two of its positive elements. We see from 1.5.12 that the elements in $C_c(X)^m$ are lower semicontinuous functions (with extended values, though), and it follows from (the proof of) 1.5.13 that $C_c(X)^m$ contains every function in $C^{1/2}(X)_+$. (Here and in the sequel M_+ will denote the set of functions f in a class M, for which $f \geq 0$.) Note though that the function -1 does not belong to $C_c(X)^m$, unless X is compact. In fact,

$$C_c(X)^m \cap C_c(X)_m = C_c(X),$$

a result we shall, however, not need.

Finally, we note that if $f \in C_c(X)^m$, the set $\{f < 0\}$ (note: in this chapter this will be the abbreviated notation for the set $\{x \in X \mid f(x) < 0\}$) has compact closure, since f majorizes an element from $C_c(X)$. Likewise, the set $\{f > 0\}^-$ is compact for every f in $C_c(X)_m$.

6.1.2. A *Radon integral* is a (linear) functional $\int : C_c(X) \to \mathbb{R}$ that is positive, i.e. $f \geq 0$ implies $\int f \geq 0$. In this section $\int$ will be fixed, and we shall show how it extends from $C_c(X)$ to a much larger class of functions.

We define the *upper integral* $\int^*$ as an extended-valued function on $C_c(X)^m$, by

$$\int^* f = \sup\left\{ \int g \mid g \in C_c(X), g \leq f \right\}, \quad f \in C_c(X)^m.$$

Analogously, we define the *lower integral* $\int_*$ on functions in $C_c(X)_m$ by

$$\int_* f = \inf\left\{ \int g \mid g \in C_c(X), g \geq f \right\}.$$

Thus, $\int^* f \in \mathbb{R} \cup \{\infty\}$, whereas $\int_* f \in \mathbb{R} \cup \{-\infty\}$. Furthermore, $\int_* f = -\int^*(-f)$ if $f \in C_c(X)_m$, and $\int_* f = \int^* f = \int f$ if $f \in C_c(X)$.

6.1.3. Lemma. *If $(f_\lambda)_{\lambda \in \Lambda}$ is a monotone decreasing net of upper semicontinuous, positive functions with compact supports, such that $f_\lambda(x) \searrow 0$ for each x in X, then $\|f_\lambda\|_\infty \searrow 0$.*

PROOF. It follows from 1.5.10 that the sets $\{f_\lambda \geq \varepsilon\}$ are closed for every $\varepsilon > 0$, and therefore compact since $\{f_\lambda > 0\}^-$ is compact by assumption. Since $(f_\lambda)_{\lambda \in \Lambda}$ tends pointwise to zero, we have $\bigcap \{f_\lambda \geq \varepsilon\} = \emptyset$, and thus $\{f_\lambda \geq \varepsilon\} = \emptyset$ eventually, by 1.6.2(ii). This means that $\|f_\lambda\|_\infty \leq \varepsilon$, eventually, whence $\|f_\lambda\|_\infty \searrow 0$. $\qquad\square$

6.1.4. Lemma. If $(f_\lambda)_{\lambda \in \Lambda}$ is a monotone decreasing net in $C_c(X)_m$ and $f_\lambda \searrow 0$, then $\int_* f_\lambda \searrow 0$.

PROOF. Every function in $C_c(X)_m$ is upper semicontinuous by 1.5.12, and the positive ones have compact supports (cf. 6.1.1). Thus, $\|f_\lambda\|_\infty \searrow 0$ by 6.1.3. We select a decreasing sequence (f_n) from the net such that $\int_* f_n \searrow \lim \int_* f_\lambda$ and $\|f_n\|_\infty < 2^{-n}$ for every n. For each n we can find g_n in $C_c(X)$ with $f_n \leq g_n$ and, replacing if necessary g_n by $g_n \wedge 2^{-n}$, we may assume that $g_n \leq 2^{-n}$. Furthermore, replacing if necessary g_n by $g_n \wedge g_1$, we may assume that all g_n have support inside the same compact set. Thus, if $g = \sum g_n$, we have $g \in C_c(X)_+$. Since $\int$ is positive, it follows that $\sum \int g_n \leq \int g < \infty$. Consequently, $\sum \int_* f_n \leq \sum \int g_n < \infty$, whence $\int_* f_n \searrow 0$. $\qquad\square$

6.1.5. Lemma. If $(f_\lambda)_{\lambda \in \Lambda}$ is a monotone increasing net in $C_c(X)$, and $f_\lambda \nearrow f$ for some f in $C_c(X)^m$, then $\int f_\lambda \nearrow \int^* f$.

PROOF. If $g \in C_c(X)$ and $g \leq f$, then $f_\lambda \wedge g \nearrow g$, i.e. $g - f_\lambda \wedge g \searrow 0$. By 6.1.4 this implies that

$$\int g = \lim \int f_\lambda \wedge g \leq \lim \int f_\lambda.$$

Since this holds for all minorants g of f, we have $\int^* f \leq \lim \int f_\lambda$. The converse inequality is obvious. $\qquad\square$

6.1.6. Corollary. If f and g belong to $C_c(X)^m$ and $t \geq 0$, then

$$\int^* (tf + g) = t \int^* f + \int^* g.$$

6.1.7. Lemma. If $(f_\lambda)_{\lambda \in \Lambda}$ is a monotone increasing net in $C_c(X)^m$ and $f_\lambda \nearrow f$ [so that $f \in C_c(X)^m$], then $\int^* f_\lambda \nearrow \int^* f$.

PROOF. Clearly $\lim \int^* f_\lambda \leq \int^* f$. On the other hand, if $g \in C_c(X)$ and $g \leq f$, then $g - f_\lambda \wedge g \searrow 0$; so by 6.1.6 and 6.1.4 we have

$$\int g - \int^* f_\lambda \wedge g = -\int^* (-g + f_\lambda \wedge g) = \int_* (g - f_\lambda \wedge g) \searrow 0.$$

Consequently,

$$\int g = \lim \int^{*} f_{\lambda} \wedge g \leq \lim \int^{*} f_{\lambda},$$

and since g was arbitrary, $\int^{*} f \leq \lim \int^{*} f_{\lambda}$.   □

6.1.8. Lemma. *If $f \in C_c(X)^m$ and $g \in C_c(X)_m$ such that $g \leq f$, then $\int_{*} g \leq \int^{*} f$.*

PROOF. Since $0 \leq f - g$ we have by 6.1.6 that

$$0 \leq \int^{*} (f - g) = \int^{*} f + \int^{*} (-g) = \int^{*} f - \int_{*} g.$$ □

6.1.9. For every real-valued function f on X we define the *upper* and the *lower* *integral* of f as

$$\int^{*} f = \inf \left\{ \int^{*} g \,\middle|\, g \in C_c(X)^m, g \geq f \right\};$$

$$\int_{*} f = \sup \left\{ \int_{*} g \,\middle|\, g \in C_c(X)_m, g \leq f \right\}.$$

Note that although 6.1.8 asserts that $\int_{*} f \leq \int^{*} f$, both the upper and the lower integral can have extended values in $\mathbb{R} \cup \{\pm\infty\}$. We say that a function f is *integrable* if $\int_{*} f = \int^{*} f \in \mathbb{R}$, and we denote by $\mathscr{L}^1(X)$ the class of integrable functions on X. The following rephrasing of this definition is very useful:

(∗) A function f is integrable if for every $\varepsilon > 0$ there are functions g in $C_c(X)^m$ and h in $C_c(X)_m$, such that $h \leq f \leq g$ and $\int^{*} g - \int_{*} h < \varepsilon$.

In particular, we see that a real-valued function f in $C_c(X)^m$ is integrable iff $\int^{*} f < \infty$. Indeed, we can use f itself as a majorant in (∗) and choose elements in $C_c(X)$ as minorants.

If $f \in \mathscr{L}^1(X)$, we define $\int f = \int^{*} f \,(= \int_{*} f)$. This leads to the formulation of *Daniell's extension theorem* for a Radon integral.

6.1.10. Theorem. *Given a Radon integral $\int$ on a locally compact Hausdorff space X, the class $\mathscr{L}^1(X)$ of integrable functions is a vector space containing $C_c(X)$, which is closed under the lattice operations $\bigvee$ and $\bigwedge$. Moreover, $\int: \mathscr{L}^1(X) \to \mathbb{R}$ is a positive functional that extends the original integral on $C_c(X)$.*

PROOF. Simple manipulations with the definition of integrability, in connection with 6.1.6 (and the corresponding result for lower integrals), show that if f and g both belong to $\mathscr{L}^1(X)$ and $t \geq 0$, then $tf + g \in \mathscr{L}^1(X)$ with

$$\int (tf + g) = t \int f + \int g.$$

Since $-C_c(X)_m = C_c(X)^m$, it is also easy to verify that if $f \in \mathscr{L}^1(X)$, then $-f \in \mathscr{L}^1(X)$ with $\int -f = -\int f$. Taken together this means that $\mathscr{L}^1(X)$ is a real vector space, and that $\int$ is a functional on it. Since $\int^* f = \int_* f = \int f$ if $f \in C_c(X)$, cf. 6.1.2, it follows that $C_c(X) \subset \mathscr{L}^1(X)$ and that $\int$ extends the original integral on $C_c(X)$. Moreover, if $f \geq 0$, then $\int^* f \geq 0$, so that $\int$ is a positive functional on $\mathscr{L}^1(X)$.

Take f_1 and f_2 in $\mathscr{L}^1(X)$ and for $\varepsilon > 0$ choose g_1, g_2 in $C_c(X)^m$, h_1, h_2 in $C_c(X)_m$, such that $h_k \leq f_k \leq g_k$ and $\int^* g_k - \int_* h_k < \varepsilon$ for $k = 1, 2$; cf. $(*)$ in 6.1.9. Note that

$$g_1 \vee g_2 - h_1 \vee h_2 \leq (g_1 - h_1) + (g_2 - h_2),$$

so that $\int^* g_1 \vee g_2 - \int_* h_1 \vee h_2 < 2\varepsilon$. It follows that $f_1 \vee f_2 \in \mathscr{L}^1(X)$. Similarly, or by exploiting the identity $f_1 \wedge f_2 + f_1 \vee f_2 = f_1 + f_2$, we see that $f_1 \wedge f_2 \in \mathscr{L}^1(X)$. $\qquad \square$

6.1.11. Corollary. *If* $f \in \mathscr{L}^1(X)$, *then* $|f| \in \mathscr{L}^1(X)$ *and* $|\int f| \leq \int |f|$.

6.1.12. Remark. As we shall see (in 6.2) the new domain of definition for the integral, viz. $\mathscr{L}^1(X)$, contains all the functions one may wish to integrate (and a good many more). The primary reason for the extension is that limit processes under the integral sign now can be handled in the generality they deserve. The fundamental results in this direction are Lebesgue's theorems on *monotone convergence* (6.1.13) and on *dominated convergence* (6.1.15). The result in 6.1.14 is known as *Fatou's lemma.*

6.1.13. Theorem. *If a function* f *on* X *is the pointwise limit of an increasing sequence* (f_n) *in* $\mathscr{L}^1(X)$ *such that* $\sup \int f_n < \infty$, *then* $f \in \mathscr{L}^1(X)$ *and* $\int f = \lim \int f_n$.

PROOF. Replacing if necessary f_n and f by $f_n - f_1$ and $f - f_1$, we may assume that $f_1 = 0$. Now put $g_n = f_{n+1} - f_n$, so that $g_n \in \mathscr{L}^1(X)_+$. Given $\varepsilon > 0$, there is therefore an h_n in $C_c(X)^m$, such that $g_n \leq h_n$ and $\int^* h_n < \int g_n + 2^{-n}\varepsilon$. Put $h = \sum h_n$, so that $h \in C_c(X)^m$ and $f \leq h$. By 6.1.6 and 6.1.7 this implies that

$$\int f_m \leq \int_* f \leq \int^* f \leq \int^* h = \sum \int^* h_n$$

$$\leq \sum \int g_n + \sum 2^{-n}\varepsilon = \lim \int f_n + \varepsilon$$

for every m, from which we conclude that $f \in \mathscr{L}^1(X)$ and that $\int f = \lim \int f_n$. $\qquad \square$

6.1.14. Lemma. *Suppose that* (f_n) *is a sequence in* $\mathscr{L}^1(X)_+$ *such that* $\liminf f_n(x) < \infty$ *for every* x *in* X *and* $\liminf \int f_n < \infty$. *Then* $\liminf f_n \in \mathscr{L}^1(X)$ *and*

$$\int \liminf f_n \leq \liminf \int f_n.$$

PROOF. Define $g_n = f_n \wedge f_{n+1} \wedge \cdots$ and note from 6.1.10 and 6.1.13 that $g_n \in \mathcal{L}^1(X)$. The sequence (g_n) is monotone increasing toward $\liminf f_n$, and since $g_n \le f_n$, we conclude from 6.1.13 that $\liminf f_n \in \mathcal{L}^1(X)$ with

$$\int \liminf f_n = \lim \int g_n \le \liminf \int f_n. \qquad \square$$

6.1.15. Theorem. *If a function f on X is a pointwise limit of a sequence (f_n) in $\mathcal{L}^1(X)$, and if $|f_n| \le g$ for some g in $\mathcal{L}^1(X)_+$ and all n, then $f \in \mathcal{L}^1(X)$ and $\int f = \lim \int f_n$.*

PROOF. Since $0 \le f_n + g \le 2g$ and $f_n + g \to f + g$, it follows from 6.1.14 that $f + g \in \mathcal{L}^1(X)$, whence $f \in \mathcal{L}^1(X)$. Furthermore, $|f - f_n| \le 2g$, so we can apply 6.1.14 to the sequence $(2g - |f - f_n|)$, to obtain

$$2 \int g \le \liminf \int (2g - |f - f_n|) = 2 \int g - \limsup \int |f - f_n|.$$

Thus, $\limsup \int |f - f_n| \le 0$, which implies that $\int |f - f_n| \to 0$. We conclude from 6.1.11 that $\int f_n \to \int f$. $\qquad \square$

6.1.16. We shall finally indicate a classical construction of Radon integrals on the real line.

Let m be a monotone increasing function on $\mathbb{R}$. Although m need not be continuous, we see that for every $\varepsilon > 0$ and every bounded interval I there are at most finitely many discontinuities of m in I with a discontinuity "jump" larger than ε. In particular, the discontinuity points for m on $\mathbb{R}$ is a countable set. We shall assume that m is lower semicontinuous, cf. 1.5.11, i.e. m is continuous from the left. This is a mild normalization condition on an arbitrary increasing function n on $\mathbb{R}$, and is always satisfied by the function m on $\mathbb{R}$ obtained by

$$m(x) = \sup\{n(y) | y < x\}.$$

For each f in $C_c(\mathbb{R})$, choose an interval $[a, b]$ such that the support of f is contained in $[a, b]$. Then consider partitions $\lambda = \{x_0, \ldots, x_n\}$ such that $a = x_0 < x_1 < \cdots < x_n = b$. Let

$$(Sf)_k = \sup\{f(x) | x_{k-1} \le x < x_k\},$$
$$(If)_k = \inf\{f(x) | x_{k-1} \le x < x_k\},$$

and use these numbers to define upper and lower sums of f with respect to m as follows:

$$\sum_{\lambda}^* f = \sum (Sf)_k (m(x_k) - m(x_{k-1})),$$
$$\sum_{\lambda *} f = \sum (If)_k (m(x_k) - m(x_{k-1})).$$

With the convention $\lambda \le \mu$ if $\lambda \subset \mu$, we see that the family Λ of finite partitions of $\mathbb{R}$ as specified above has an upward filtering order, so that we obtain the two nets $(\sum_{\lambda}^* f)_{\lambda \in \Lambda}$ and $(\sum_{*\lambda} f)_{\lambda \in \Lambda}$. Clearly the first net is monotone decreasing and the second monotone increasing. Moreover, $\sum_{*\lambda} f \leq \sum_{\lambda}^* f$ for every λ, and

$$\sum_{\lambda}^* f - \sum_{*\lambda} f = \sum_{\lambda} ((Sf)_k - (If)_k)(m(x_k) - m(x_{k-1})).$$

Since f is uniformly continuous on $[a, b]$, there is for each $\varepsilon > 0$ a $\delta > 0$ such that $(Sf)_k - (If)_k < \varepsilon$ if $x_k - x_{k-1} < \delta$. Thus, if λ_0 is an equidistant partition of $[a, b]$ of length $n > (b - a)\delta^{-1}$, it follows that for $\lambda \geq \lambda_0$ we have $\sum_{\lambda}^* f - \sum_{*\lambda} f < \varepsilon(m(b) - m(a))$. Consequently, the nets of upper and lower sums converge to the same number, called the *Stieltjes integral* of f with respect to m and denoted by $\int f \, dm$.

It is clear that the function $f \to \sum_{\lambda}^* f$ on $C_c(\mathbb{R})$ is positive homogeneous and subadditive, and takes positive values on positive functions. Also, $f \to \sum_{*\lambda} f$ is superadditive and $-\sum_{*\lambda} f = \sum_{\lambda}^* - f$. It follows that in the limit the Stieltjes integral $f \to \int f \, dm$ is a positive, linear functional on $C_c(\mathbb{R})$, i.e. a Radon integral.

We claim that *every* Radon integral on $\mathbb{R}$ can be realized as a Stieltjes integral with respect to some function m. Indeed, if $\int$ is a Radon integral on $\mathbb{R}$ we extend it to the class $\mathscr{L}^1(\mathbb{R})$ as described in 6.1.10, and we note that $\mathscr{L}^1(\mathbb{R})$ certainly contains all characteristic functions of bounded intervals. Let 1_x^y denote the characteristic function corresponding to the half-open interval $[x, y[$. Then define $m(x) = \int 1_0^x$ if $x \geq 0$ and $m(x) = -\int 1_x^0$ if $x < 0$. Clearly, m is a monotone increasing function on $\mathbb{R}$, and it follows from the monotone convergence theorem (6.1.13) that m is lower semicontinuous. If $f \in C_c(X)$ with support in an interval $[a, b]$, and $\lambda = \{x_0, \ldots, x_n\}$ is a partition as described before, then $f \leq \sum (Sf)_k 1_{x_{k-1}}^{x_k}$, whence

$$\int f \leq \sum \int (Sf)_k 1_{x_{k-1}}^{x_k} = \sum (Sf)_k (m(x_k) - m(x_{k-1})) = \sum_{\lambda}^* f.$$

Likewise, $\int f \geq \sum_{*\lambda} f$ for every λ, and we conclude that $\int f = \int f \, dm$, as desired.

6.1.17. Remarks. Stieltjes' construction is evidently patterned after the *Riemann integral*, obtained by taking m to be the identical function $m(x) = x$ in 6.1.16. It is also clear that the Riemann–Stieltjes' construction will define an integral on more functions that just those in $C_c(X)$. In fact, a function f with compact support is Riemann integrable iff its discontinuity points form a set of Lebesgue measure zero. However, the class of Riemann integrable functions is not stable under monotone sequential limits in any reasonable sense, so that Lebesgue's all important convergence theorems cannot be formulated within this class.

Finally, we remark (with regret) that the Riemann–Stieltjes' construction is intimately connected with the total order on $\mathbb{R}$, and does not generalize to higher dimensions.

6.2. Measurability

Synopsis. Sequentially complete function classes. σ-rings and σ-algebras. Borel sets and functions. Measurable sets and functions. Integrability of measurable functions.

6.2.1. A class $\mathscr{F}$ of real-valued functions on a set X is *(monotone) sequentially complete*, if every function on X, which is the pointwise limit of a (monotone increasing or decreasing) sequence of functions from $\mathscr{F}$, itself belongs to $\mathscr{F}$. It is clear that for every family of functions on X there is a smallest (monotone) sequentially complete class containing it, viz. the intersection of all (monotone) sequentially complete classes containing the family. This is only an existence result (however convenient), and it is not possible, in general, to give a constructive description of *all* the functions in the sequential completion.

6.2.2. Lemma. *If $\mathscr{A}$ is an algebra of real functions on X that is stable under the lattice operations $\bigvee$ and $\bigwedge$, then the monotone sequential completion $\mathscr{B}(\mathscr{A})$ of $\mathscr{A}$ is a sequentially complete algebra of functions, stable under $\bigvee$ and $\bigwedge$.*

PROOF. Take f in $\mathscr{A}$ and let $\mathscr{B}_1$ denote the class of functions g in $\mathscr{B}(\mathscr{A})$ such that $f + g \in \mathscr{B}(\mathscr{A})$. Since $\mathscr{A} \subset \mathscr{B}_1$ and $\mathscr{B}_1$ is monotone sequentially complete, it follows from the minimality of $\mathscr{B}(\mathscr{A})$ that $\mathscr{B}_1 = \mathscr{B}(\mathscr{A})$. Thus, $\mathscr{A} + \mathscr{B}(\mathscr{A}) \subset \mathscr{B}(\mathscr{A})$. Now take g in $\mathscr{B}(\mathscr{A})$, and let $\mathscr{B}_2$ denote the class of functions f in $\mathscr{B}(\mathscr{A})$ such that $f + g \in \mathscr{B}(\mathscr{A})$. We just proved that $\mathscr{A} \subset \mathscr{B}_2$, and since evidently $\mathscr{B}_2$ is monotone sequentially complete, it follows that $\mathscr{B}_2 = \mathscr{B}(\mathscr{A})$. Thus, $\mathscr{B}(\mathscr{A}) + \mathscr{B}(\mathscr{A}) \subset \mathscr{B}(\mathscr{A})$. The proof that $\mathbb{R}\mathscr{B}(\mathscr{A}) \subset \mathscr{B}(\mathscr{A})$ is similar (but simpler), and we see that $\mathscr{B}(\mathscr{A})$ is a vector space.

With $+$ replaced by $\bigvee$ or by $\bigwedge$, the argument above shows that $\mathscr{B}(\mathscr{A})$ is stable under maximum and minimum.

To show that $\mathscr{B}(\mathscr{A})$ is sequentially complete, let (f_n) be a sequence in $\mathscr{B}(\mathscr{A})$ that converges pointwise to some function f on X. Since $f_n \vee 0 \to f \vee 0$ and $f_n \wedge 0 \to f \wedge 0$, it suffices to consider the case where $f_n \geq 0$ for all n. For fixed n and some $m > n$ define

$$g_{nm} = f_n \wedge f_{n+1} \wedge \cdots \wedge f_m.$$

Then $g_{nm} \in \mathscr{B}(\mathscr{A})$, and since the sequence (g_{nm}) is monotone decreasing (in m) and positive, it has a pointwise limit g_n, and $g_n \in \mathscr{B}(\mathscr{A})$. Moreover, the sequence (g_n) is monotone increasing (since $g_{n_1 m} \leq g_{n_2 m}$ if $n_1 \leq n_2 < m$) and converges to f since

$$\lim g_n(x) = \liminf f_n(x) = \lim f_n(x) = f(x)$$

for every x in X. Thus $f \in \mathscr{B}(\mathscr{A})$, as claimed

Finally, to show that $\mathscr{B}(\mathscr{A})$ is an algebra, let $\mathscr{B}_3$ denote the class of functions f in $\mathscr{B}(\mathscr{A})$ such that $f^2 \in \mathscr{B}(\mathscr{A})$. Clearly, $\mathscr{A} \subset \mathscr{B}_3$ and if (f_n) is a sequence from $\mathscr{B}_3$ converging pointwise to some function f, then $f_n^2 \to f^2$, pointwise. Since

$\mathscr{B}(\mathscr{A})$ is sequentially complete, $f^2 \in \mathscr{B}(\mathscr{A})$. This proves that $\mathscr{B}_3$ is sequentially complete, and therefore $\mathscr{B}_3 = \mathscr{B}(\mathscr{A})$. For every pair f, g in $\mathscr{B}(\mathscr{A})$ we therefore have

$$fg = \tfrac{1}{2}((f + g)^2 - f^2 - g^2) \in \mathscr{B}(\mathscr{A}),$$

which proves that $\mathscr{B}(\mathscr{A})$ is an algebra. $\hfill\square$

6.2.3. A system $\mathscr{S}$ of subsets of a set X is called a *σ-ring* if it is stable under the formation of differences and countable unions. Since $\bigcap A_n = A_1 \backslash (\bigcup A_1 \backslash A_n)$, it follows that a σ-ring is also stable under countable intersections. If $X \in \mathscr{S}$, we say that $\mathscr{S}$ is a *σ-algebra*.

This terminology is based on the easily proved fact that a σ-ring $\mathscr{S}$ with the operations $+$ and $\cdot$ defined by

$$A + B = (A \cup B) \backslash (A \cap B) = (A \backslash B) \cup (B \backslash A) \qquad \text{(symmetric difference)},$$

$$A \cdot B = A \cap B \qquad \text{(intersection)},$$

is a (Boolean) ring, which is unital iff $\mathscr{S}$ is a σ-algebra. The prefix σ as usual indicates a countable process, cf. σ-compact set, σ-finite measure, σ-weak topology.

Just as in topology (1.2.11), there is for every family of subsets of X a smallest σ-ring/σ-algebra containing the family. But, contrary to the topological case (1.2.12), there is no procedure for constructing all the sets in the generated σ-system.

If $\mathscr{F}$ is a monotone sequentially complete algebra of functions on X, the class $\mathscr{S}$ of subsets $A \subset X$, such that $[A] \in \mathscr{F}$, is a σ-ring. {Here and in the rest of the chapter we shall constantly use $[A]$ to denote the characteristic function for A; i.e. $[A](x) = 1$ if $x \in A$, otherwise $[A](x) = 0$, cf. 1.5.10.} Indeed, we have

$$[A \backslash B] = [A] - [A][B], \qquad [A \cup B] = [A] + [B] - [A][B],$$

$$\left[\bigcup A_n \right] = \lim[A_1 \cup A_2 \cup \cdots \cup A_n].$$

The following result in the opposite direction is less obvious.

6.2.4. Lemma. *Let $\mathscr{S}$ be a σ-algebra of subsets of a set X, and denote by $\mathscr{F}$ the class of functions f on X such that $\{f > t\} \in \mathscr{S}$ for each t in $\mathbb{R}$. Then $\mathscr{F}$ is a sequentially complete, unital algebra of functions on X, stable under $\bigvee$ and $\bigwedge$. Moreover, $|f|^p \in \mathscr{F}$ for every f in $\mathscr{F}$ and $p > 0$.*

PROOF. If f and g belong to $\mathscr{F}$, then

$$\{f + g > t\} = \bigcup (\{f > r_n\} \cap \{g > t - r_n\}) \in \mathscr{S},$$

for each t in $\mathbb{R}$, if (r_n) is an enumeration of the rational numbers. Thus, $f + g \in \mathscr{F}$. If $\alpha > 0$, then $\{\alpha f > t\} = \{f < \alpha^{-1}t\} \in \mathscr{S}$, so that $\alpha f \in \mathscr{F}$. Moreover,

$$\{-f > t\} = \{f < -t\}$$
$$= X\backslash\{f \geq -t\} = X\backslash \bigcap \{f > -t - n^{-1}\} \in \mathscr{S},$$

so that $-f \in \mathscr{F}$. Taken together it shows that $\mathscr{F}$ is a vector space.

The expressions

$$\{f \vee g > t\} = \{f > t\} \cup \{g > t\},$$
$$\{f \wedge g > t\} = \{f > t\} \cap \{g > t\},$$

show that $\mathscr{F}$ is stable under the lattice operations $\bigvee$ and $\bigwedge$.

If $f \in \mathscr{F}$ and $t \geq 0$, then $\{|f|^p > t\} = \{|f| > t^{1/p}\} \in \mathscr{S}$, because $|f| = f \vee 0 - f \wedge 0 \in \mathscr{F}$. If $t < 0$, then $\{|f|^p > t\} = X \in \mathscr{S}$. It follows that $|f|^p \in \mathscr{F}$ for every $p > 0$. As

$$fg = \tfrac{1}{2}((f + g)^2 - f^2 - g^2),$$

this implies (with $p = 2$) that $fg \in \mathscr{F}$ for all f and g in $\mathscr{F}$, so that $\mathscr{F}$ is an algebra.

To show that $\mathscr{F}$ is sequentially complete, it suffices by 6.2.2 to show that $\mathscr{F}$ is monotone sequentially complete. Therefore, take a monotone increasing sequence (f_n) in $\mathscr{F}$ that converges pointwise to some function f. Then for each t in $\mathbb{R}$

$$\{f > t\} = \bigcup \{f_n > t\} \in \mathscr{S},$$

so that $f \in \mathscr{F}$, completing the proof.   □

6.2.5. For a topological space X the system $\mathscr{B}$ of *Borel sets* is defined as the smallest σ-algebra of subsets of X that contains all open sets (equivalently, contains all closed sets). The theory of Borel sets is due to Baire and Lebesgue.

Taking $X = \mathbb{R}^n$ we know that every open set is the countable union of open boxes, and each such is the intersection of $2n$ open half-spaces. Thus, the system of Borel subsets of $\mathbb{R}^n$ is the σ-ring generated by all half-spaces of the form $\{x \in \mathbb{R}^n | x_k > t\}$, where $1 \leq k \leq n$ and $t \in \mathbb{R}$.

6.2.6. A real-valued function f on a topological space X is a *Borel function*, if $\{f > t\}$ is a Borel subset of X for every t in $\mathbb{R}$ (cf. the definition of a continuous function in 1.4.1). Since the system of subsets $B \subset \mathbb{R}$ for which $f^{-1}(B) \in \mathscr{B}_X$ is a σ-algebra, it follows from the second half of 6.2.5 that for a Borel function f on X we have $f^{-1}(B) \in \mathscr{B}_X$ for every Borel subset B of $\mathbb{R}$.

As a natural continuation of the above we say that a function $f: X \to Y$ between topological spaces X and Y is a *Borel map* if $f^{-1}(B) \in \mathscr{B}_X$ for every B in $\mathscr{B}_Y$. As in the case of continuous functions, we see that composition of Borel maps again produces a Borel map.

6.2.7. Proposition. *For a topological space X the class $\mathscr{B}(X)$ of Borel functions on X is a sequentially complete, unital algebra, which is stable under the lattice operations $\bigvee$ and $\bigwedge$. Moreover, $|f|^p \in \mathscr{B}(X)$ for every f in $\mathscr{B}(X)$.*

PROOF. Direct application of 6.2.4. □

6.2.8. Lemma. *If X is a locally compact Hausdorff space, there is for each compact subset C of X a monotone decreasing net $(f_\lambda)_{\lambda \in \Lambda}$ in $C_c(X)$, such that $f_\lambda \searrow [C]$. If X is second countable, the net map be taken as a sequence.*

PROOF. Consider the set Λ of functions f in $C_c(X)_+$, such that $f|C = 1$. Since Λ is stable under infimum, it constitutes a monotone decreasing net if we use the usual pointwise order of functions. For each x in $X \setminus C$ there is by 1.7.5 an element f in Λ such that $f(x) = 0$. Consequently, the net $(f_\lambda)_{\lambda \in \Lambda}$ converges pointwise to $[C]$.

If X is second countable, we can write $C = \bigcap A_n$ for some sequence A_n of open sets, and then apply 1.7.5 to obtain a sequence as desired. □

6.2.9. Proposition. *If X is a second countable, locally compact Hausdorff space, the class $\mathscr{B}(X)$ of Borel functions on X is the monotone sequential completion of $C_c(X)$. Moreover, the class $\mathscr{B}_b(X)$ of bounded Borel functions is the monotone sequential completion of $C_c(X)$ inside the system of bounded functions on X.*

PROOF. Let $\mathscr{B}(C_c(X))$ denote the monotone sequential completion of $C_c(X)$, and note from 6.2.2 that $\mathscr{B}(C_c(X))$ is a sequentially complete algebra of functions, stable under $\bigvee$ and $\bigwedge$. Since $C_c(X) \subset \mathscr{B}(X)$, it follows from 6.2.7 that $\mathscr{B}(C_c(X)) \subset \mathscr{B}(X)$.

To prove the converse inclusion, note from 6.2.8 that $[C] \in \mathscr{B}(C_c(X))$ for every compact subset C of X. Since every open set in X is the countable union of compact subsets this implies that $[A] \in \mathscr{B}(C_c(X))$ for every open set A, and thus (cf. the remark in the end of 6.2.3) $[B] \in \mathscr{B}(C_c(X))$ for every Borel set $B \subset X$.

Now take f in $\mathscr{B}(X)$. To prove that $f \in \mathscr{B}(C_c(X))$ it suffices to consider the case $f \geq 0$. For n in $\mathbb{N}$ and $1 \leq k \leq n2^n$ define

$$B_{nk} = \{(k-1)2^{-n} < f \leq k2^{-n}\} \subset \mathscr{B};$$
$$f_n = \sum (k-1)2^{-n}[B_{nk}].$$

From the first part of the proof we see that $[B_{nk}] \in \mathscr{B}(C_c(X))$, whence $f_n \in \mathscr{B}(C_c(X))$ for every n. Since $f_n \nearrow f$, we conclude that $f \in \mathscr{B}(C_c(X))$.

To prove the second half of the proposition, let $\mathscr{B}_b(C_c(X))$ denote the smallest class of bounded functions on X, containing $C_c(X)$, that is monotone sequentially complete. This means that if (f_n) is a monotone (increasing or decreasing) sequence from $\mathscr{B}_b(C_c(X))$, converging pointwise to some bounded function f, then $f \in \mathscr{B}_b(C_c(X))$. For each n in $\mathbb{N}$ let

$$\mathscr{B}_n(X) = \{f \in \mathscr{B}(X)| -n \leq f \leq n\},$$

and let $\mathscr{B}_n(C_c(X))$ denote the monotone sequential completion of the class $\{f \in C_c(X)| -n \leq f \leq n\}$. Clearly, $\mathscr{B}_n(C_c(X)) \subset \mathscr{B}_n(X)$. On the other hand, the set $\mathscr{B}'_n(X)$ of functions f in $\mathscr{B}(X)$, for which $(-n \vee f) \wedge n \in \mathscr{B}_n(C_c(X))$, is monotone sequentially complete and contains $C_c(X)$. It follows that $\mathscr{B}'_n(X) = \mathscr{B}(X)$, from which we deduce that $\mathscr{B}_n(X) \subset \mathscr{B}_n(C_c(X))$. Therefore $\mathscr{B}_n(X) = \mathscr{B}_n(C_c(X))$. Since $\mathscr{B}_b(X) = \bigcup \mathscr{B}_n(X)$ and since $\bigcup \mathscr{B}_n(C_c(X))$ is monotone sequentially complete inside $\mathscr{B}_b(C_c(X))$, we finally conclude that

$$\mathscr{B}_b(C_c(X)) = \bigcup \mathscr{B}_n(C_c(X)) = \mathscr{B}_b(X). \qquad \square$$

6.2.10. Remarks. For an arbitrary (large) locally compact Hausdorff space X, the (monotone) sequential completion $\mathscr{B}(C_c(X))$ of $C_c(X)$ may be strictly smaller than $\mathscr{B}(X)$. This class is known as the *Baire functions*, and we see from 6.2.2 that they constitute a sequentially complete algebra of functions, stable under $\bigvee$ and $\bigwedge$. Moreover, 1 is a Baire function iff X is σ-compact. The class of subsets $B \subset X$, such that $[B]$ is a Baire function, is the *Baire sets*, and they form a σ-ring contained in $\mathscr{B}_X$. It is not difficult to prove that the class of Baire sets is the σ-ring generated by the compact G_δ-sets in X. The salient fact is that a subset C of X is a compact G_δ-set iff $C = \{f \geq \varepsilon\}$ for some f in $C_c(X)$ and $\varepsilon > 0$.

In the converse direction, if we insist on having all Borel functions available, we might ask for a reasonably small class $\mathscr{F}$ of functions on X, whose sequential completion will be $\mathscr{B}(X)$. One such is obtained by taking $C_b^{1/2}(X)$ as the family of bounded, lower semicontinuous functions on X (cf. 1.5.10), and then define

$$\mathscr{F} = C_b^{1/2}(X) - C_b^{1/2}(X).$$

Evidently $\mathscr{F}$ is a vector space (actually it is an algebra), and since it contains all characteristic functions $[A]$, where A is an open set, it is not hard to show that $\mathscr{B}(X)$ is the monotone sequential completion of $\mathscr{F}$.

6.2.11. Consider now a locally compact Hausdorff space X and denote by $\mathscr{C}$ the class of compact subsets of X. Given a Radon integral $\int$ on $C_c(X)$ we consider its extension to $\mathscr{L}^1(X)$ as described in 6.1.10. We denote by $\mathscr{M}^1$ the class of subsets $B \subset X$ such that $[B] \in \mathscr{L}^1(X)$, and we define $\mathscr{M}$—the *measurable sets*—as the class of subsets $A \subset X$ such that $A \cap C \in \mathscr{M}^1$ for every C in $\mathscr{C}$.

6.2.12. Proposition. *The system $\mathscr{M}$ of measurable sets for a Radon integral on a locally compact Hausdorff space X is a σ-algebra that contains the class $\mathscr{B}$ of Borel sets. Moreover, $\mathscr{C} \subset \mathscr{M}^1$.*

PROOF. If $C \in \mathscr{C}$, then $[C] \in C_c(X)_m$ by 6.2.8. Evidently $\int_* [C] \geq 0$, so by the remarks in 6.1.8 it follows that $[C] \in \mathscr{L}^1(X)$, i.e. $C \in \mathscr{M}^1$.

If $A \in \mathscr{M}$ and $C \in \mathscr{C}$, then $(X \setminus A) \cap C = C \setminus (A \cap C)$, whence

$$[(X \setminus A) \cap C] = [C] - [A \cap C] \in \mathscr{L}^1(X),$$

since $[C] \in \mathscr{L}^1(X)$. Consequently, $X \setminus A \in \mathscr{M}$. From 6.1.10 we see that $\mathscr{M}^1$ is stable under finite unions and intersections, and it then follows from Lebesgue's monotone convergence theorem, 6.1.13, that $\mathscr{M}$ is stable under countable unions and intersections. Thus $\mathscr{M}$ is a σ-algebra.

Since $\mathscr{C} \subset \mathscr{M}^1$, every closed subset of X belongs to $\mathscr{M}$, and as $\mathscr{M}$ is a σ-algebra, it follows that $\mathscr{B} \subset \mathscr{M}$. $\square$

6.2.13. A real-valued function f on X is said to be *measurable* (with respect to a given Radon integral $\int$) if $\{f > t\} \in \mathscr{M}$ for every t in $\mathbb{R}$. As in 6.2.6 this implies that $f^{-1}(B) \in \mathscr{M}$ for every Borel set B in $\mathbb{R}$. Also, $f \circ g$ is measurable for every Borel function g on $\mathbb{R}$. We denote by $\mathscr{L}(X)$ the class of measurable functions on X, and since $\mathscr{B} \subset \mathscr{M}$ we see immediately that $\mathscr{B}(X) \subset \mathscr{L}(X)$. Moreover, by 6.2.4 we have the following result.

6.2.14. Proposition. *The class $\mathscr{L}(X)$ of measurable functions with respect to a Radon integral on a locally compact Hausdorff space X is a sequentially complete, unital algebra, which is stable under the lattice operations $\bigvee$ and $\bigwedge$. Moreover, $|f|^p \in \mathscr{L}(X)$ for every f in $\mathscr{L}(X)$.*

6.2.15. Lemma. *For every subset B of X we have*

$$\int_* [B] = \sup\left\{ \int [C] \,\middle|\, C \subset B, C \in \mathscr{C} \right\};$$

$$\int^* [B] = \inf\left\{ \int^* [A] \,\middle|\, B \subset A, A \text{ open} \right\}.$$

Moreover, $B \in \mathscr{M}^1$ iff $B \in \mathscr{M}$ and $\int^* [B] < \infty$.

PROOF. For every $h \le [B]$ in $C_c(X)_m$ and every n, the set $C_n = \{h \ge n^{-1}\}$ is compact, since h is upper semicontinuous (1.5.12) and dominated by functions with compact supports. Furthermore,

$$h \le [\{h > 0\}] \le [B].$$

Since $\bigcup C_n = \{h > 0\}$ and $\{C_n\} \subset \mathscr{M}^1$ (6.2.12), it follows from Lebesgue's monotone convergence theorem (6.1.13) that $\int h \le \lim \int [C_n]$. By definition $\int_* [B]$ is the supremum of numbers $\int h$ (cf. 6.1.9), and it follows that $\int_* [B]$ is also the supremum of numbers $\int [C]$, where $C \in \mathscr{C}$ and $C \subset B$.

To prove the second equation, we may assume that $\int^* [B] < \infty$. There is then for each $\varepsilon > 0$ a g in $C_c(X)^m$ with $g \ge [B]$, such that $\int^* g \le \int^* [B] + \varepsilon$. Taking $A = \{g > 1 - \varepsilon\}$ we see that A is open because g is lower semicontinuous. Moreover, $B \subset A$ and

$$\int^* [A] \le (1 - \varepsilon)^{-1} \int^* g \le (1 - \varepsilon)^{-1} \left(\int^* [B] + \varepsilon \right).$$

This shows that $\int^* [B]$ is the infimum of numbers $\int^* [A]$, where A is open and $B \subset A$.

If $B \in \mathscr{M}$ and $\int^* [B] < \infty$, then, as we just proved, $\int^* [A] < \infty$ for some open set $A \supset B$. Since $[A] \in C_c(X)^m$ (1.5.13), we noted in 6.1.9 that $[A] \in \mathscr{L}^1(X)$, i.e. $A \in \mathscr{M}^1$. From the first equation we see that there is a C in

$\mathscr{C}$ with $C \subset A$, such that

$$\int^* [A] = \int_* [A] \leq \int [C] + \varepsilon.$$

We then have

$$B = (B \cap C) \cup (B \backslash C) \subset (B \cap C) \cup (A \backslash C),$$

and since both $B \cap C$ and $A \backslash C$ belong to $\mathscr{M}^1$ we conclude that

$$\int^* [B] \leq \int [B \cap C] \vee [A \backslash C]$$

$$\leq \int [B \cap C] + [A \backslash C] \leq \int_* [B] + \varepsilon.$$

Since ε is arbitrary, it follows that $[B] \in \mathscr{L}^1(X)$, i.e. $B \in \mathscr{M}^1$.   □

6.2.16. Theorem. *If $\int$ is a Radon integral on a locally compact Hausdorff space X, every integrable function is measurable. Conversely, a measurable function f on X is integrable iff $\int^* |f| < \infty$.*

PROOF. First note that if $f \in \mathscr{L}^1(X)_+$, then $f \wedge 1 \in \mathscr{L}^1(X)_+$. Indeed, the classes $C_c(X)_+, (C_c(X)^m)_+$, and $(C_c(X)_m)_+$ are all stable under the operation $f \rightarrow f \wedge 1$; and if $0 \leq h \leq f \leq g$, as in (*) in 6.1.9, then

$$h \wedge 1 \leq f \wedge 1 \leq g \wedge 1 \quad \text{and} \quad g \wedge 1 - h \wedge 1 \leq g - h.$$

Thus, for each f in $\mathscr{L}^1(X)_+$ and $t > 0$ we can define

$$f_n = (n(f - f \wedge t)) \wedge 1 \in \mathscr{L}^1(X).$$

Since $f_n \nearrow [\{f > t\}]$ and $f_n \leq t^{-1}f$, it follows from 6.1.13 that $\{f > t\} \in \mathscr{M}^1$.

In the general case, where $f \in \mathscr{L}^1(X)$, we see that if $t \geq 0$, then

$$\{f > t\} = \bigcup_n \{f \vee 0 > t + n^{-1}\} \in \mathscr{M},$$

since $\mathscr{M}$ is a σ-algebra. If $t < 0$, we use that

$$\{f > t\} = X \backslash \{-f \geq -t\} = X \backslash \bigcap \{-f > -t - n^{-1}\} \in \mathscr{M}.$$

Assume now that f is a measurable function with $\int^* |f| < \infty$. Then the same is true for $f \vee 0$ and $(-f) \vee 0$, since these functions are dominated by $|f|$. Since $f = f \vee 0 - (-f) \vee 0$, we may therefore assume that $f \geq 0$. For n in $\mathbb{N}$ and $1 \leq k \leq n2^n$ define

$$A_{nk} = \{(k-1)2^{-n} < f \leq k2^{-n}\};$$

$$f_n = \sum (k-1)2^{-n}[A_{nk}].$$

By assumption $A_{nk} \in \mathscr{M}$ for every k, and since $f_n \leq f$, we have $\int^* [A_{nk}] < 2^n(k-1)^{-1} \int^* f < \infty$ for all $k > 1$. By 6.2.15 this implies that $A_{nk} \in \mathscr{M}^1$, whence $f_n \in \mathscr{L}^1(X)$. As $f_n \nearrow f$ we conclude from the monotone convergence theorem (6.1.13) that $f \in \mathscr{L}^1(X)$.   □

6.3. Measures

Synopsis. Radon measures. Inner and outer regularity. The Riesz representation theorem. Essential integral. The σ-compact case. Extended integrability.

6.3.1. A *measure* on a σ-ring $\mathscr{S}$ of subsets of a set X is a function $\mu \colon \mathscr{S} \to [0, \infty]$ that is σ-*additive* in the sense that

$$\mu\left(\bigcup A_n \right) = \sum \mu(A_n)$$

for every sequence (A_n) of pairwise disjoint sets in $\mathscr{S}$. If X is a locally compact Hausdorff space, and $\mathscr{S}$ contains all Borel sets in X, we say that μ is a *Radon measure* if

(i) $\mu(C) < \infty$ for every C in $\mathscr{C}$.
(ii) $\mu(A) = \sup\{\mu(C) | C \subset A, C \in \mathscr{C}\}$ for every A in $\mathscr{S}$.

Condition (ii) is called *inner regularity* of μ. *Outer regularity*, which we shall encounter later (6.3.6), is expressed by the condition

(iii) $\mu(A) = \inf\{\mu(B) | A \subset B, B \text{ open in } X\}$ for every A in $\mathscr{S}$.

6.3.2. Proposition. *To each Radon integral* $\int$ *on a locally compact Hausdorff space* X *corresponds two measures on the* σ-*algebra* $\mathscr{M}$ *of measurable subsets of* X: *the outer measure* μ^* *and the inner measure* μ_*, *defined by*

(i) $\mu^*(A) = \int^* [A], A \in \mathscr{M}$;
(ii) $\mu_* A = \int_* [A], A \in \mathscr{M}$.

Of these, μ_* *is a Radon measure and* μ^* *is an outer regular measure dominating* μ_*. *Moreover,* $\mu_*(A) = \mu^*(A)$ *whenever* A *is a countable union of sets from* $\mathscr{M}^1$.

PROOF. From 6.2.15 it follows that if $\mu^*(A) < \infty$, then $A \in \mathscr{M}^1$, so that $\mu_*(A) = \mu^*(A)$. This immediately implies that μ^* is a measure, because the identity $\mu^*(\bigcup A_n) = \sum \mu^*(A_n)$ is only nontrivial if $\sum \mu^*(A_n) < \infty$, and in that case it follows from the monotone convergence theorem (6.1.13). Furthermore, we see from 6.2.15 that μ^* is outer regular and μ_* is inner regular. To show that μ_* is a measure (and thus a Radon measure), take a sequence (A_n) of pairwise disjoint sets in $\mathscr{M}$, and compute for each compact subset C of $\bigcup A_n$ that

$$\mu_*(C) = \int [C] = \int \sum [C \cap A_n]$$

$$= \sum \int [C \cap A_n] = \sum \mu_*(C \cap A_n) \leq \sum \mu_*(A_n),$$

since $C \cap A_n \in \mathcal{M}^1$. The inner regularity of μ_* now shows that $\mu_*(\bigcup A_n) = \sum \mu_*(A_n)$. Since μ_* and μ^* agree on $\mathcal{M}^1$, they agree on countable unions from $\mathcal{M}^1$ by σ-additivity. $\square$

6.3.3. Lemma. Let μ be a measure on a σ-algebra $\mathcal{S}$ of subsets of a set X, and denote by $\mathcal{F}$ the class of $\mathcal{S}$-measurable functions as defined in 6.2.4. There is then a unique positive homogeneous, additive function $\Phi: \mathcal{F}_+ \to [0, \infty]$, satisfying the conditions

(i) $\Phi([A]) = \mu(A)$ if $A \in \mathcal{S}$.
(ii) $\Phi(f) = \lim \Phi(f_n)$ if $f_n \nearrow f$ in $\mathcal{F}_+$.

PROOF. Consider the class $\mathcal{F}_s$ of simple functions in $\mathcal{F}_+$ of the form $f = \sum \alpha_n[A_n]$, where $\alpha_n \geq 0$ and $A_n \in \mathcal{S}$. Define Φ on $\mathcal{F}_s$ [as we must by (i)] by setting

$$\Phi(f) = \sum \alpha_n \mu(A_n).$$

Straightforward (but rather lengthy) manipulations, involving only the finite additivity of μ on the algebra $\mathcal{S}$, show that Φ is a well-defined, positive homogeneous, additive function on the subcone $\mathcal{F}_s$ of $\mathcal{F}_+$.

Suppose that $(f_n) \subset \mathcal{F}_s$ and $f_n \nearrow [A]$ for some A in $\mathcal{S}$. For $\varepsilon > 0$ let $B_n = \{f_n \geq 1 - \varepsilon\}$, so that $(1 - \varepsilon)[B_n] \leq f_n$. As $f_n \nearrow [A]$ we must have $\bigcup B_n = A$, whence $\mu(B_n) \nearrow \mu(A)$ by the σ-additivity of μ. Consequently,

$$\mu(A) \geq \lim \Phi(f_n) \geq (1 - \varepsilon)\lim \Phi([B_n])$$
$$= (1 - \varepsilon)\lim \mu(B_n) = (1 - \varepsilon)\mu(A).$$

Since ε is arbitrary, $\Phi(f_n) \nearrow \Phi([A])$. From this result it is easy to show that $\Phi(f_n) \nearrow \Phi(f)$, whenever $f_n \nearrow f$ in $\mathcal{F}_s$.

Now if (f_n) and (g_m) are increasing sequences in $\mathcal{F}_s$ such that $f_n \nearrow f$ and $g_m \nearrow f$ for some f in $\mathcal{F}_+$, then $f_n \wedge g_m \in \mathcal{F}_s$ for all n and m, and $f_n \wedge g_m \nearrow f_n$ for each fixed n. By the previous result this means that

$$\lim \Phi(g_m) \geq \lim \Phi(f_n \wedge g_m) = \Phi(f_n),$$

whence $\lim \Phi(g_m) \geq \lim \Phi(f_n)$. Exchanging the roles of f_n and g_m we see that

$$\lim \Phi(f_n) = \lim \Phi(g_m). \qquad (*)$$

Every element f in $\mathcal{F}_+$ is the pointwise limit of an increasing sequence (f_n) from $\mathcal{F}_s$. Indeed, as in the proofs of 6.2.9 and 6.2.16 let

$$A_{nk} = \{(k - 1)2^{-n} < f \leq k2^{-n}\}$$

for $1 \leq k \leq n2^n$, and define $f_n = \sum (k - 1)2^{-n}[A_{nk}]$. We extend Φ from $\mathcal{F}_s$ to $\mathcal{F}_+$ by setting $\Phi(f) = \lim \Phi(f_n)$, and we note from $(*)$ that $\Phi(f) = \lim \Phi(g_m)$ for every increasing sequence (g_m) in $\mathcal{F}_s$ with $g_m \nearrow f$. Thus, the definition of $\Phi(f)$ does not depend on any particular sequence (as long as it increases to f), and it follows easily that Φ is positive homogeneous and additive on $\mathcal{F}_+$.

Finally, to show that Φ satisfies condition (ii), let $f_n \nearrow f$ in $\mathscr{F}_+$. For each n there is an increasing sequence (g_{nm}) in $\mathscr{F}_s$ with $g_{nm} \nearrow f_n$. But then, taking

$$h_n = g_{1n} \vee g_{2n} \vee \cdots \vee g_{nn}$$

we obtain an increasing sequence (h_n) in $\mathscr{F}_s$ with $h_n \nearrow f$. Since $h_n < f_n$, we conclude, using (*), that

$$\lim \Phi(f_n) \le \Phi(f) = \lim \Phi(h_n) \le \lim \Phi(f_n);$$

whence $\Phi(f) = \lim \Phi(f_n)$, as desired. The unicity of Φ, given μ and the conditions (i) and (ii), follows from the construction. $\square$

6.3.4. Theorem. *There is a bijective correspondence between Radon integrals on the locally compact Hausdorff space X and Radon measures on the σ-algebra $\mathscr{B}$ of Borel sets of X, given by*

$$\mu(A) = \int_* [A], \quad A \in \mathscr{B}.$$

PROOF. If $\int$ is a Radon integral on X, then $\mu_* | \mathscr{B}$ is a Radon measure on $\mathscr{B}$ by 6.3.2.

Conversely, if μ is a Radon measure on $\mathscr{B}$, we use 6.3.3 to define a positive homogeneous, additive function $\int_\mu$ on the class $\mathscr{B}(X)_+$ of positive Borel functions, taking values in $[0, \infty]$. If $f \in C_c(X)$ and $f \ge 0$, then $f \le \alpha[C]$ for some $\alpha > 0$ and C in $\mathscr{C}$. Since $\mu(C) < \infty$, it follows that $\int_\mu f < \infty$. Thus $\int_\mu$ is finite on $C_c(X)_+$, so that $\int = \int_\mu | C_c(X)_+$ is a Radon integral on X.

To show that the two maps are the inverse of each other, take a Radon integral $\int$ on X and define the inner measure μ_* as in 6.3.2. From this measure we obtain the integral $\int_{\mu_*}$ by 6.3.3. If f is a positive, simple function with compact support of the form $f = \sum \alpha_n[B_n]$, $B_n \in \mathscr{B}$, then

$$\int_{\mu_*} f = \sum \alpha_n \mu_*(B_n) = \sum \alpha_n \int_* [B_n] = \int_* f.$$

For each f in $C_c(X)_+$ we can choose an increasing sequence (f_n) of positive, simple functions with compact supports such that $f_n \nearrow f$. It follows that

$$\int_{\mu_*} f = \lim \int_{\mu_*} f_n = \lim \int_* f_n = \int f$$

by 6.3.3 and 6.1.13, whence $\int_{\mu_*} = \int$.

Conversely, if we start with a Radon measure μ, extend it to an integral $\int_\mu$ using 6.3.3, take $\int = \int_\mu | C_c(X)_+$, and finally define μ_* relative to $\int$ as in 6.3.2, we must show that $\mu = \mu_*$ on $\mathscr{B}$. Since both measures are inner regular, it suffices to show that $\mu(C) = \mu_*(C)$ for each C in $\mathscr{C}$. By 6.2.8 there is a decreasing net $(f_\lambda)_{\lambda \in \Lambda}$ in $C_c(X)$ such that $f_\lambda \searrow [C]$, whence by 6.1.5

$$\mu_*(C) = \int_* [C] = \lim \int f_\lambda \ge \mu(C).$$

From the inner regularity we conclude that $\mu_*(B) \geq \mu(B)$ for every Borel set B. Fixing C as above, we choose f in $C_c(X)$, $0 \leq f \leq 1$, such that $f \geq [C]$ (cf. 1.7.5). Thus $f^n \searrow [C_1]$, where $C_1 = \{f = 1\} \supset C$. By (ii) in 6.3.3 (applied to the case $f - f^n \nearrow f - [C_1]$)

$$\mu_*(C_1) = \int_* [C_1] = \lim \int_* f^n$$

$$= \lim \int_\mu f^n = \int_\mu [C_1].$$

Since $\mu(C) \leq \mu_*(C)$ and $\mu(C_1 \setminus C) \leq \mu_*(C_1 \setminus C)$, the equality above forces $\mu(C) = \mu_*(C)$, as desired.   □

6.3.5. *Riesz' representation theorem* (6.3.4) (F. Riesz 1909) shows indirectly that Daniell's extension theorem is not best possible, if the goal is to make the class of integrable functions as large as possible. (It isn't.) If in the class $\mathscr{L}(X)$ of measurable functions on X we define

$$\mathscr{L}^1_{ess}(X) = \left\{ f \in \mathscr{L}(X) \,\middle|\, \int_* |f| < \infty \right\},$$

then the *essential integral* $\int_{ess}$, defined on $\mathscr{L}^1_{ess}(X)$ by

$$\int_{ess} f = \int_* f \vee 0 + \int^* f \wedge 0,$$

is a positive linear functional on $\mathscr{L}^1_{ess}(X)$ extending $\int$ and satisfying Lebesgue's convergence theorems. We have a strict inclusion $\mathscr{L}^1(X) \subset \mathscr{L}^1_{ess}(X)$, whenever the two measures μ_* and μ^* in 6.3.2 are different. However, under the innocent assumption (with regards to the applications) that the space X is a *countable* union of compact subsets—a σ-compact space—the dichotomy between μ_* and μ^* disappears.

6.3.6. Proposition. *If X is a locally compact, σ-compact Hausdorff space, every Radon measure is outer regular. Moreover, if $\int$ is a Radon integral, a measurable function f is integrable with respect to $\int$ iff $\int_* |f| < \infty$.*

PROOF. By assumption $X = \bigcup C_n$, where (C_n) is a sequence in $\mathscr{C}$. This means that if $\int$ is a Radon integral and $\mathscr{M}$ denotes the class of measurable subsets of X, then for each B in $\mathscr{M}$ we have $B = \bigcup (B \cap C_n)$. Since $B \cap C_n \in \mathscr{M}^1$, it follows from 6.3.2 that $\mu^*(B) = \mu_*(B)$. Since μ^* is outer regular and μ_* is the prototype of a Radon measure by 6.3.4, we have proved the first half of the proposition. The second follows in the same manner: If $f \geq 0$ is measurable, define $f_n = f \wedge n[C_n]$. Then $f_n \in \mathscr{L}^1(X)$ by 6.2.16 and since $f_n \nearrow f$, it follows from the monotone convergence theorem (6.1.13) that if $\int_* f < \infty$ then $\lim \int f_n < \infty$, whence $f \in \mathscr{L}^1(X)$.   □

6.3.7. It is sometimes convenient that the integral, like the measure, can take infinite values. We say that a measurable function $f \geq 0$ (with respect to a Radon integral $\int$ on a locally compact Hausdorff space X) is *extended integrable* if

$$\int_* f = \int^* f \qquad \left(\in \mathbb{R} \cup \{\infty\} \right).$$

Furthermore, we say that an arbitrary measurable function f is extended integrable if either $f \vee 0$ or $-(f \wedge 0)$ belongs to $\mathscr{L}^1(X)$ while the other is extended integrable. We then define

$$\int_{ext} f = \int f \vee 0 - \int (-f \wedge 0) \qquad \left(\in \mathbb{R} \cup \{\pm\infty\} \right).$$

Since the symbol $\infty - \infty$ has no useful definition, the extended integrable functions do not form a vector space. The positive extended integrable functions, however, is a positive cone, on which the integral acts as a positive homogeneous, σ-additive function.

If X is σ-compact, every positive, measurable function is extended integrable (6.3.6), whereas an arbitrary measurable function is extended integrable except in the case $\int f \vee 0 = \infty, \int f \wedge 0 = -\infty$.

6.3.8. Remark. Riesz' representation theorem (6.3.4) shows that the discussion about what is more important, the measure or the integral, is not of mathematical nature. It may, however, be waged on aestetical, proof–technological, yea even pedagogical, assumptions. There is little doubt (in this author) that the Lebesgue measure on $\mathbb{R}$ is more basic than the Lebesgue integral. This is marked by using the standard notation $\int f(x)dx$ for the Lebesgue integral of a function f. This symbolism (going back to Leibniz) is versatile and intuitive [derived from $\sum f(x)\Delta x$]. In particular, it is well suited to describe a change of variables and to distinguish the variables in multiple integrals. It is much more doubtful (for this author), whether general (Radon) measures are more basic than general (Radon) integrals. Moreover, the integral notation is clearly superior to measure notation when we have to derive one measure/integral from another (cf. 6.5.4 and 6.5.6). For these reasons the reader will not find the traditional notation $\int f(x)d\mu(x)$ [or the slightly more "logical" $\int f(x)\mu(dx)$; not to mention the rather masochistic $\int d\mu(x)f(x)$] for the integral of a function with respect to a measure. The exception being the case of translation invariant, "Lebesgue-like" measures; see 6.6.15.

6.4. L^p-Spaces

Synopsis. Null functions and the almost everywhere terminology. The Hölder and Minkowski inequalities. Egoroff's theorem. Lusin's theorem. The Riesz–Fischer theorem. Approximation by continuous functions. Complex spaces. Interpolation between L^p-spaces.

6.4.1. In this section we consider a fixed Radon integral $\int$ on a locally compact Hausdorff space X. We shall study various subspaces of the vector lattice $\mathscr{L}(X)$ of measurable functions (6.2.14). Of fundamental importance is the class of *null functions* defined by

$$\mathscr{N}(X) = \left\{ f \in \mathscr{L}(X) \,\middle|\, \int |f| = 0 \right\}.$$

From this we derive the class of *null sets*

$$\mathscr{N} = \{ N \in \mathscr{M} \mid [N] \in \mathscr{N}(X) \}.$$

A statement about points in X, which is true except at the points of a null set, is said to hold *almost everywhere*. For example, $f \in \mathscr{N}(X)$ iff $f = 0$ almost everywhere. Indeed,

$$\{ f \neq 0 \} = \bigcup \{ |f| > n^{-1} \} \quad \text{and} \quad [\{ |f| > n^{-1} \}] \leq n|f|.$$

It is easy to verify that $\mathscr{N}(X)$ is a sequentially complete ideal in $\mathscr{L}(X)$, stable under $\bigvee$ and $\bigwedge$. Likewise $\mathscr{N}$ is a σ-ring (and an ideal in $\mathscr{M}$ for the Boolean ring structure defined in 6.2.3). It follows from the definition that *every* positive function dominated by a null function is again a null function. Similarly, *every* subset of a null set is again a null set. We can therefore expect the wildest behavior, set-theoretically, of null functions and null sets; but since the effect on the integral is negligible, this will cause no trouble.

6.4.2. Having established the "almost everywhere" terminology, we can now allow functions on X with values in the extended real line $\mathbb{R} \cup \{ \pm \infty \}$. Thus, we say from now on that a function $f \colon X \to \mathbb{R} \cup \{ \pm \infty \}$ belongs to $\mathscr{L}(X)$ if there is a null set N such that $f(X) \in \mathbb{R}$ for $x \notin N$ and such that $[X \backslash N]f$ is a measurable function in the old sense. The definition is not ideal, since the algebraic and lattice rules now only hold almost everywhere. Passing to the quotient space $\mathscr{L}(X)/\mathscr{N}(X)$ will restore the operations, but at the price that we no longer work with function spaces. The reason why we allow functions with infinite values is that many limit operations give functions of this type, and it is cumbersome everytime to recall how to redefine them (from $\pm\infty$ back to $\mathbb{R}$) on an irrelevant null set. As an example of this phenomenon we present a version of the monotone convergence theorem, known as *Beppo Levi's theorem.*

6.4.3. Proposition. *If (f_n) is a sequence in $\mathscr{L}^1(X)$ such that $f_n(x) \leq f_{n+1}(x)$ for almost all x and every n, and if $\lim \int f_n < \infty$, there is an element f in $\mathscr{L}^1(X)$ such that $\int f = \lim \int f_n$ and $f(x) = \lim f_n(x)$ almost everywhere.*

PROOF. For each n there is a null set N_n such that $f_n(x) \leq f_{n+1}(x)$ for $x \notin N_n$. With $N = \bigcup N_n$ we have a null set N and an extended-valued function f such that $f_n(x) \nearrow f(x)$ for $x \notin N$. Replacing if necessary f_n by $f_n - f_1$ we may assume that $f_n(x) \geq 0$ for $x \notin N$. Since

$$\int [\{f_n > m\}] \leq \int m^{-1} f_n$$

for every n and m, we see from 6.1.13 that

$$\int [\{f > m\}] \leq m^{-1} \lim \int f_n.$$

It follows that $N_\infty = \{f = \infty\}$ is a null set, so that 6.1.13 can be applied to the restriction of (f_n) and f to $X \backslash (N_\infty \cup N)$.   □

6.4.4. It is important to keep in mind that the "almost everywhere" terminology depends on the chosen integral. We say that a Radon integral $\int$ is *continuous* (or *diffuse*) if $\int [\{x\}] = 0$ for every point x in X, i.e. if all one-point sets are null sets. For a continuous integral a statement can therefore be true almost everywhere, and yet fail on a countable set of points. (Almost all real numbers are irrational.) The name derives from the fact that if $X = \mathbb{R}$, a Radon ($=$Stieltjes) integral (cf. 6.1.16) with respect to an increment function m is continuous iff m is continuous. The opposite case is an *atomic* Radon integral, characterized by having a set $S \subset X$, such that $X \backslash S$ is a null set and $S \cap C$ is countable for every compact subset C of X. Thus, S itself is countable if X is σ-compact. Particular cases of atomic integrals are the Dirac integrals (Dirac measures) δ_x, $x \in X$, given by $\delta_x(f) = f(x)$; cf. 2.5.7. Note that for a Dirac integral δ_x a statement on X is true almost everywhere, provided that it holds at x.

It is easy to show that every Radon integral $\int$ can be decomposed as a sum $\int = \int_c + \int_a$, where $\int_c$ is continuous and $\int_a$ is atomic. Just set

$$S = \left\{ x \in X \,\middle|\, \int [\{x\}] > 0 \right\}.$$

Then $S \cap C$ is countable for every C in $\mathscr{C}$ (since $\int [C] < \infty$). Thus, $S \in \mathscr{M}$ by 6.2.11, and we can define

$$\int_a f = \int [S] f, \qquad \int_c f = \int [X \backslash S] f, \quad f \in C_c(X).$$

Evidently $\int_a$ is atomic, $\int_c$ is continuous, and $\int_a + \int_c = \int$.

6.4.5. For $1 \leq p < \infty$ we define the Lebesgue space of order p as

$$\mathscr{L}^p(X) = \left\{ f \in \mathscr{L}(X) \,\middle|\, |f|^p \in \mathscr{L}^1(X) \right\}.$$

Since for all real s and t we have

$$(s + t)^p \leq 2^p(|s|^p \vee |t|^p) \leq 2^p(|s|^p + |t|^p),$$

it follows from 6.2.14 and 6.2.16 that $\mathscr{L}^p(X)$ is a vector space. For each f in $\mathscr{L}^p(X)$ we define

$$\|f\|_p = \left(\int |f|^p \right)^{1/p}$$

Then the map $f \to \|f\|_p$ is positive homogeneous on $\mathscr{L}^p(X)$, and for $p = 1$ it is obviously a seminorm. That it also is a seminorm for $p > 1$ will follow from the two fundamental inequalities of Hölder and Minkowski (6.4.6 and 6.4.7).

If $f \in \mathscr{L}(X)$, we define the *essential supremum* of f as

$$\operatorname{ess\,sup} f = \inf \left\{ t \in \mathbb{R} \,\middle|\, \int^* [\{f > t\}] = 0 \right\}$$

$$= \inf \left\{ t \in \mathbb{R} \,\middle|\, \int^* (f - f \wedge t) = 0 \right\}.$$

We denote by $\mathscr{L}^\infty(X)$ the vector space of essentially bounded, measurable functions, equipped with the seminorm

$$\|f\|_\infty = \operatorname{ess\,sup} |f|.$$

6.4.6. Lemma. *If $f \in \mathscr{L}^p(X)$ and $g \in \mathscr{L}^q(X)$, where p and q are conjugate exponents, i.e. $p^{-1} + q^{-1} = 1$ (by definition 1 and ∞ are conjugate), then $fg \in \mathscr{L}^1(X)$ and*

$$\|fg\|_1 \leq \|f\|_p \|g\|_q.$$

PROOF. The case $p = 1, q = \infty$ is easy, since $|fg| \leq |f| \|g\|_\infty$ almost everywhere. Therefore, assume that both p and q are finite.

The inequality $a^x + x \leq ax + 1$, for $0 \leq x \leq 1$ and $a > 0$ is easily verified, since $x \to a^x$ is a convex function. From this we get

$$(st^{-1})^{1/p} + p^{-1} \leq p^{-1} st^{-1} + 1$$

for all positive s and t. Multiplying with t this is transformed into

$$s^{1/p} t^{1/q} \leq p^{-1} s + q^{-1} t.$$

Inserting $s = |f|^p \alpha^{-p}$ and $t = |g|^q \beta^{-q}$, where $\alpha = \|f\|_p$ and $\beta = \|g\|_q$, we obtain

$$\alpha^{-1} \beta^{-1} |f| |g| \leq p^{-1} \alpha^{-p} |f|^p + q^{-1} \beta^{-q} |g|^q. \qquad (*)$$

Since $fg \in \mathscr{L}(X)$ (6.2.14), it follows from 6.2.16 that $fg \in \mathscr{L}^1(X)$, and we see from $(*)$ that

$$\alpha^{-1} \beta^{-1} \|fg\|_1 \leq p^{-1} \alpha^{-p} \|f\|_p^p + q^{-1} \beta^{-q} \|g\|_q^q = p^{-1} + q^{-1} = 1,$$

which is the desired inequality. $\qquad\square$

6.4.7. Lemma. *If $1 \leq p \leq \infty$ and f and g belong to $\mathscr{L}^p(X)$, then $f + g \in \mathscr{L}^p(X)$ and*

$$\|f + g\|_p \leq \|f\|_p + \|g\|_p.$$

PROOF. The cases $p = 1$ and $p = \infty$ are evident. If $1 < p < \infty$, we put $h = |f + g|^{p-1}$, and note that $h \in \mathscr{L}^q(X)$ when $p^{-1} + q^{-1} = 1$ [because $f + g \in \mathscr{L}^p(X)$ by 6.4.5]. By the Hölder inequality (6.4.6) we therefore have

$$(\|f + g\|_p)^p \leq \int h|f| + \int h|g|$$

$$\leq \|h\|_q(\|f\|_p + \|g\|_p) = (\|f + g\|_p)^{p-1}(\|f\|_p + \|g\|_p),$$

from which Minkowski's inequality is immediate. $\qquad\square$

6.4.8. The space $\mathscr{N}(X)$ of null functions is contained in $\mathscr{L}^p(X)$ for every p $(1 \leq p \leq \infty)$. Moreover,

$$\mathscr{N}(X) = \{f \in \mathscr{L}^p(X) | \|f\|_p = 0\}$$

again for $1 \leq p \leq \infty$. It follows from Minkowski's inequality (6.4.7) that the quotient spaces

$$L^p(X) = \mathscr{L}^p(X)/\mathscr{N}(X)$$

are normed spaces for every p, with $\|\cdot\|_p$ as the norm. That these spaces are complete (the *Riesz–Fischer theorem*, 6.4.10) is a consequence of the next result, known as *Egoroff's theorem*.

6.4.9. Proposition. *Given a Cauchy sequence in $\mathscr{L}^p(X)$, for $1 \leq p < \infty$, there is for each $\varepsilon > 0$ a subsequence (f_n), an open set A with $\int [A] < \varepsilon$ and a null set N, such that the sequence (f_n) converges uniformly on $X \backslash A$ and converges pointwise on $X \backslash N$.*

PROOF. Choose the subsequence such that $\|f_{n+1} - f_n\|_p \leq (\varepsilon 2^{-n(p+1)})^{1/p}$ for every n. With

$$B_n = \{|f_{n+1} - f_n| \geq 2^{-n}\}$$

we therefore have

$$\int [B_n] \leq 2^{np} \int |f_{n+1} - f_n|^p < 2^{np}\varepsilon 2^{-n(p+1)} = \varepsilon 2^{-n}.$$

Taking $A_n = \bigcup_{m>n} B_m$ we obtain the estimate $\int [A_n] < \varepsilon 2^{-n}$. In particular, $\int [A_0] < \varepsilon$ and $\int [N] = 0$, where $N = \bigcap A_n$. By 6.2.15 we can choose an open set $A \supset A_0$ such that $\int [A] < \varepsilon$.

If $x \notin A_n$, we have $|f_{m+1}(x) - f_m(x)| < 2^{-m}$ as soon as $m > n$. In particular, (f_n) is uniformly convergent on $X \backslash A_0$, hence on $X \backslash A$. Moreover, since $X \backslash N = \bigcup (X \backslash A_n)$, it follows that (f_n) converges pointwise on $X \backslash N$. $\qquad\square$

6.4.10. Theorem. *Each space $L^p(X)$, $1 \leq p \leq \infty$, is a Banach space.*

PROOF. Let $Q: \mathscr{L}^p(X) \to L^p(X)$ denote the quotient map. If $p < \infty$ and (Qf_n) is a Cauchy sequence in $L^p(X)$, then (f_n) is a Cauchy sequence in $\mathscr{L}^p(X)$ (for the seminorm $\| \cdot \|_p$), and we can find a subsequence as described in 6.4.9, which we shall continue to call (f_n). Define $f(x) = \lim f_n(x)$ if $x \notin N$ and $f(x) = 0$ if $x \in N$. Taking $g = \sum |f_{n+1} - f_n|$, we have $\|g\|_p \leq \sum \|f_{n+1} - f_n\|_p < \infty$ so g^p is an integrable function that majorizes $|f - f_n|^p$ on $X \setminus N$. By Lebesgue's dominated convergence theorem (6.1.15) it follows that $(\|f - f_n\|_p)^p \to 0$; whence $f \in \mathscr{L}^p(X)$, and $Qf_n \to Qf$ in L^p.

If (Qf_n) is a Cauchy sequence in $L^\infty(X)$ we set

$$N_{nm} = \{|f_n - f_m| > \|f_n - f_m\|_\infty\} \quad \text{and} \quad N_n' = \{|f_n| > \|f_n\|_\infty\}.$$

All these sets are null sets and since $\mathscr{N}$ is a σ-ring, also their union $N = (\bigcup N_{nm}) \cup (\bigcup N_n')$ is a null set. On $X \setminus N$ the sequence (f_n) converges uniformly to a bounded function f. Defining $f = 0$ on N we have that $f \in \mathscr{L}^\infty(X)$ and that $Qf_n \to Qf$ in $L^\infty(X)$. $\square$

6.4.11. Proposition. *For every* $p < \infty$ *(the image of)* $C_c(X)$ *is dense in* $L^p(X)$.

PROOF. If $f \in \mathscr{L}^p(X)$, we have the decomposition $f = f \vee 0 + f \wedge 0$ in $\mathscr{L}^p(X)$. It therefore suffices to approximate a positive function f in $\mathscr{L}^p(X)$.

By 6.1.9 there is for each $\varepsilon > 0$ an h in $C_c(X)^m$ such that $f^p \leq h$ and $\int h \leq \int f^p + \varepsilon^p$. Since $p \geq 1$, we have the inequality $s^p + t^p \leq (s + t)^p$ for all positive s and t, and therefore

$$\|h^{1/p} - f\|_p^p = \int (h^{1/p} - f)^p \leq \int h - f^p \leq \varepsilon^p.$$

By definition of the upper integral (6.1.2) there is now a g in $C_c(X)_+$ such that $g \leq h$ and $\int h - g \leq \varepsilon^p$. As above this implies that $\|h^{1/p} - g^{1/p}\|_p \leq \varepsilon$. Taken together we have $\|f - g^{1/p}\|_p \leq 2\varepsilon$, and since $g^{1/p} \in C_c(X)$, the proof is complete. $\square$

6.4.12. Corollary. *For each* f *in* $\mathscr{L}^p(X)$, $1 \leq p < \infty$, *and* $\varepsilon > 0$ *there is an open set* A *with* $\int [A] < \varepsilon$, *such that* $f | X \setminus A$ *belongs to* $C_0(X \setminus A)$.

PROOF. Choose by 6.4.11 a sequence (f_n) in $C_c(X)$ such that $\|f - f_n\|_p \to 0$. Then use 6.4.9 to find the open set A. Since $(f_n | X \setminus A) \subset C_0(X \setminus A)$, the result follows from the fact that $C_0(X \setminus A)$ is uniformly closed. This is *Lusin's theorem*. $\square$

6.4.13. Proposition. *If* $p < \infty$, *there is for every measurable function* f *in* $\mathscr{L}^p(X)$ *a Borel function* g *such that* $f - g$ *is a null function. The same is true for any measurable function* f *that is zero almost everywhere outside a* σ-*compact subset of* X.

PROOF. If $f \in \mathscr{L}^p(X)$, there is by 6.4.11 a sequence (f_n) in $C_c(X)$ such that $\|f - f_n\|_p \to 0$. Applying 6.4.9 we may assume that $f_n(x) \to f(x)$ for every x in $X \setminus N$, where N is a null set. Furthermore, we may assume that N is a Borel set, replacing it if necessary with $\bigcap A_n$, where (A_n) is a sequence of open sets containing N such that $\int[A] < n^{-1}$ (cf. 6.2.15). This means that $([X \setminus N]f_n)$ is a sequence of Borel functions and the pointwise limit g is therefore also a Borel function (6.2.7). Clearly, $f - g \in \mathcal{N}(X)$.

If $f \in \mathcal{L}(X)$ and $\{f \neq 0\} \subset \bigcup C_n$, where (C_n) is a sequence of compact sets in X, we first use the decomposition $f = f \vee 0 + f \wedge 0$ to reduce the problem to the case $f \geq 0$. Moreover, assuming that $C_n \subset C_{n+1}$ for all n, we let $f_n = f \wedge n[C_n]$. Then $\int^* f_n \leq n \int[C_n] < \infty$, so that $f_n \in \mathcal{L}^1(X)$ by 6.2.16. From the first part of the proof there is a Borel function g_n and a Borel null set N_n such that $f_n(x) = g_n(x)$ if $x \in X \setminus N_n$. With $N = \bigcup N_n$ we see that

$$g_n(x) = f_n(x) \nearrow f(x),$$

if $x \in X \setminus N$. This shows that $g = [X \setminus N]f$ is a Borel function with $f - g$ in $\mathcal{N}(X)$. $\square$

6.4.14. Even though the integration theory is at heart concerned only with real-valued functions, there is no difficulty in extending it to complex functions.

We say that a function $f: X \to \mathbb{C}$ is Borel/measurable provided that the same holds for the two functions $\mathrm{Re}\, f$ and $\mathrm{Im}\, f$. In fact, this is equivalent with the demand that $f^{-1}(B) \in \mathcal{B}$ (respectively $\mathcal{M}$) for every Borel subset B in $\mathbb{C}$ (because it suffices to consider sets B of the form $\{\alpha \in \mathbb{C} \,|\, \mathrm{Re}\, \alpha > t\}$ and $\{\alpha \in \mathbb{C} \,|\, \mathrm{Im}\, \alpha > t\}$, cf. 6.2.5, and the counterimages of such sets depend only on $\mathrm{Re}\, f$ or on $\mathrm{Im}\, f$).

In the same manner we define the complex $\mathcal{L}^p$-spaces and the complex null functions as the complexifications of the corresponding real spaces. This means that the integral $\int$ becomes a complex linear functional on the vector space of complex, integrable functions [which we still, with an absolute disregard for consistency, shall denote by $\mathcal{L}^1(X)$]. This extension process is painless, but the following complex version of 6.1.11 and 6.2.16 requires a special proof.

6.4.15. Proposition. *A measurable function $f: X \to \mathbb{C}$ is integrable iff $\int^* |f| < \infty$, in which case $|\int f| \leq \int |f|$.*

PROOF. With $g = \mathrm{Re}\, f$ and $h = \mathrm{Im}\, f$ we have $f = g + ih$ and $|f|^2 = g^2 + h^2$. Thus

$$|g| \vee |h| \leq |f| \leq |g| + |h|. \tag{*}$$

In conjunction with 6.2.16 this proves the first half of the proposition.

To show the integral estimate we note that all functions $g|f|^{-1/2}$, $h|f|^{-1/2}$, and $|f|^{1/2}$ belong to (real) $\mathcal{L}^2(X)$ so that we may apply the Cauchy–Schwarz inequality ($=$ Hölder's inequality with $p = 2$). This yields

$$\left| \int f \right|^2 = \left(\int g \right)^2 + \left(\int h \right)^2 = \left(\int g|f|^{-1/2}|f|^{1/2} \right)^2 + \left(\int h|f|^{-1/2}|f|^{1/2} \right)^2$$

$$\leq \left(\int g^2|f|^{-1} \right) \int |f| + \left(\int h^2|f|^{-1} \right) \int |f| = \left(\int |f| \right)^2.   \square$$

6.4.16. The inequalities in (∗) in 6.4.15 show that a measurable function $f: X \to \mathbb{C}$ belongs to $\mathscr{L}^p(X)$ for $1 \le p \le \infty$ iff $|f| \in \mathscr{L}^p(X)$. At the same time 6.4.15 shows that the Hölder and Minkowski inequalities also hold for complex functions, so that we can define the complex normed spaces $L^p(X)$ for $1 \le p \le \infty$. That these spaces are Banach spaces follows from the fact that (f_n) is a Cauchy sequence in (complex) $\mathscr{L}^p(X)$ iff both $(\operatorname{Re} f_n)$ and $(\operatorname{Im} f_n)$ are Cauchy sequences in (real) $\mathscr{L}^p(X)$.

The next results on the relative position of various $\mathscr{L}^p$-spaces are valid both in the real and the complex case (and with identical proofs).

6.4.17. Proposition. *If* $1 \le p < r < q \le \infty$, *then*

$$\mathscr{L}^p(X) \cap \mathscr{L}^q(X) \subset \mathscr{L}^r(X),$$

and for each f *in the intersection we have*

$$\|f\|_r \le \|f\|_p \vee \|f\|_q.$$

PROOF. Assume first that $q < \infty$. There is then a t in $]0, 1[$ such that $r = tp + (1 - t)q$. If $f \in \mathscr{L}^p(X) \cap \mathscr{L}^q(X)$, then $|f|^{tp} \in \mathscr{L}^{1/t}(X)$ and $|f|^{(1-t)q} \in \mathscr{L}^{1/1-t}(X)$. Hölder's inequality (6.4.6) shows immediately that $|f|^r \in \mathscr{L}^1(X)$, i.e. $f \in \mathscr{L}^r(X)$, and gives the estimate

$$\int |f|^r = \int |f|^{tp} |f|^{(1-t)q} \le \left(\int |f|^p \right)^t \left(\int |f|^q \right)^{1-t}$$

This by reformulation shows that

$$\|f\|_r^r \le \|f\|_p^{pt} \|f\|_q^{(1-t)q} \le (\|f\|_p \vee \|f\|_q)^r.$$

If $q = \infty$ and $f \in \mathscr{L}^p(X) \cap \mathscr{L}^\infty(X)$, we have directly that

$$|f|^r = |f|^p |f|^{r-p} \le |f|^p \|f\|_\infty^{r-p}$$

(almost everywhere); which shows that $f \in \mathscr{L}^r(X)$ and that

$$\|f\|_r^r \le \|f\|_p^p \|f\|_\infty^{r-p} \le (\|f\|_p \vee \|f\|_\infty)^r. \qquad \square$$

6.4.18. It follows from the line above that

$$\limsup \|f\|_r \le \|f\|_\infty.$$

That, in fact, we have $\lim \|f\|_r \to \|f\|_\infty$ for every f in some $\mathscr{L}^p(X) \cap \mathscr{L}^\infty(X)$, is seen from the estimate

$$\int |f|^r \ge \left(\|f\|_\infty - \varepsilon \right)^r \int [B_\varepsilon],$$

where $B_\varepsilon = \{|f| \ge \|f\|_\infty - \varepsilon\}$, which shows that for every $\varepsilon > 0$ we have

$$\|f\|_r \ge \left(\|f\|_\infty - \varepsilon \right) \left(\int [B_\varepsilon] \right)^{1/r},$$

whence $\liminf \|f\|_r \ge \|f\|_\infty - \varepsilon$.

6.4.19. Most often $\mathscr{L}^p(X) \not\subset \mathscr{L}^q(X)$ if $p \neq q$. This is so in particular for the $\mathscr{L}^p$-spaces arising from the Lebesgue integral on $\mathbb{R}^n$. There are, however, two cases for which we do have inclusions: If the integral $\int$ is finite, i.e. $\int 1 < \infty$; or if the integral is atomic with a smallest atom, i.e. for some $\varepsilon > 0$ we either have $\int [A] = 0$ or $\int [A] \geq \varepsilon$ for every A in $\mathscr{M}$. In the first case we may assume without loss of generality that $\int 1 = 1$ (so that $\int$ is a probability distribution). The second case means (when X is σ-compact) that the integral is concentrated on a countable subset of X, and that the weight of every atom is $\geq \varepsilon$. Then the L^p-spaces are isomorphic to the sequence spaces ℓ^p, cf. 2.1.18, and we are justified in treating only these.

6.4.20. Corollary. *If $\int 1 \doteq 1$ and $1 \leq p < q \leq \infty$, then*

$$\mathscr{L}^q(X) \subset \mathscr{L}^p(X) \quad and \quad \|\cdot\|_p \leq \|\cdot\|_q.$$

PROOF. We just have to show that $\mathscr{L}^q(X) \subset \mathscr{L}^1(X)$ and that $\|\cdot\|_1 \leq \|\cdot\|_q$ for every q, then 6.4.17 gives the rest. If $q = \infty$, this is evident. For $q < \infty$ we put $t = q(q-1)^{-1}$, and apply Hölder's inequality to a function f in $\mathscr{L}^q(X)$ and the function 1 in $\mathscr{L}^t(X)$ to obtain that $f \in \mathscr{L}^1(X)$ and that

$$\|f\|_1 = \int |f| \cdot 1 \leq \|f\|_q \|1\|_t = \|f\|_q. \qquad \square$$

6.4.21. Corollary. *For the sequence spaces ℓ^p, $1 \leq p \leq \infty$, we have for $1 \leq p < q \leq \infty$ that $\ell^p \subset \ell^q$ and $\|\cdot\|_q \leq \|\cdot\|_p$.*

PROOF. We always have $\ell^p \subset \ell^\infty$ and $\|\cdot\|_\infty \leq \|\cdot\|_p$, and thus 6.4.17 applies. $\qquad \square$

6.5. Duality Theory

Synopsis. σ-compactness and σ-finiteness. Absolute continuity. The Radon–Nikodym theorem. Radon charges. Total variation. The Jordan decomposition. The duality between L^p-spaces.

6.5.1. To prove the Radon–Nikodym theorem (6.5.4) (which is the key result in this section) in the regie of abstract measure theory, it is by and large necessary that the measures involved are σ-*finite*. This means that there is a countable family $\{A_n\}$ of measurable sets with finite measure, such that $X = \bigcup A_n$. In topological measure theory (our theory) one *may* treat the general case, using the inner regularity of the involved measures. This would entail the use of the essential integral defined in 6.3.5. In the interest of brevity (and without sacrificing any good applications) we shall avoid this complication and stick to the σ-finite case. Thus, we assume throughout this section that X is a locally compact, σ-*compact* Hausdorff space. Read in 1.7.7–1.7.13 about the nice properties of such spaces.

Note that we would gain no generality by working with σ-finite Radon integrals/measures on an arbitrary locally compact Hausdorff space X. Indeed, if $\{A_n\} \subset \mathcal{M}^1$ and $\bigcup A_n = X$, we may assume by 6.2.15 that all the A_n are open. Then by an inductive application of 6.2.15 we can find for each A_n a sequence of open, relatively compact subsets B_{nm}, such that $\overline{B_{nm}} \subset B_{nm+1}$ for all m and $A_n \backslash \bigcup B_{nm} \in \mathcal{N}$. Thus, the union $Y = \bigcup_{n,m} B_{nm}$ is an open, σ-compact subset of X and $X \backslash Y \in \mathcal{N}$. Replacing X with Y we are back at the σ-compact case.

6.5.2. Proposition. *If $\int_0$ and $\int$ are Radon integrals on a locally compact Hausdorff space X, the following conditions are equivalent:*

(i) *For every monotone decreasing sequence (f_n) in $C_c(X)_+$ the condition* $\lim \int f_n = 0$ *implies that* $\lim \int_0 f_n = 0$.
(ii) *For every Borel set N in X, $\int [N] = 0$ implies that $\int_0 [N] = 0$.*
(iii) *For every Borel function $f \geq 0$, $\int f = 0$ implies that $\int_0 f = 0$.*

PROOF. (i) $\Rightarrow$ (ii). By 6.2.8 there is for every compact subset C of N a decreasing net $(f_\lambda)_{\lambda \in \Lambda}$ in $C_c(X)$ such that $f_\lambda \searrow [C]$. Choose a decreasing sequence $(f_n) \subset (f_\lambda)$, such that

$$\int f_n \searrow \lim \int f_\lambda \quad \text{and} \quad \int_0 f_n \searrow \lim \int_0 f_\lambda.$$

Since $C \subset N$ we have $\int [C] = 0$, whence $\int f_n \searrow 0$. By assumption this implies that $\int_0 f_n \searrow 0$, and we conclude that $\int_0 [C] = 0$. Since this holds for every $C \subset N$, it follows from the inner regularity of Radon measures (6.3.1) that $\int_0 [N] = 0$.

(ii) $\Rightarrow$ (iii). If $\int f = 0$, then $\{f > 0\}$ is a null set for $\int$, hence also for $\int_0$, whence $\int_0 f = 0$.

(iii) $\Rightarrow$ (i). If (f_n) is a monotone decreasing sequence in $C_c(X)_+$, it has a pointwise limit $f \geq 0$ in $C_c(X)_m$. If $\int f_n \searrow 0$, then $\int f = 0$. By assumption this implies that $\int_0 f = 0$, and since $\int_0 f_n \searrow \int_0 f$, we are done. $\qquad \square$

6.5.3. If $\int_0$ and $\int$ satisfy the conditions in 6.5.2, we say that $\int_0$ is *absolutely continuous* with respect to $\int$, in symbols $\int_0 \ll \int$. If both $\int_0 \ll \int$ and $\int \ll \int_0$, we say that $\int$ and $\int_0$ are *equivalent*, in symbols $\int_0 \sim \int$. From 6.5.2(iii) we see that this is equivalent with $\mathcal{N}(X) = \mathcal{N}_0(X)$. Thus, $\int_0 \sim \int$ iff $L_0^\infty(X) = L^\infty(X)$.

To formulate the Radon–Nikodym theorem we need the concept of a *locally integrable* function. This means a function f (necessarily measurable) such that $[C]f \in \mathcal{L}^1(X)$ for every C in $\mathscr{C}$ (cf. the definition of $\mathcal{M}$ from $\mathcal{M}^1$ in 6.2.11). Clearly, the set $\mathcal{L}_{\text{loc}}^1(X)$ of locally integrable functions on X is a vector space, stable under $\vee$ and $\wedge$.

6.5.4. Theorem. *If $\int_0$ and $\int$ are Radon integrals on a locally compact, σ-compact Hausdorff space X, then $\int_0$ is absolutely continuous with respect to $\int$ iff there is* *a Borel function m ≥ 0, locally integrable with respect to ∫, such that*

$$\int_0 f = \int fm$$

for every Borel function f on X. The function m is uniquely determined modulo null functions for ∫.

PROOF. Evidently the definition $\int_0 f = \int fm$ gives for every m in $\mathcal{L}^1_{loc}(X)_+$ a Radon integral $\int_0$ with $\int_0 \ll \int$; and it is also clear that $\int_0$ determines m up to null functions for $\int$.

To show the interesting implication we first assume that $\int_0 \le \int$ and that $\int 1 < \infty$. For each f in $C_c(X)$ we then have by the Cauchy–Schwarz inequality that

$$\left| \int_0 f \right|^2 \le \left(\int_0 |f| \right)^2 \le \left(\int |f| \right)^2 \le \left(\int 1 \right)\left(\int |f|^2 \right).$$

Since $C_c(X)$ is dense in $L^2(X)$ (6.4.11) and $\int_0$ is bounded with respect to the 2-norm, it extends to a bounded functional on $L^2(X)$. Each such is given by a vector in $L^2(X)$ by 3.1.9, and by 6.4.13 there is therefore a real-valued Borel function m such that

$$\int_0 f = (f|m) = \int fm, \quad f \in C_c(X). \tag{*}$$

Since $C_c(X)$ is dense in both $\mathcal{L}^1_0(X)$ and $\mathcal{L}^2(X)$, and $\mathcal{L}^2(X) \subset \mathcal{L}^1(X) \subset \mathcal{L}^1_0(X)$, it follows that (*) holds for every f in $\mathcal{L}^2(X)$. Taking $f = [C]$, where C is a compact subset of $\{m \le -\varepsilon\}$, we have

$$0 \le \int_0 [C] = \int [C]m \le -\varepsilon \int [C].$$

Thus, $\int [C] = 0$; and since C and ε are arbitrary, it follows that $m \ge 0$ almost everywhere. Now take $f = [D]$ in (*), where D is a compact subset of $\{m \ge 1 + \varepsilon\}$, to obtain

$$\int [D] \ge \int_0 [D] = \int [D]m \ge (1 + \varepsilon) \int [D].$$

Thus $\int [D] = 0$; and since D and ε are arbitrary, we conclude that $0 \le m \le 1$ almost everywhere. Since both $\int_0$ and $\int \cdot m$ are Radon integrals on X, and coincide on $C_c(X)$ by (*), it follows that (*) holds for every positive Borel function f on X (in the sense that if one side is finite then so is the other, with the same value).

Let us now drop the assumption that $\int_0 \le \int$, but retain the assumptions $\int_0 1 < \infty$ and $\int 1 < \infty$. Defining $\int_1 = \int_0 + \int$ we have $\int_0 \le \int_1$, and by the previous argument there is therefore a Borel function m_0, with $0 \le m_0 \le 1$, such that

$$\int_0 f = \int_1 fm_0 = \int_0 fm_0 + \int fm_0 \qquad (**)$$

for every Borel function $f \geq 0$ on X. Taking $f = [C]$, where C is a compact subset of $\{m_0 = 1\}$ we see from $(**)$ that

$$\int_0 [C] = \int_0 [C] + \int [C],$$

whence $\int [C] = 0$. Since this holds for every C we know that $\{m_0 = 1\}$ is a null set for $\int$. By assumption $\{m_0 = 1\}$ is therefore also a null set for $\int_0$. Iteration of the formula $(**)$ gives

$$\int_0 f = \int_0 fm_0^n + \int \sum_{k=1}^n fm_0^k. \qquad (***)$$

Since $m_0^n \searrow 0$ almost everywhere (with respect to both $\int$ and $\int_0$) we have $\int_0 fm_0^n \searrow 0$ for each f in $\mathscr{L}_0^1(X)_+$ by 6.1.13. If we therefore define $m = m_0(1 - m_0)^{-1}$ as an almost everywhere finite Borel function, we see from $(***)$ that

$$\int_0 f = \int fm$$

for every f in $C_c(X)$, and therefore for every Borel function $f \geq 0$ on X.

With no assumptions on $\int_0$ and $\int$ (except $\int_0 \ll \int$) we choose an increasing sequence (C_n) in $\mathscr{C}$ such that $\bigcup C_n = X$. With $X_n = C_n \backslash C_{n-1}$ we see that X is the disjoint union of the locally compact spaces X_n, and that $\int_0 [X_n] < \infty$, $\int [X_n] < \infty$ for every n. Replacing X with X_n in the previous argument we can therefore find a Borel function $m_n \geq 0$ on X_n such that $\int_0 f[X_n] = \int fm_n$ for every Borel function $f \geq 0$ on X. Taking $m = \sum m_n$ it follows from 6.1.13 that $\int_0 f = \int fm$ for every f in $\mathscr{B}(X)_+$, as desired. $\qquad \square$

6.5.5. We wish to consider the complex linear span of Radon integrals on X. This will be referred to as the space of Radon charges on X, and consists of certain linear functionals on $C_c(X)$. Equipped with the weak topology τ induced by the seminorms $f \to |\int f|$, where $\int$ ranges over all Radon integrals on X, it follows from 2.4.1 and 2.4.4 that the Radon charges constitute the dual space of the topological vector space $(C_c(X), \tau)$. Clearly, a more constructive description of these elements is desirable. We therefore define a *Radon charge* to be a functional Φ on $C_c(X)$, which for every f in $C_c(X)_+$ satisfies

$$\sup\{|\Phi(g)| \,|\, g \in C_c(X), |g| \leq f\} < \infty.$$

6.5.6. Theorem. *To each Radon charge Φ on a locally compact, σ-compact Hausdorff space X there is a Radon integral $\int$ and a Borel function u such that $|u| = 1$ almost everywhere (with respect to $\int$) and $\Phi = \int \cdot u$.*

PROOF. For each f in $C_c(X)_+$ define

$$\int f = \sup\{|\Phi(g)|\,|\,g \in C_c(X), |g| \leq f\}.$$

Then $0 \leq \int f < \infty$ and $f \to \int f$ is a positive homogeneous function. Evidently, we may replace $|\Phi(g)|$ with $\operatorname{Re}\Phi(g)$ in the definition above, from which it follows that $\int$ is superadditive $[\int f_1 + \int f_2 \leq \int(f_1 + f_2)]$, since $|g_1| \leq f_1$ and $|g_2| \leq f_2$ imply $|g_1 + g_2| \leq f_1 + f_2$. To show that $\int$ is additive it therefore suffices to show that it is subadditive. Given f_1 and f_2 and $\varepsilon > 0$, choose g in $C_c(X)$ such that $|g| \leq f_1 + f_2$ and

$$\int f_1 + f_2 \leq \varepsilon + |\Phi(g)|.$$

Put $g_j = gf_j(f_1 + f_2)^{-1}$ for $j = 1, 2$, and note that these are *continuous* functions. Moreover, $g_1 + g_2 = g$ and

$$|g_j| = |g|f_j(f_1 + f_2)^{-1} \leq f_j, \quad j = 1, 2.$$

Thus,

$$|\Phi(g)| \leq |\Phi(g_1)| + |\Phi(g_2)| \leq \int f_1 + \int f_2.$$

In conjunction with the previous estimate this shows that $\int$ is subadditive. Thus, $\int$ is a positive homogeneous, additive function on $C_c(X)_+$, and therefore extends uniquely to a positive functional on $C_c(X)$, i.e. a Radon integral on X.

Assume first that $\int 1 < \infty$. Then for each f in $C_c(X)$ we have

$$|\Phi(f)|^2 \leq \left(\int |f|\right)^2 \leq \left(\int 1\right)\left(\int |f|^2\right).$$

As in the proof of 6.5.4 this produces a Borel function u in $\mathcal{L}^2(X)$ such that

$$\Phi(f) = (f|\bar{u}) = \int fu, \quad f \in C_c(X).$$

It follows that

$$\left|\int fu\right| = |\Phi(f)| \leq \int |f|$$

for all f in $C_c(X)$. On the other hand, if $f \geq 0$,

$$\int f = \sup\left\{\left|\int gu\right|\,\Big|\,|g| \leq f\right\} \leq \int f|u|$$

by 6.4.15. Using 6.4.11 we see that the two inequalities above hold for every f in $\mathscr{L}^1(X)$. In particular, replacing in the first one f by fu_n, where $f \in \mathscr{L}^1(X)_+$ and $u_n = \bar{u}(|u| + n^{-1})^{-1}$, we get in the limit

$$\int f|u| \leq \int f.$$

It follows that $\int f|u| = \int f$ for every f in $\mathscr{L}^1(X)_+$, whence $|u| = 1$ almost everywhere.

To prove the general case, use the σ-compactness of X to find a function h in $C_0(X)_+$ such that $\int h < \infty$ but $h(x) > 0$ for every x in X. Then consider the Radon charge $\Phi(\cdot h)$ and the finite Radon integral $\int \cdot h$, and note that $\int \cdot h$ is obtained from $\Phi(\cdot h)$ as in the first part of the proof. There is therefore a Borel function u on X, with $|u| = 1$ almost everywhere (with respect to $\int \cdot h$) such that $\Phi(\cdot h) = \int \cdot hu$. Since $h > 0$, the integral $\int \cdot h$ and $\int$ have the same null sets, so that $|u| = 1$ almost everywhere with respect to $\int$. Moreover, $hC_c(X) = C_c(X)$, so $\Phi(f) = \int fu$ for every f in $C_c(X)$.   □

6.5.7. Corollary. *To each real Radon charge Φ there are two Radon integrals $\int_+$ and $\int_-$, concentrated on disjoint Borel subsets of X, such that $\Phi = \int_+ - \int_-$.*

PROOF. We have $\Phi = \int \cdot u$, and since Φ is a real functional, u must be real-valued (almost everywhere). Taking $A_+ = \{u = 1\}$ and $A_- = \{u = -1\}$ we have $X \setminus (A_+ \cup A_-) \in \mathscr{N}$, and we now define $\int_+$ and $\int_-$ as the restriction of $\int$ to A_+ and A_-, respectively.   □

6.5.8. The decomposition of a Radon charge in 6.5.6 serves the same purpose as the polar decomposition of an operator on a Hilbert space (3.2.17 and 5.1.14). The integral obtained from the charge is called the *total variation* of Φ, and denoted by $|\Phi|$. It is easy to verify that $|\Phi|$ is the smallest Radon integral on X (in the order of positive functionals) that satisfies

$$|\Phi(f)| \leq \int |f|, \quad f \in C_c(X).$$

Furthermore, it is evident from the proof of 6.5.6 that when Φ (and therefore also $|\Phi|$) is given, the "sign" u is uniquely determined up to null functions for $|\Phi|$.

The result in 6.5.7 is called the *Jordan decomposition* of Φ, and the equation $X = A_+ \cup A_- \cup N$ is known as the *Hahn decomposition*. Note that for a real charge Φ, the total variation is just $\Phi_+ + \Phi_-$; cf. 4.4.10.

A (real or complex) Radon charge Φ is *finite* if the total variation $|\Phi|$ is a finite Radon integral. Denoting by $M(X)$ the vector space of finite Radon charges, and defining $\|\Phi\| = |\Phi|(1)$ for every Φ in $M(X)$, we have the following result.

6.5.9. Proposition. *The space $M(X)$ of finite Radon charges on X, equipped with the norm $\|\Phi\| = |\Phi|(1)$, is isometrically isomorphic to the Banach space $(C_0(X))^*$.*

PROOF. From the definition of $|\Phi|$ in 6.5.6 it follows that

$$\|\Phi\| = |\Phi|(1) = \sup\{|\Phi(g)|\,\big|\,g \in C_c(X), \|g\|_\infty \le 1\}.$$

Since $C_c(X)$ is uniformly dense in $C_0(X)$, we see that $\|\Phi\|$ is just the norm of Φ as a bounded functional on $C_0(X)$. We therefore have an isometric injection of $M(X)$ into $(C_0(X))^*$ (and we have verified that $\|\cdot\|$ really is a norm).

Conversely, if $\Phi \in (C_0(X))^*$, then for each f in $C_c(X)_+$ we have

$$\sup\{|\Phi(g)|\,\big|\,g \in C_c(X), |g| \le f\} \le \sup\{\|\Phi\|\,\|g\|_\infty\,\big|\,g \in C_c(X), |g| \le f\}$$

$$= \|\Phi\|\,\|f\|_\infty.$$

Thus Φ is a Radon charge, and taking an increasing net $(f_\lambda)_{\lambda \in \Lambda}$ in $C_c(X)_+$ such that $f_\lambda \nearrow 1$, it follows that $|\Phi|(1) = \lim |\Phi|(f_\lambda) \le \|\Phi\|$, so that $\Phi \in M(X)$. □

6.5.10. Proposition. *For a fixed Radon integral $\int$ on X, the space of finite Radon charges Φ, such that $|\Phi| \ll \int$, is isometrically isomorphic to $L^1(X)$, via the map $f \to \Phi_f$ from $L^1(X)$ into $M(X)$ given by*

$$\Phi_f(g) = \int gf, \quad f \in \mathscr{L}^1(X), g \in C_c(X).$$

PROOF. If $f \in \mathscr{L}^1(X)$, then Φ_f is a Radon charge with

$$|\Phi_f|(h) = \sup\{|\Phi_f(g)|\,\big|\,g \in C_c(X), |g| \le h\}$$

$$= \sup\left\{\left|\int gf\right|\,\Big|\,g \in C_c(X), |g| \le h\right\} = \sup\left\{\left|\int gf\right|\,\Big|\,g \in \mathscr{L}^1(X), |g| \le h\right\}$$

for every h in $C_c(X)_+$. Evidently $|\Phi_f|(h) \le \int h|f|$. On the other hand, we can insert $g = \bar{f}(|f| + \varepsilon)^{-1}h$ to obtain $|\Phi_f|(h) \ge \int h|f|^2(|f| + \varepsilon)^{-1}$ for every $\varepsilon > 0$. Thus $|\Phi_f| = \int \cdot |f|$. In particular, $|\Phi_f| \ll \int$ and $\|\Phi_f\| = |\Phi_f|(1) = \|f\|_1$.

In the converse direction, if Φ is a finite Radon charge with $|\Phi| \ll \int$, we know from the Radon–Nikodym theorem (6.5.4) that $|\Phi| = \int \cdot m$ for some positive Borel function m in $\mathscr{L}^1(X)$. Moreover, $\Phi = |\Phi|(\cdot u)$ by 6.5.6, with $|u| = 1$, almost everywhere. Combining these results we have $\Phi = \Phi_f$, with $f = um$. □

6.5.11. Theorem. *Let $\int$ be a Radon integral on a locally compact, σ-compact Hausdorff space X. If $1 < p \le \infty$ and $1 \le q < \infty$, such that $p^{-1} + q^{-1} = 1$ (where $\infty^{-1} = 0$ by definition), the bilinear form*

$$\langle f, g \rangle = \int fg, \quad f \in \mathscr{L}^p(X), g \in \mathscr{L}^q(X),$$

determines an isometric isomorphism of $L^p(X)$ onto $(L^q(X))^*$.

PROOF. From Hölder's inequality (6.4.6) we see that the bilinear form gives a norm decreasing linear map of $L^p(X)$ into $(L^q(X))^*$.

Assume now that Φ is a bounded functional on $L^q(X)$. By 6.4.11 we can regard Φ as a functional on $C_c(X)$, and we see that for h in $C_c(X)_+$ and g in $C_c(X)$ with $|g| \le h$ we have

$$|\Phi(g)| \le \|\Phi\| \, \|g\|_q \le \|\Phi\| \, \|h\|_q. \tag{*}$$

This inequality shows that Φ is a Radon charge on X (6.5.5), and by 6.5.6 it therefore has the form $\Phi = \int_0 \cdot u$ for some Radon integral $\int_0$ on X and some Borel function u, which we can choose such that $|u| = 1$. From $(*)$ we also see that $\int_0 h \le \|\Phi\| \, \|h\|_q$, and thus by 6.5.2(i) we have $\int_0 \ll \int$ [because if (h_n) is decreasing and $\int h_n \searrow 0$, then $h_n^q \le h_n h_1^{q-1}$, so $\int h_n^q \searrow 0$]. Thus $\int_0 = \int \cdot m$ for some m in $\mathscr{B}(X)_+$, and we have the inequality

$$\int hm \le \|\Phi\| \left(\int h^q \right)^{1/q} \tag{**}$$

for every h in $C_c(X)_+$. This inequality is preserved under monotone limits, and it therefore holds also if $h \in C_c(X)_+^m$. From the criterion of integrability given in 6.1.9 $(*)$ it follows that the inequality $(**)$ holds for every $h \ge 0$ in $\mathscr{L}^q(X)$.

If $q = 1$, we see from $(**)$ that $m \le \|\Phi\|$ almost everywhere, whence $\|m\|_\infty \le \|\Phi\|$. If $q > 1$, we insert $h = m^{p-1}[C]$, where $p^{-1} + q^{-1} = 1$ and $C \in \mathscr{C}$ (remember m is locally integrable), to obtain

$$\int [C]m^p \le \|\Phi\| \left(\int [C]m^p \right)^{1/q},$$

and thus

$$\left(\int [C]m^p \right)^{1/p} \le \|\Phi\|.$$

Since C is arbitrary, we conclude that $m \in \mathscr{L}^p(X)$ with $\|m\|_p \le \|\Phi\|$. Taking $f = um$ we know that $f \in \mathscr{L}^p(X)$ with $\|f\|_p \le \|\Phi\|$ for every $p \le \infty$, and, moreover,

$$\Phi(g) = \int_0 gu = \int gf = \langle g, f \rangle, \quad g \in C_c(X).$$

But since $C_c(X)$ is dense in $L^q(X)$ and both Φ and $\langle \cdot, f \rangle$ are continuous on $L^q(X)$, it follows that $\Phi = \langle \cdot, f \rangle$, as desired. Since both maps $\Phi \to f$ and $f \to \Phi_f$ are norm decreasing, and the inverse of each other, they are both isometries, and the proof is complete. $\qquad\square$

6.5.12. With 6.5.11 at hand we can now extend the density results in 6.4.11 to the case $p = \infty$. Of course it is not true that $C_c(X)$ is norm dense in $L^\infty(X)$, but identifying $L^\infty(X)$ with $(L^1(X))^*$ equipped with the w^*-topology, we claim that (the image of) $C_c(X)$ is w^*-dense in $L^\infty(X)$. If not, there is by 2.4.10 a nonzero g in $\mathscr{L}^1(X)$, such that $\int fg = 0$ for every f in $C_c(X)$, in contradiction with 6.5.10.

6.6. Product Integrals

Synopsis. Product integral. Fubini's theorem. Tonelli's theorem. Locally compact groups. Uniqueness of the Haar integral. The modular function. The convolution algebras $L^1(G)$ and $M(G)$.

6.6.1. Given locally compact Hausdorff spaces X and Y, consider $X \times Y$ with the product topology (1.4.8). Then $X \times Y$ is a locally compact Hausdorff space because the set of products from $\mathscr{C}_X \times \mathscr{C}_Y$ contain a basis for the product topology. Moreover, we see that $X \times Y$ is σ-compact iff both X and Y are σ-compact.

For each pair of functions f on X and g on Y we denote by $f \otimes g$ (f tensor g) the function on $X \times Y$ given by $f \otimes g(x, y) = f(x)g(y)$. Note that if $f \in C_c(X)$ and $g \in C_c(Y)$, then $f \otimes g \in C_c(X \times Y)$. Similarly, $C_0(X) \otimes C_0(Y) \subset C_0(X \times Y)$ and $C_b(X) \otimes C_b(Y) \subset C_b(X \times Y)$.

For each of the three topological spaces X, Y, and $X \times Y$ one may consider the class of Borel sets (6.2.5) and the space of Borel functions (6.2.6). The next lemma deals with the relations among these sets.

6.6.2. Lemma. *For a real function f on $X \times Y$ and y in Y, consider the function $f(\cdot, y)$ on X. If $f \in C_c(X \times Y)$, then $f(\cdot, y) \in C_c(X)$. If $f \in C_c(X \times Y)^m$, then $f(\cdot, y) \in C_c(X)^m$. If $f \in \mathscr{B}(X \times Y)$, then $f(\cdot, y) \in \mathscr{B}(X)$. Similar results hold for each function $f(x, \cdot)$, $x \in X$.*

PROOF. Let $\iota_y: X \to X \times Y$ be the continuous function given by $\iota_y(x) = (x, y)$. Then $f(\cdot, y) = f \circ \iota_y$. From this we see immediately that if $f \in C_c(X \times Y)$, then $f(\cdot, y)$ is continuous. Furthermore, it is clear that the support of $f(\cdot, y)$ is contained in the projection from $X \times Y$ onto X of the support of f, so that $f(\cdot, y) \in C_c(X)$. It follows from the definitions that $f(\cdot, y) \in C_c(X)^m$ if $f \in C_c(X \times Y)^m$. Finally, since composition of Borel functions again produce Borel functions, we see from the identity $f(\cdot, y) = f \circ \iota_y$ that $f(\cdot, y) \in \mathscr{B}(X)$ if $f \in \mathscr{B}(X \times Y)$.   □

6.6.3. Proposition. *If $\int_x$ and $\int_y$ are Radon integrals on the locally compact Hausdorff spaces X and Y, respectively, there is a unique Radon integral $\int_x \otimes \int_y$ on $X \times Y$ such that*

$$\int_x \otimes \int_y f \otimes g = \left(\int_x f \right)\left(\int_y g \right), \quad f \in C_c(X), g \in C_c(Y).$$

PROOF. Let $C_c(X) \otimes C_c(Y)$ denote the linear span in $C_c(X \times Y)$ of elements of the form $f \otimes g$, and note from the equation $(f_1 \otimes g_1)(f_2 \otimes g_2) = f_1 f_2 \otimes g_1 g_2$, that $C_c(X) \otimes C_c(Y)$ is an algebra.

If $h = \sum f_k \otimes g_k \in C_c(X) \otimes C_c(Y)$, define

$$\int_x \otimes \int_y h = \sum \left(\int_x f_k \right)\left(\int_y g_k \right) \tag{*}$$

(as we must). Note that if $h = 0$, then $h(\cdot, y) = 0$ for each y in Y, whence $\sum (\int_x f_k) g_k(y) = 0$. Since this holds for every y, we conclude that $\sum (\int_x f_k)(\int_y g_k) = 0$, so that the formula $(*)$ gives a well-defined functional on $C_c(X) \otimes C_c(Y)$. The same argument, with $= 0$ replaced by ≥ 0, shows that $\int_x \otimes \int_y$ is a positive functional.

Suppose that $A \subset X$ and $B \subset Y$ are open, relatively compact subsets. We then have

$$C_0(A) \subset C_c(X), \quad C_0(B) \subset C_c(Y), \quad C_0(A \times B) \subset C_c(X \times Y).$$

Note now that $C_0(A) \otimes C_0(B)$ is a self-adjoint subalgebra of $C_0(A \times B)$ that separates the points in $A \times B$ and does not vanish identically at any point. By the Stone–Weierstrass theorem (4.3.5) it follows that $C_0(A) \otimes C_0(B)$ is uniformly dense in $C_0(A \times B)$. Choose by 1.7.5 functions d in $C_c(X)_+$ and e in $C_c(Y)_+$ such that $d|A = 1$ and $e|B = 1$. Then $|h| \leq \|h\|_\infty d \otimes e$ for every h in $C_0(A \times B)$. In particular,

$$\left| \int_x \otimes \int_y h \right| \leq \|h\|_\infty \left(\int_x d \right) \left(\int_y e \right)$$

for every h in $C_c(A) \otimes C_c(B)$. Consequently, $\int_x \otimes \int_y$ extends uniquely to a bounded, positive functional on $C_0(A \times B)$ (cf. 2.1.11).

Let $(A_\lambda)_{\lambda \in \Lambda}$ and $(B_\mu)_{\mu \in M}$ be the nets (ordered under inclusion) of open, relatively compact subsets of X and Y, respectively. Then $X = \bigcup A_\lambda$ and $Y = \bigcup B_\mu$, whence $X \times Y = \bigcup A_\lambda \times B_\mu$, so that

$$C_c(X \times Y) = \bigcup C_0(A_\lambda \times B_\mu).$$

It follows from the argument above that $\int_x \otimes \int_y$ has a unique extension to a positive functional on $C_c(X \times Y)$, i.e. a Radon integral on $X \times Y$.   □

6.6.4. Lemma. *If $h \in C_c(X \times Y)$, the function $x \to \int_y h(x, \cdot)$ belongs to $C_c(X)$ and $\int_x \int_y h(\cdot, \cdot) = \int_x \otimes \int_y h$.*

PROOF. Choose open, relatively compact subsets $A \subset X$ and $B \subset Y$ such that the support of h is contained in $A \times B$. With notations and arguments as in the proof of 6.6.3, there is then a sequence (h_n) in $C_0(A) \otimes C_0(B)$ converging uniformly to h. Thus

$$\left| \int_y h(x, \cdot) - \int_y h_n(x, \cdot) \right| \leq \int_y \|h - h_n\|_\infty d \otimes e$$

$$\leq \|h - h_n\|_\infty \|d\|_\infty \int_y e,$$

for every x in X. Since the functions $x \to \int_y h_n(x, \cdot)$ evidently belong to $C_0(A)$, we conclude that the same is true of the limit function. Furthermore, we see from the construction of $\int_x \otimes \int_y$ that

$$\left| \int_x \int_y h(\cdot, \cdot) - \int_x \otimes \int_y h \right| = \lim \left| \int_x \int_y h(\cdot, \cdot) - \int_x \otimes \int_y h_n \right|$$

$$= \lim \left| \int_x \int_y h(\cdot, \cdot) - h_n(\cdot, \cdot) \right|$$

$$\leq \lim \|h - h_n\|_\infty \left(\int_x d \right) \left(\int_y e \right) = 0. \qquad \square$$

6.6.5. Lemma. *If* $h \in C_c(X \times Y)^m$, *the function* $f: x \to \int_y^* h(x, \cdot)$ *belongs to* $C_c(X)^m$. *Moreover, if* $h \in \mathcal{L}^1(X \times Y)$, *then* $f \in \mathcal{L}^1(X)$ *and*

$$\int_x f = \int_x \int_y^* h(\cdot, \cdot) = \int_x \otimes \int_y h.$$

PROOF. Take an increasing net $(h_\lambda)_{\lambda \in \Lambda}$ in $C_c(X \times Y)$ such that $h_\lambda \nearrow h$ and put $f_\lambda(x) = \int_y h_\lambda(x, \cdot)$. Then $f_\lambda \in C_c(X)$ by 6.6.4 and $f_\lambda \nearrow f$, whence $f \in C_c(X)^m$ (but maybe often with the value $+\infty$). Now if $h \in \mathcal{L}^1(X \times Y)$, then, again by 6.6.4,

$$\int_x f_\lambda = \int_x \int_y h_\lambda = \int_x \otimes \int_y h_\lambda \nearrow \int_x \otimes \int_y h < \infty.$$

It follows from this that $f(x) < \infty$ almost everywhere, and (by 6.1.5) that

$$\int_x f = \lim \int_x f_\lambda = \int_x \otimes \int_y h. \qquad \square$$

6.6.6. Theorem. *Let* $\int_x$ *and* $\int_y$ *be Radon integrals on the locally compact Hausdorff spaces* X *and* Y. *If* h *is a Borel function on* $X \times Y$, *which is integrable with respect to the product integral* $\int_x \otimes \int_y$, *then the Borel function* $y \to h(x, y)$ *belongs to* $\mathcal{L}^1(Y)$ *for almost all* x *in* X, *and the almost everywhere defined function* $x \to \int_y h(x, \cdot)$ *belongs to* $\mathcal{L}^1(X)$ *with*

$$\int_x \int_y h(\cdot, \cdot) = \int_x \otimes \int_y h.$$

PROOF. If $h \in \mathcal{L}^1(X \times Y)$, there are by 6.1.9 functions h_n in $C_c(X \times Y)^m$ and $h'_n \in C_c(X \times Y)_m$ such that $h'_n \leq h \leq h_n$ and $\int_x \otimes \int_y (h_n - h'_n) < n^{-3}$. In particular, both h_n and h'_n belong to $\mathcal{L}^1(X \times Y)$. By 6.6.5 there is a null set $N_n \subset X$ such that the functions $y \to h_n(x, y)$ and $y \to h'_n(x, y)$ belong to $\mathcal{L}^1(Y)$ for $x \notin N_n$. Furthermore, the integrated functions belong to $\mathcal{L}^1(X) \cap C_c(X)^m$ and to $\mathcal{L}^1(X) \cap C_c(X)_m$, respectively. Set

$$A_n = \left\{ \int_y (h_n(x, \cdot) - h'_n(x, \cdot)) > n^{-1} \right\} \subset X.$$

Then A_n is open, and by 6.6.5

$$\int_x [A_n] \leq n \int_x \int_y (h_n(\cdot, \cdot) - h'_n(\cdot, \cdot)) = n \int_x \otimes \int_y (h_n - h'_n) < n^{-2}.$$

Consequently, the set

$$N = (\bigcup N_n) \cup \left(\bigcap_m \bigcup_{n > m} A_n \right)$$

is a null set in X. We see that if $x \notin N$, then for some m we have $x \notin A_n$ for all $n > m$, whence $h'_n(x, \cdot) \leq h(x, \cdot) \leq h_n(x, \cdot)$ with $\int_y (h_n(x, \cdot) - h'_n(x, \cdot)) \leq n^{-1}$. Since $h_n(x, \cdot) \in C_c(Y)^m$ and $h'_n(x, \cdot) \in C_c(Y)_m$, it follows from $(*)$ in 6.1.9 that $h(x, \cdot) \in \mathcal{L}^1(Y)$ if $x \notin N$. Moreover,

$$\int_y h'_n(x, \cdot) \leq \int_y h(x, \cdot) \leq \int_y h_n(x, \cdot),$$

where the first function belongs to $C_c(X)_m$ and the third to $C_c(X)^m$, and

$$\int_x \left(\int_y h_n(\cdot, \cdot) - \int_y h'_n(\cdot, \cdot) \right) < n^{-3}.$$

Again from 6.1.9 we conclude that the function $x \to \int_y h(x, \cdot)$ belongs to $\mathcal{L}^1(X)$ with integral

$$\int_x \int_y h(\cdot, \cdot) = \lim \int_x \int_y h_n(\cdot, \cdot) = \int_x \otimes \int_y h. \qquad \square$$

6.6.7. The result above is known as *Fubini's theorem*. Since the definition of the product integral (see 6.6.3) is symmetric in x and y, it follows from 6.6.6 that if $h \in \mathcal{L}^1(X \times Y)$, the iterated integrals below both exist and are equal. Thus,

$$\int_x \int_y h(\cdot, \cdot) = \int_x \otimes \int_y h = \int_y \int_x h(\cdot, \cdot).$$

It is well known that this exchange of the integration order is the most useful part of Fubini's theorem. Therefore, it is a bit inconvenient that one has to check the integrability with respect to a somewhat nebulous product integral *before* computing the double integrals. This problem is solved in the σ-compact case by the following variation of Fubini's theorem, known as *Tonelli's theorem*.

6.6.8. Corollary. *Let h be a Borel function on $X \times Y$ that vanishes outside a σ-compact subset of $X \times Y$. If the Borel function $y \to |h(x, y)|$ belongs to $\mathcal{L}^1(Y)$ for almost all x in X, and if the almost everywhere defined function $x \to \int_y |h(x, \cdot)|$ belongs to $\mathcal{L}^1(X)$, then $h \in \mathcal{L}^1(X \times Y)$, whence*

$$\int_x \int_y h(\cdot, \cdot) = \int_x \otimes \int_y h = \int_y \int_x h(\cdot, \cdot).$$

PROOF. Let $h_n = |h| \wedge n[C_n]$, where $C_n \in \mathscr{C}$ and $h|(X \setminus \bigcup C_n) = 0$ almost everywhere. Then $h_n \in \mathscr{L}^1(X \times Y)$ and $h_n \nearrow |h|$. By assumption

$$\int_x \otimes \int_y h_n = \int_x \int_y h_n(\cdot, \cdot) \leq \int_x \int_y |h(\cdot, \cdot)| < \infty.$$

Thus, from the monotone convergence theorem (6.1.13) it follows that $|h| \in \mathscr{L}^1(X \times Y)$, and 6.2.16 in conjunction with Fubini's theorem (6.6.6) completes the proof. $\qquad\square$

6.6.9. A *topological group* is a group G equipped with a topology in which the group operations are continuous. Thus, $x, y \to xy$ is continuous from $G \times G$ to G, and $x \to x^{-1}$ is continuous from G to G (and thus a homeomorphism); cf. E 1.5.6, E 1.6.14, E 1.6.15, E 1.7.4, and E.1.7.7. We shall not assume that the group is abelian, and therefore use the multiplicative notation for product and inverse, just as we let 1 denote the unit in G. The most interesting topological groups are all *locally compact Hausdorff*, and we shall limit ourselves exclusively to such groups.

If f is a (real or complex) function on G, we define for each x in G the left (respectively right) translated function $_x f$ (respectively $^x f$) by $_x f(y) = f(x^{-1}y)$ $[^x f(y) = f(yx)]$. The definitions are chosen such that $_{xy} f = {}_x({}_y f)$ and $^{xy} f = {}^x({}^y f)$.

A (left) *Haar integral* on a locally compact group G is a Radon integral $\int \neq 0$ that is (left) translation invariant, i.e. $\int_x f = \int f$ for every x in G and f in $C_c(G)$. Similarly, one may define a right Haar integral. It is clear that the translation invariance will extend to every function in $\mathscr{L}^1(G)$. In particular, we see that $\int [xA] = \int [A]$ for every x in G and every Borel subset A in G (because $[xA] = {}_x[A]$). If A is open and nonempty, then for each compact subset C of G we have $C \subset \bigcup x_n A$ for some finite subset $\{x_n\}$ of G. Since $\int \neq 0$, we see from the inner regularity of Radon measures (6.3.1) that $\int [A] > 0$ for every open, nonempty subset A of G.

6.6.10. Lemma. *If $f \in C_0(G)$, the maps $x \to {}_x f$ and $x \to {}^x f$ are uniformly continuous from G into $C_0(G)$.*

PROOF. Given $\varepsilon > 0$ there is a compact set $C \subset G$, such that $|f(y)| < \varepsilon$ for $y \notin C$. For each y in C there is a symmetric neighborhood $A(y)$ of 1 [i.e. $x \in A(y)$ iff $x^{-1} \in A(y)$] such that $|f(x^{-1}y) - f(y)| < \varepsilon$ when $x \in A(y)$. A standard compactness argument now produces a symmetric neighborhood $A [= A(y_1) \cap \cdots \cap A(y_n)]$ of 1 such that $|f(x^{-1}y) - f(y)| < \varepsilon$ for every y in C and x in A. If $x^{-1}y \in C$ (and $x \in A$), it follows from the symmetry that

$$|f(x^{-1}y) - f(y)| = |f(x^{-1}y) - f(x(x^{-1}y))| < \varepsilon.$$

If both $x^{-1}y \notin C$ and $y \notin C$, then $|f(x^{-1}y) - f(y)| < 2\varepsilon$. Consequently, $\|_x f - f\|_\infty \leq 2\varepsilon$ for every x in A, so that we have uniform continuity at 1. But since $_{xy} f = {}_x({}_y f)$, this will imply uniform continuity at every point. The argument for the map $x \to {}^x f$ is symmetrical. $\qquad\square$

6.6.11. Lemma. *For each Radon integral $\int$ on G and $1 \leq p < \infty$, the maps $x \to {}_x f$ and $x \to {}^x f$ are uniformly continuous from G into $\mathscr{L}^p(G)$ for every f in $C_c(G)$.*

PROOF. Choose a compact set C such that $f(y) = 0$ if $y \notin C$. Choose then a compact neighborhood C_1 of 1. Now use 6.6.10 with a given $\varepsilon > 0$ to find a neighborhood $A \subset C_1$ of 1, such that $\|{}_x f - f\|_p^p = \int |{}_x f - f|^p \leq \varepsilon^p \int [C_1 C]$. $\square$

6.6.12. Theorem. *On a locally compact topological group G, the Haar integral is unique up to a positive scalar factor.*

PROOF. By 6.2.8 there is a monotone decreasing net $(g_\lambda)_{\lambda \in \Lambda}$ in $C_c(G)_+$ such that $\bigcap \{g_\lambda > 0\} = \{1\}$. Replacing, if necessary, g_λ by the function $x \to g_\lambda(x)g_\lambda(x^{-1})$, we may assume that the net consists of symmetric functions.

If $\int_x$ and $\int_y$ both are (left) Haar integrals on G consider the numbers $\gamma_\lambda = (\int_x g_\lambda)(\int_y g_\lambda)^{-1}$. Since $\gamma_\lambda > 0$ (cf. the last lines in 6.6.9), we may assume, passing if necessary to a subnet, that either $\gamma_\lambda \to \gamma$ or $\gamma_\lambda^{-1} \to \gamma$ for some $\gamma \geq 0$. Interchanging $\int_x$ and $\int_y$ transforms the second case to the first, which we may therefore assume to hold.

For f in $C_c(G)$ we now obtain from Fubini's theorem (6.6.6) applied to the product integral $\int_x \otimes \int_y$ and the functions

$$(x, y) \to f(x)g_\lambda(x^{-1}y) \quad \text{and} \quad (x, y) \to f(yx)g_\lambda(x),$$

together with the translation invariance of $\int_x$ and $\int_y$, that

$$\int_x f \int_y g_\lambda = \int_x \int_y f(x)g_\lambda(y) = \int_x \int_y f(x)g_\lambda(x^{-1}y)$$

$$= \int_y \int_x f(x)g_\lambda(x^{-1}y) = \int_y \int_x f(yx)g_\lambda(x^{-1}) = \int_x \int_y f(yx)g_\lambda(x).$$

For each $\varepsilon > 0$ there is by 6.6.11 a neighborhood A of 1 such that $\int_y |{}^x f - f| < \varepsilon$ for every x in A. Since $\{g_\lambda > 0\} \subset A$ for large λ, we see from above that

$$\left| \int_x f - \gamma \int_y f \right| = \lim \left(\int_y g_\lambda \right)^{-1} \left| \int_x f \int_y g_\lambda - \int_x g_\lambda \int_y f \right|$$

$$= \lim \left(\int_y g_\lambda \right)^{-1} \left| \int_x \int_y (f(yx) - f(y))g_\lambda(x) \right|$$

$$\leq \limsup \left(\int_y g_\lambda \right)^{-1} \int_x \|{}^x f - f\|_1 g_\lambda(x) \leq \varepsilon\gamma.$$

As ε and f are arbitrary, $\int_x = \gamma \int_y$. $\square$

6.6.13. In 1933 A. Haar proved that every locally compact topological group has a translation invariant integral/measure. For the classical groups ($\mathbb{R}$, $\mathbb{Z}$, $\mathbb{T}$, and various groups of regular or unitary $n \times n$ matrices) these integrals are

well known, and the uniqueness result in 6.6.12 is more important—Haar's
amazing result notwithstanding.

6.6.14. We have only mentioned the left Haar integral, but of course there is
also (for symmetry reasons) a (unique) right Haar integral. These two need
not coincide. The easiest available (counter) example obtains by taking G to
be the group of 2×2 matrices of the form

$$x = \begin{pmatrix} a & b \\ 0 & 1 \end{pmatrix}, \quad a > 0, b \in \mathbb{R}.$$

This group can also be visualized as the group of affine transformations of $\mathbb{R}$
onto itself, where $x(t) = at + b$ for t in $\mathbb{R}$. Define

$$\int_l f = \int\int f(a, b) a^{-2} da db,$$

$$\int_r f = \int\int f(a, b) a^{-1} da db,$$

where the right-hand sides are ordinary Lebesgue integrals on $\mathbb{R}_+ \times \mathbb{R}$, and
$f \in C_c(G)$. Elementary computations show that $\int_l$ and $\int_r$ are, respectively, a left
and a right Haar integral on G.

6.6.15. We now fix a left Haar integral on the locally compact group G, and
we use the standard notation $\int f(x)\,dx$ to denote the value of the integral on
a function f, because this notation is well suited to describe the constant
change of variables, that is the trademark of harmonic analysis.

For each x in G the Radon integral on G given by $f \to \int f(yx)\,dy, f \in C_c(G)$,
is evidently left invariant. By the uniqueness theorem (6.6.12) there is therefore
a number $\Delta(x) > 0$ such that

$$\Delta(x) \int f(yx)\,dy = \int f(y)\,dy, \quad f \in \mathscr{L}^1(G). \tag{*}$$

The function $\Delta: G \to \,]0, \infty[$ so obtained is called the *modular function*. If
$\Delta = 1$, we say that G is *unimodular*. Clearly, this happens iff the Haar integral
is right invariant.

The choice between Δ and Δ^{-1} to be named the modular function is
determined by tradition (and favors the measure over the integral). Thus we
see from (*) that $\int [Ax](y)\,dy = \Delta(x) \int [A](y)\,dy$ for every Borel subset A of G
(because $[Ax^{-1}] = {}^x[A]$), and that, symbolically, $d(yx) = \Delta(x)\,dy$.

6.6.16. Proposition. *The modular function Δ on G is a continuous homomorphism of G into the multiplicative group $\exp \mathbb{R}$ of positive real numbers. Moreover, G is unimodular whenever G is abelian, or discrete, or compact.*

PROOF. Choose some f in $C_c(G)$ with $\int f(x)\,dx = 1$. By 6.6.11 the map $x \to {}^x f$
is continuous from G into $L^1(G)$. Consequently, the function $x \to \int {}^x f(y)\,dy =$

$\Delta(x)^{-1}$ is continuous. Furthermore,

$$\Delta(xy)^{-1} = \int {}^{xy}f(z)\,dz = \int {}^{x}({}^{y}f(z))\,dz$$

$$= \Delta(x)^{-1} \int {}^{y}f(z)\,dz = \Delta(x)^{-1}\Delta(y)^{-1},$$

which shows that Δ is multiplicative.

If G is abelian, ${}^{x}f = {}_{x^{-1}}f$, and, consequently, $\Delta = 1$. If G is discrete, the Haar integral is simply the counting measure on G [i.e. $\int f(x)\,dx = \sum f(x)$], and this is clearly both left and right invariant. If G is compact, the function 1 is integrable, whence

$$\Delta(x) \int 1 = \Delta(x) \int {}^{x}1 = \int 1$$

for every x in G, so that $\Delta = 1$.   □

6.6.17. Lemma. *For each f in $C_c(G)$ we have*

$$\int f(x^{-1})\Delta(x)^{-1}\,dx = \int f(x)\,dx.$$

PROOF. Define $\check{f}(x) = f(x^{-1})$. For each x in G we then have

$$\int f(x^{-1}y^{-1})\Delta(y^{-1})\,dy = \Delta(x) \int \check{f}(yx)\Delta(yx)^{-1}\,dy$$

$$= \int \check{f}(y)\Delta(y)^{-1}\,dy = \int f(y^{-1})\Delta(y^{-1})\,dy.$$

This shows that the left-hand side of the formula in the lemma defines a left invariant Radon integral. By 6.6.12 there is therefore a $\gamma > 0$ such that

$$\int f(x^{-1})\Delta(x)^{-1}\,dx = \gamma \int f(x)\,dx, \quad f \in C_c(G).$$

Given $\varepsilon > 0$ we can find a neighborhood A of 1 in G such that $|\Delta(x)^{-1} - 1| \le \varepsilon$ for x in A. Choosing f as a symmetric function with support inside A, we see that $|1 - \gamma| \le \varepsilon$; and since ε is arbitrary, $\gamma = 1$.   □

6.6.18. With $\check{f}(x) = f(x^{-1})$ as in 6.6.17 we observe that the integral $f \to \int \check{f}(x)\,dx$ is right invariant. Using 6.6.17 we see how the modular function describes the connection between this integral and the left Haar integral, viz.

$$\int \check{f}(x)\,dx = \int f(x^{-1})\Delta(x)\Delta(x)^{-1}\,dx = \int f(x)\Delta(x)^{-1}\,dx.$$

Furthermore, it follows from 6.6.17 that if we define

$$f*(x) = \overline{f(x^{-1})}\Delta(x)^{-1}, \tag{*}$$

then

$$\|f*\|_1 = \int |f(x^{-1})|\,\Delta(x)^{-1}\,dx = \int |f(x)|\,dx = \|f\|_1$$

for every f in $C_c(G)$, and therefore also for f in $\mathscr{L}^1(G)$, so that $f \to f*$ is an isometric involution in the Banach space $L^1(G)$.

6.6.19. Proposition. *If $f \in \mathscr{L}^p(G)$ for $1 \le p < \infty$, the maps $x \to {}_x f$ and $x \to {}^x f$ are uniformly continuous from G into $L^p(G)$.*

PROOF. Given $\varepsilon > 0$ we can by 6.4.11 choose g in $C_c(G)$ such that $\|f - g\|_p \le \varepsilon$. By 6.6.11 there is then a neighborhood A of 1 in G such that $\|g - {}_x g\|_p \le \varepsilon$ and $\|g - {}^x g\|_p \le \varepsilon$ for every x in A. Since Δ is continuous (6.6.16), we may also assume that $\Delta(x)^{-1/p} \le 2$ for all x in A. Now $\|{}_x f - {}_x g\|_p = \|f - g\|_p$ by the invariance of the Haar integral, whereas $\|{}^x f - {}^x g\|_p = \Delta(x)^{-1/p}\|f - g\|_p$ by (*) in 6.6.15. Combining these estimates we have

$$\|{}_x f - f\|_p \le 3\varepsilon, \qquad \|{}^x f - f\|_p \le 4\varepsilon$$

for every x in A. $\qquad\qquad\qquad\qquad\qquad\qquad\qquad\qquad\qquad\qquad\qquad\square$

6.6.20. Proposition. *If f and g are Borel functions in $\mathscr{L}^1(G)$ and $\mathscr{L}^p(G)$, respectively (where $1 \le p < \infty$), the function $y \to f(y)g(y^{-1}x)$ belongs to $\mathscr{L}^1(G)$ for almost all x, and the almost everywhere defined function $x \to \int f(y)g(y^{-1}x)\,dy$ belongs to $\mathscr{L}^p(G)$, with*

$$\left\| \int f(y)g(y^{-1}\cdot)\,dy \right\|_p \le \|f\|_1\|g\|_p.$$

PROOF. Since f is the limit in $L^1(G)$ of a sequence from $C_c(G)$ by 6.4.11, there is a σ-compact subset B of G such that $f|(G\backslash B) = 0$ almost everywhere. Similarly, $g|(G\backslash A) = 0$ almost everywhere for some σ-compact subset A. The function h on $G \times G$ given by $h(x, y) = f(y)g(y^{-1}x)$ is the composition of the continuous function $(x, y) \to (y^{-1}x, y)$ and the (product) Borel function $g \otimes f$ and, consequently, is a Borel function. Moreover, $h = 0$ almost everywhere outside the σ-compact subset $D = BA \times B$ in $G \times G$. For each k in $\mathscr{L}^q(BA)$, where $p^{-1} + q^{-1} = 1$, we can therefore apply Tonelli's theorem (6.6.8) to the product hk, and since by Hölder's inequality (6.4.6) we have

$$\int\int |h(x, y)k(x)|\,dx\,dy = \int\int |f(y)g(y^{-1}x)k(x)|\,dx\,dy$$

$$\le \int |f(y)|\,\|{}_y g\|_p\|k\|_q\,dy = \|f\|_1\|g\|_p\|k\|_q,$$

we conclude that the function $y \to f(y)g(y^{-1}x)k(x)$ belongs to $\mathscr{L}^1(G)$ for almost all x, and that the almost everywhere defined function $x \to \int f(y)g(y^{-1}x)\,dy\,k(x)$ belongs to $\mathscr{L}^1(G)$, with a 1-norm dominated by $\|f\|_1 \|g\|_p \|k\|_q$. Since we can choose k strictly positive on BA, it follows that the function $x \to \int f(y)g(y^{-1}x)\,dy$ is well-defined almost everywhere, and we conclude from 6.5.11 that it belongs to $\mathscr{L}^p(BA) \subset \mathscr{L}^p(G)$, with a p-norm dominated by $\|f\|_1 \|g\|_p$. $\square$

6.6.21. Theorem. *For a locally compact group G with Haar integral $\int$, the space $L^1(G)$ is a Banach algebra with an isometric involution. Product and involution in $L^1(G)$ are given by the formulas*

$$f \times g(x) = \int f(y)g(y^{-1}x)\,dy, \qquad f^*(x) = \overline{f(x^{-1})}\Delta(x)^{-1}.$$

PROOF. It follows from 6.6.20 that $f \times g \in L^1(G)$ with $\|f \times g\|_1 \leq \|f\|_1 \|g\|_1$. It is elementary to check that the convolution product is distributive with respect to the sum, but it requires Fubini's theorem to show that the product is associative. Indeed, if f, g, and h belong to $\mathscr{L}^1(G)$, both functions $(f \times g) \times h$ and $f \times (g \times h)$ belong to $\mathscr{L}^1(G)$; and they are equal almost everywhere because the function

$$(x, y, z) \to f(z)g(z^{-1}y)h(y^{-1}x)$$

belongs to $\mathscr{L}^1(G \times G \times G)$.

The operation $f \to f^*$ is conjugate linear and isometric with period two by 6.6.18. Moreover, we have

$$g^* \times f^*(x) = \int g^*(y)f^*(y^{-1}x)\,dy$$

$$= \int \overline{f(x^{-1}y)}\Delta(x^{-1}y)\overline{g(y^{-1})}\Delta(y^{-1})\,dy$$

$$= \int \overline{f(y)}\overline{g(y^{-1}x^{-1})}\,dy\,\Delta(x^{-1}) = (f \times g)^*(x),$$

which shows that $*$ is antimultiplicative, and thus is an involution on $L^1(G)$. $\square$

6.6.22. The *convolution algebra* $L^1(G)$ defined in 6.6.21 is only a C^*-algebra (4.3.7) when G is finite. But the Hilbert space structure is not far away. Indeed, applying 6.6.20 with $p = 2$ we obtain a norm decreasing, $*$-preserving homomorphism $f \to F_f$ of $L^1(G)$ into $\mathbf{B}(L^2(G))$ given by $F_f g = f \times g$, $g \in L^2(G)$. This *regular representation* of $L^1(G)$ is faithful, i.e. $F_f = 0$ implies $f = 0$.

If G is abelian and $\hat{G}$ denotes the dual group [$\hat{G} = \hom(G, \mathbb{T})$, see 4.2.8], the restriction of the Fourier transformation $f \to \hat{f}$ to $\mathscr{L}^1(G) \cap \mathscr{L}^2(G)$ extends uniquely to an isometry F of $L^2(G)$ onto $L^2(\hat{G})$ (when the Haar integral on $\hat{G}$ is suitably normalized). For $G = \mathbb{R}$ the construction of this *Plancherel isomor-*

phism is quite elementary; see E 3.1.16. Combining this with the regular representation we see that for f in $L^1(G)$ and g in $L^2(G)$ we have

$$FF_f g = F(f \times g) = \hat{f} \cdot Fg = M_{\hat{f}} F_g,$$

so that the convolution operator F_f is transformed by the Plancherel isomorphism into the multiplication operator $M_{\hat{f}}$, cf. 4.7.6.

6.6.23. The convolution product can be defined for other classes of functions. Thus $f \times g$ exists as an element in $C_0(G)$ whenever $f \in \mathscr{L}^p(G)$ and $\check{g} \in \mathscr{L}^q(G)$ with $p^{-1} + q^{-1} = 1$ [because $f \times g(x) = \int f(y)_x \check{g}(y) \, dy$]. In the case $p = 1$, $q = \infty$, however, we only have $f \times g$ as a uniformly continuous, bounded function on G.

We wish to define the convolution product of finite Radon charges (cf. 6.5.8). If $\Phi, \Psi \in M(G)$, we define $\Phi \otimes \Psi$ as a finite Radon charge on $G \times G$, either by mimicking the proof of 6.6.3 or by taking polar decompositions $\Phi = |\Phi|(u \cdot)$ and $\Psi = |\Psi|(v \cdot)$ as in 6.5.6 and 6.5.8 and then setting

$$(\Phi \otimes \Psi)h = (|\Phi| \otimes |\Psi|)((u \otimes v)h), \quad h \in C_c(G \times G). \tag{$*$}$$

Having done this, we define the product in $M(G)$ by the formula

$$(\Phi \times \Psi)f = (\Phi \otimes \Psi)(f \circ \pi), \quad f \in C_c(G), \tag{$**$}$$

where $\pi: G \times G \to G$ is the product map $\pi(x, y) = xy$. Note that although $f \circ \pi$ is a bounded continuous function on $G \times G$ it does not belong to $C_c(G \times G)$ (if $f \neq 0$), so that we need the assumption that Φ and Ψ are *finite* charges.

It follows from $(*)$ and $(**)$ that $\Phi \times \Psi \in M(G)$, with

$$\|\Phi \times \Psi\| \le \|\Phi \otimes \Psi\| = \|\Phi\| \, \|\Psi\|.$$

Given a third charge Ω it is easy to see that

$$((\Phi \times \Psi) \times \Omega)f = (\Phi \otimes \Psi \otimes \Omega)(f \circ \tau) = (\Phi \times (\Psi \times \Omega))f,$$

where $\tau(x, y, z) = xyz$, so that the product is associative. In conjunction with 6.5.9 this shows that $M(G)$ is a unital Banach algebra (the point measure δ_1 at 1 being the unit). Defining

$$\Phi^* f = \overline{\Phi \tilde{f}}, \quad f \in C_c(G),$$

where $\tilde{f}(x) = \overline{f(x^{-1})}$, we see that

$$(\Psi^* \otimes \Phi^*)(f \otimes g) = \overline{\Psi(\tilde{f})\Phi(\tilde{g})} = \overline{(\Phi \otimes \Psi)(\tilde{g} \otimes \tilde{f})} = \overline{(\Phi \otimes \Psi)(f \otimes g)^s},$$

where $h^s(x, y) = \overline{h(y^{-1}, x^{-1})}$ for every function on $G \times G$. It follows from $(**)$ that for each f in $C_c(G)$ we have

$$(\Psi^* \times \Phi^*)f = (\Psi^* \otimes \Phi^*)(f \circ \pi) = \overline{(\Phi \otimes \Psi)(f \circ \pi)^s}$$

$$= \overline{(\Phi \otimes \Psi)(\tilde{f} \circ \pi)} = (\Phi \times \Psi)^* f,$$

so that $*$ is an isometric involution on $M(G)$.

6.6.24. Proposition. *The isometry* $f \to \Phi_f$ *defined in 6.5.10 is a $*$-isomorphism of the convolution algebra* $L^1(G)$ *onto a closed, $*$-invariant ideal of* $M(G)$.

PROOF. Elementary computations show that the map $f \to \Phi_f$ is a $*$-isomorphism, i.e. $\Phi_f^* = \Phi_{f^*}$ and $\Phi_{f \times g} = \Phi_f \times \Phi_g$.

To show that $L^1(G)$ is an ideal in $M(G)$ it suffices to prove that $\Phi \times \Phi_f \in \mathcal{L}^1(G)$, when Φ is a finite Radon integral and f belongs to $\mathcal{L}^1(G)_+$. Choose a σ-compact subset A of G such that $\Phi([A]) = \Phi(1)$. Then apply Tonelli's theorem (6.6.8) to the product integral $\Phi \otimes \int$ and the function $h(x, y) = [A](x)f(x^{-1}y)$ (which has σ-compact support on $G \times G$). We get

$$\Phi\left(\int h(\cdot, y)\, dy\right) = \Phi([A]\, \|f\|_1) = \Phi(1)\, \|f\|_1,$$

from which we conclude that the function

$$y \to \Phi(h(\cdot, y)) = \Phi(\,_y\check{f})$$

exists almost everywhere and belongs to $\mathcal{L}^1(G)$ with $\int \Phi(\,_y\check{f})\, dy = \Phi(1)\, \|f\|_1$. Now take g in $C_c(G)$ and compute, again employing Fubini's theorem, that

$$\int \Phi(\,_y\check{f})g(y)\, dy = \Phi\left(\int \,_y\check{f}g(y)\, dy\right)$$

$$= \Phi\left(\int \check{f}(y^{-1}\cdot)g(y)\, dy\right) = \Phi\left(\int \check{f}(y^{-1})g(\cdot\, y)\, dy\right)$$

$$= \Phi\left(\int g(\cdot\, y)f(y)\, dy\right) = \Phi \otimes \Phi_f(g \circ \pi) = \Phi \times \Phi_f g.$$

Thus $\Phi \times \Phi_f$ equals the element $y \to \Phi(\,_y\check{f})$ in $L^1(G)$. $\square$

Bibliography

N.I. Ahiezer and I.M. Glazman, *Theory of Linear Operators in Hilbert Space.* Ungar, New York, 1961 (Russian original, 1950).

L. Alaoglu, *Weak topologies of normed linear spaces.* Ann. of Math. **41** (1940), 252–267.

P. Alexandroff and H. Hopf, *Topologie.* Springer-Verlag, Berlin, 1935.

P. Alexandroff and P. Urysohn, *Mémoire sur les espaces topologiques compactes.* Verh. Akad. Wetensch. Amsterdam **14** (1929), 1–29.

W.B. Arveson, *An Invitation to C*-algebras.* Springer-Verlag, Heidelberg, 1976.

G. Ascoli, *Sugli spazi lineari metrici e le loro varietà lineari.* Ann. Mat. Pura Appl. **10** (1932), 33–81, 203–232.

L. Asimow and A.J. Ellis, *Convexity Theory and Its Applications in Functional Analysis.* Academic Press, London/New York, 1980.

S. Banach, *Théorie des opérations linéaires.* Monografje Matematyczne, Warszawa, 1932.

S. Banach, Oeuvres, I–II. Akad. Pol. Sci. Warszawa, 1967, 1979.

T. Bonnesen and W. Fenchel, *Theorie der konvexen Körper.* Springer-Verlag, Berlin, 1934. Reprinted 1974.

E. Borel, *Lecons sur les fonctions de variables réelles.* Gauthier-Villars, Paris, 1905.

E. Borel, *Théorie de fonctions.* Gauthier-Villars, Paris, 1921.

N. Bourbaki, *Topologie générale.* Herman et Cie, Paris, 1940.

N. Bourbaki, *Intégration.* Herman et Cie, Paris, 1952.

N. Bourbaki, *Espaces vectoriels topologiques.* Herman et Cie, Paris, 1955.

J.W. Calkin, *Two-sided ideals and congruences in the ring of bounded operators in Hilbert space.* Ann. of Math. **42** (1941), 839–873.

C. Carathéodory, *Vorlesungen über reelle Funktionen,* second edition. Teubner, Leipzig, 1927.

H. Cartan, *Sur le mesure de Haar.* C. R. Acad. Sci. Paris, **211** (1940), 759–762.

E. Čech, *On bicompact spaces.* Ann. of Math. **38** (1937), 823–844.

G. Choquet, *Cours D'Analyse,* II (Topologie). Masson et Cie, Paris, 1964.

P.J. Cohen, *The independence of the continuum hypothesis,* I–II. Proc. Nat. Acad. Sci. USA **50** (1963), 1143–1148; **51** (1964), 105–110.

R. Courant and D. Hilbert, *Methoden der mathematischen Physik,* I–II. Springer-Verlag, Berlin, 1924, 1937.

P.J. Daniell, *A general form of integral.* Ann. of Math. **19** (1917–18), 279–294.

E.B. Davies, *One-Parameter Semigroups*. Academic Press, London/New York, 1980.

J. Dieudonné, *Treatise on Analysis*, I–V. Academic Press, New York, 1969, 1976, 1972, 1974, 1977.

U. Dini, *Fondamenti per la teorica delle funzioni di variabili reali*. Pisa, 1878. German translation published by Teubner, Berlin, 1892.

J. Dixmier, *Les algèbres d'opérateurs dans l'espace hilbertien*. Gauthier-Villars, Paris, 1957 (second edition 1969).

R.G. Douglas, *Banach Algebra Techniques in Operator Theory*. Academic Press, New York, 1972.

J. Dugundji, *Topology*. Allyn & Bacon, Boston, 1966.

N. Dunford and J.T. Schwartz, *Linear Operators*, I–III. Interscience, New York, 1958, 1963, 1971.

R.E. Edwards, *Functional Analysis*. Holt, Rinehart & Winston, New York, 1965.

R. Engelkind, *Outline of General Topology*. North-Holland, Amsterdam, 1968.

E. Fischer, *Sur la convergence en moyenne*. C. R. Acad. Sci. Paris, **144** (1907), 1022–1024.

E. Fischer, *Applications d'un théorème sur la convergence en moyenne*. C. R. Acad. Sci. Paris, **144** (1907), 1148–1151.

M. Fréchet, *Les espaces abstraits*. Gauthier-Villars, Paris, 1928.

I. Fredholm, *Sur une classe d'équations fonctionelles*. Acta Math. **27** (1903), 365–390.

K.O. Friedrichs, *Spektraltheorie halbbeschränkter Operatoren*, I–III. Math. Ann. **109–10** (1934–35), 465–487, 685–713, 777–779.

B. Fuglede, *A commutativity theorem for normal operators*. Proc. Nat. Acad. Sci. USA **36** (1950), 35–40.

I.M. Gelfand, *Normierte Ringe*. Mat. Sbornik **9** (1941), 3–24.

I.M. Gelfand and M.A. Naimark, *On the imbedding of normed rings into the ring of operators in Hilbert space*. Mat. Sbornik **12** (1943), 197–213.

I.M. Gelfand and G.E. Shilov, *Generalized Functions*. Academic Press, New York, 1964 (Russian original, 1958).

K. Gödel, *The Consistency of the Continuum Hypothesis*. Princeton Univ. Press, Princeton, 1940.

A. Haar, *Der Massbegriff in der Theorie der kontinuierlichen Gruppen*, Ann. of Math. **34** (1933), 147–169.

P.R. Halmos, *Measure Theory*. Van Nostrand, Princeton, 1950.

P.R. Halmos, *Introduction to Hilbert Space and the Theory of Spectral Multiplicity*. Chelsea, New York, 1951.

P.R. Halmos, *What does the spectral theorem say?* Amer. Math. Monthly **70** (1963), 241–247.

P.R. Halmos, *A Hilbert Space Problem Book*. Van Nostrand, New York, 1967.

P.R. Halmos and V.S. Sunder, *Bounded Integral Operators on L^2-Spaces*. Springer-Verlag, Heidelberg, 1978.

H. Hahn, *Reelle Funktionen*. Akad. Verlag, Leipzig, 1932.

F. Hansen and G.K. Pedersen, *Jensens inequality for operators and Löwners theorem*. Math. Ann. **258** (1982), 229–241.

F. Hausdorff, *Mengenlehre*, third edition. W. de Gruyter, Berlin/Leipzig, 1935.

E. Hellinger, *Neue Begründung der Theorie quadratischer Formen von unendlichvielen Veränderlichen*. J. Reine Angew. Math. **136** (1909), 210–271.

E. Hellinger and O. Toeplitz, *Grundlagen für einer Theorie der unendlichen Matrizen*. Math. Ann. **69** (1910), 289–330.

E. Hellinger and O. Toeplitz, *Integralgleichungen und Gleichungen mit unendlichvielen Unbekannten*. Encyklop. d. math. Wiss. II C 13, 1335–1616. Teubner, Leipzig, 1927.

E. Hewitt and K.A. Ross, *Abstract Harmonic Analysis*, I–II. Springer-Verlag, Berlin, 1963, 1970.

D. Hilbert, *Grundzüge einer allgemeinen Theorie der linearen Integralgleichungen*, I–VI. Nachr. Akad. Wiss. Göttingen, 1904–10. Published in book form by Teubner, Leipzig, 1912.

E. Hille and R.S. Phillips, *Functional Analysis and Semi-Groups*. Amer. Math. Soc. Colloq. Publ. 31, New York, 1957.

K. Hoffmann, *Banach Spaces of Analytic Functions*. Prentice-Hall, Englewood Cliffs, 1962.

O. Hölder, *Über einen Mittelwertsatz*. Nachr. Akad. Wiss. Göttingen, 38–47 (1889).

S.T. Hu, *Elements of General Topology*. Holden-Day, San Francisco, 1964.

R.V. Kadison, *A representation theory for commutative topological algebras*. Memoirs Amer. Math. Soc. 7 (1951).

R.V. Kadison and G.K. Pedersen, *Means and convex combinations of unitary operators*. Math. Scand. 57 (1985), 249–266.

R.V. Kadison and J.R. Ringrose, *Fundamentals of the Theory of Operator Algebras*, I–II. Academic Press, New York, 1983, 1986.

S. Kakutani, *Two fixed-point theorems concerning bicompact convex sets*. Proc. Imp. Acad. Tokyo, 19 (1938), 242–245.

E. Kamke, *Mengenlehre*. W. de Gruyter, Berlin/Leipzig, 1928.

I. Kaplansky, *A theorem on rings of operators*. Pacific J. Math. 1 (1951), 227–232.

T. Kato, *Perturbation Theory for Linear Operators*, second edition. Springer-Verlag, Heidelberg, 1976.

J.L. Kelley, *The Tychonoff product theorem implies the axiom of choice*. Fund. Math. 37 (1950), 75–76.

J.L. Kelley, *General Topology*. Van Nostrand, New York, 1955.

J. Kelley and I. Namioka, *Linear Topological Spaces*. Van Nostrand, Princeton, 1963.

M.G. Krein, *The theory of self-adjoint extensions of semi-bounded Hermitian operators and its applications*, I–II. Mat. Sbornik 20 (1947), 431–495; 21 (1947), 365–404.

M.G. Krein and D. Milman, *On extreme points of regularly convex sets*. Studia Math. 9 (1940), 133–138.

M.G. Krein and V. Smulian, *On regularly convex sets in the space conjugate to a Banach space*. Ann. of Math. 41 (1940), 556–583.

C. Kuratowski, *Topologie*, I–II. Monografje Matematyczne 3, 21, Warszawa, 1933, 1950.

H. Lebesgue, *Oeuvres Scientifiques*, I–V. L'Enseignement Mathematique, Genéve, 1973.

P. Lévy, *Lecons d'analyse fonctionelle*. Gauthier-Villars, Paris, 1922.

L.H. Loomis, *An Introduction to Abstract Harmonic Analysis*. Van Nostrand, New York, 1953.

E.H. Moore and H.L. Smith, *A general theory of limits*. Amer. J. Math. 44 (1922), 102–121.

F.J. Murray, *An Introduction to Linear Transformations in Hilbert Space*. Ann. of Math. Studies 4, Princeton, 1941.

M.A. Naimark, *Normed Rings*, E.P. Nordhoff, Groningen, 1960 (Russian original, 1955).

J. von Neumann, *Collected Works*, I–VI. Pergamon Press, Oxford, 1961–63.

O.M. Nikodym, *Sur une généralisation des intégrales de M.J. Radon*. Fund. Math. 15 (1930), 131–179.

O.M. Nikodym, *Remarques sur les intégrales de Stieltjes en connexion avec celles de MM. Radon et Fréchet*. Ann. Soc. Polon. Math. 18 (1945), 12–24.

G.K. Pedersen, *C*-Algebras and Their Automorphism Groups*. Academic Press, London/New York, 1979.

R.R. Phelps, *Lectures on Choquets Theorem*. Van Nostrand, Princeton, 1966.

M. Reed and B. Simon, *Methods of Modern Mathematical Physics*, I–IV, Academic Press, New York, 1972–1978.

F. Rellich, *Störungstheorie der Spektralzerlegung*. Proc. Int. Congress Math. Cambridge (MA) 1 (1950), 606–613.

C.E. Rickart, *General Theory of Banach Algebras*. Van Nostrand, Princeton, 1960.

F. Riesz, *Oeuvres Complètes*, I–II. Akadémiai Kiadó, Budapest, 1960.

F. Riesz and B. Sz-Nagy, *Lecons d'analyse fonctionelle*, sixth edition. Gauthier-Villars, Paris, 1972.

W. Rudin, *Real and Complex Analysis*. McGraw-Hill, New York, 1966.

W. Rudin, *Functional Analysis*. McGraw-Hill, New York, 1973.

C. Ryll-Nardzewski, *On the ergodic theorems*, I–II. Studia Math. 12 (1951), 65–73, 74–79.

S. Saks, *Theory of the Integral*, second edition. Monografje Matematyczne, Warszawa, 1952.

L. Schwartz, *Théorie des Distributions*, I–II. Herman et Cie, Paris, 1951.

H. Seifert and W. Threlfall, *Lehrbuch der Topologie*. Teubner, Leipzig, 1935.

W. Sierpinski, *General Topology*, second edition. Univ. of Toronto Press, Toronto, 1952.

I.M. Singer and J.A. Thorpe, *Lecture Notes on Elementary Topology and Geometry*. Scott-Foresman Co., Illinois, 1967.

L.A. Steen and J.A. Seebach, *Counterexamples in Topology*, second edition. Springer-Verlag, New York, 1978.

T.J. Stieltjes, *Recherches sur les fractions continues*. Ann. Fac. Sci. Toulouse 8 (1894), 1–22.

M.H. Stone, *On one-parameter unitary groups in Hilbert space*. Ann. of Math. 33 (1932), 643–648.

M.H. Stone, *Linear Transformations in Hilbert Space and Their Applications to Analysis*. Amer. Math. Soc. Colloq. Publ. 15, New York, 1932.

M.H. Stone, *The generalized Weierstrass approximation theorem*. Math. Mag. 21 (1947–48), 167–184, 237–254.

M.H. Stone, *On the compactification of topological spaces*. Ann. Soc. Polon. Math. 21 (1948), 153–160.

B. Sz-Nagy, *Spektraldarstellung linearer Transformationen des Hilbertschen Raumes*. Ergebnisse d. Math. 5, Springer-Verlag, Berlin, 1942.

M. Takesaki, *Theory of Operator Algebras*, I, Springer-Verlag, Heidelberg, 1979.

F. Trèves, *Topological Vector Spaces, Distributions, and Kernels*. Academic Press, New York, 1967.

J.W. Tukey, *Convergence and Uniformity in Topology*. Ann. of Math. Studies 2, Princeton, 1940.

A. Tychonoff, *Über einen Metrizationssatz von P. Urysohn*. Math. Ann. 95 (1926), 139–142.

A. Tychonoff, *Über die topologische Erweiterung von Räumen*. Math. Ann. 102 (1929), 544–561.

A. Tychonoff, *Über einen Funktionenraum*. Math. Ann. 111 (1935), 762–766.

P. Urysohn, *Über Metrization des kompakten topologischen Raumes*. Math. Ann. 92 (1924), 275–293.

P. Urysohn, *Über die Mächtigkeit der zusammenhängenden Mengen*. Math. Ann. 94 (1925), 262–295.

P. Urysohn, *Zum Metrizationsproblem*. Math. Ann. 94 (1925), 309–315.

A. Weil, *L'intégration dans les groupes topologiques et ses applications*. Herman et Cie, Paris, 1940.

H. Weyl, *Über beschränkte quadratische Formen, deren Differenz vollstetig ist*. Rend. Circ. Mat. Palermo 27 (1909), 373–392.

A. Wintner, *Spektraltheorie der unendlichen Matrizen*. Hirzel, Leipzig, 1929.

K. Yosida, *Functional Analysis*. Springer-Verlag, New York, 1968.

E. Zermelo, *Beweis, dass jede Menge wohlgeordnet werden kann*. Math. Ann. 59 (1904), 514–516.

E. Zermelo, *Neuer Beweis für die Möglichkeit einer Wohlordnung*. Math. Ann. 65 (1908), 107–128.

M. Zorn, *A remark on method in transfinite algebra*. Bull. Amer. Math. Soc. 41 (1935), 667–670.

List of Symbols

Index

Graduate Texts in Mathematics

continued from page ii